Birds of
BANGLADESH

Richard Grimmett
In loving memory of his mum, Molly Grimmett.
Happy memories shine bright and strong.

Paul Thompson
In memory of David L. Johnson,
for many happy days sharing birds and bird lore in Bangladesh.

Birds of
BANGLADESH

Richard Grimmett, Paul Thompson & Tim Inskipp

Illustrated by
Richard Allen, Adam Bowley, Clive Byers, Dan Cole, John Cox,
Gerald Driessens, Carl d'Silva, Martin Elliott, Kim Franklin, John Gale,
Alan Harris, Peter Hayman, Dave Nurney, Craig Robson,
Christopher Schmidt, Brian Small, Jan Wilczur,
Tim Worfolk & Martin Woodcock

H E L M
LONDON · OXFORD · NEW YORK · NEW DELHI · SYDNEY

HELM
Bloomsbury Publishing Plc
50 Bedford Square, London, WC1B 3DP, UK
29 Earlsfort Terrace, Dublin 2, Ireland

BLOOMSBURY, HELM and the Helm logo are trademarks of Bloomsbury Publishing Plc

First published in the United Kingdom 2021

Bloomsbury Publishing Plc does not have any control over, or responsibility for, any third-party websites
referred to in this book. All internet addresses given in this book were correct at the time of going to press.
The authors and publisher regret any inconvenience caused if addresses have changed or
sites have ceased to exist, but can accept no responsibility for any such changes.

A catalogue record for this book is available from the British Library.

Library of Congress Cataloguing-in-Publication data has been applied for.

ISBN: HB: 978-1-4729-9059-4
PB: 978-1-4729-3755-1
ePub: 978-1-4729-9248-2
ePDF: 978-1-4729-3756-8

2 4 6 8 10 9 7 5 3 1

Design by Fluke Art
Printed and bound in China by RR Donnelley Asia Printing Solutions Limited

To find out more about our authors and books visit www.bloomsbury.com and sign up for our newsletters.

CONTENTS

ACKNOWLEDGEMENTS

The authors are grateful to the artists whose work illustrates this book, including Alan Harris who executed the cover.

Richard Grimmett would like to thank George Grimmett for help preparing the text on climate and the geographical setting for the introduction, and Ella Grimmett for help to arrange and check the species texts and other assistance. A very grateful thanks also to Helen Taylor for the support and love that has underpinned our family life.

Also appreciated is the assistance given to Richard Grimmett from colleagues at BirdLife International, Ian Burfield, Adjoa Boateng and Tom Lambert, to help access information from the BirdLife Data Zone and for preparing the map on Important Bird and Biodiversity Areas in Bangladesh. The authors have great admiration and appreciation for the teams at BirdLife and Handbook of the Birds of the World for providing such an authoritative reference point on the taxonomy, nomenclature and status of the world's birds, which has been followed in this work (see opposite). What would the birding and conservation world do without such rigorous foundations?

Many people have generously shared birding locations, visits and sightings with Paul Thompson, who in particular thanks David Johnson, Enam Ul Haque and Ronald Halder for their company and pioneering publications on Bangladesh birds, and Sayam U. Chowdhury for helping compile the burgeoning records of recent rarities. We again thank Enam Ul Haque, founder of the Bangladesh Bird Club, who kindly provided Bangla names for all the species recorded.

Work by Carol Inskipp for earlier books, together with Richard Grimmettt and Tim Inskipp (e.g. *Birds of the Indian Subcontinent*), has been invaluable in preparing the introduction (especially the family summaries) and for the species accounts (with regard to habits and voice) and we express our considerable thanks for allowing us to build off these for this book.

Finally, thanks also go to our publisher Bloomsbury, its commissioning editor Jim Martin, and editors Alice Ward and Jenny Campbell for their belief in this project; to Tim Harris and Guy Kirwan for editing the text; and to Julie Dando for preparing final copy of the maps and for the wonderful layout.

INTRODUCTION

Bangladesh is one of the most densely populated and intensively cultivated countries on Earth, and remnant natural and semi-natural habitats are under huge pressure. Despite this, the country can boast the Sundarbans, the largest area of mangrove forest in the world, vast expanses of intertidal habitat and a fine number of accessible and productive sites to go birdwatching. Even the capital city Dhaka and the surrounding area has some nice spots to see birds.

Birding is not a widely practiced pastime in Bangladesh, but this is starting to change. Bird photography is becoming popular and bird clubs are springing up. Domestic tourism and an interest in nature is rapidly growing, and deshi (national) and foreign birders alike are finding birding to be rewarding, with many opportunities for new discoveries.

To aid and encourage this interest, and to support researchers and conservation practitioners, this book aims to present in a readily accessible form information on the occurrence, identification, habits and habitats of all 705 species reliably recorded in the country up to the end of 2019.

The detailed text covers all 535 bird species regularly occurring in the country or that have bred or were presumed resident in the past. Accompanying this text, there are 103 colour plates depicting each species and the variations in plumages with sex and age. There are also distribution maps to provide information on the extent and nature of each species' occurrence. Appendix 1 provides information on the 170 species that have been recorded rarely and are currently considered to be vagrants to the country, although as our knowledge improves several may prove to be regular.

In addition, Appendix 3 lists 162 species for which there are published references to Bangladesh but which are considered unproven.

TAXONOMY AND NOMENCLATURE

The taxonomy and nomenclature used follows the *Handbook of the Birds of the World* and BirdLife International digital checklist of the birds of the world (HBW and BirdLife International 2019). We have done this to improve standardisation in the region and follow what has been adopted by, and make life easier for, birders in Bangladesh (Thompson & Chowdhury 2020). This has resulted in major changes from *Birds of the Indian Subcontinent* (Grimmett *et al.* 2011), which largely followed Inskipp *et al.* (1996) and Gill & Wright (2006). The aim is to take account of (although not necessarily follow) the many recent proposals for taxonomic changes, particularly in Rasmussen & Anderton (2012), Dickinson & Remsen (2013), Dickinson & Christidis (2014) and del Hoyo & Collar (2014–16). The changes from *Birds of the Indian Subcontinent* are too many to list here, but alternative names and taxonomic notes are provided in the species accounts to help the reader understand these.

PLATES AND SPECIES ACCOUNTS

Species that occur regularly in Bangladesh are illustrated in colour and described in the plate captions, as are species that formerly occurred and are presumed to have been resident but are now feared nationally extinct (IUCN Bangladesh 2015). Vagrants/rarities (typically with fewer than ten records) are described in Appendix 1, with reference to distinguishing features from other more regularly recorded species where appropriate. The illustrations show distinctive sexual and racial variation whenever possible, as well as immature plumages. Some distinctive races are also depicted.

The accompanying text very briefly summarises the species' distribution, status and habitats, and provides information on the most important identification characters (**ID**), including **voice** and approximate body length, including bill and tail, in centimetres. Length is expressed as a range when there is marked variation within the species (e.g. as a result of sexual dimorphism or racial differences).

A general description is given of the species' status as a resident, winter visitor, summer visitor or passage migrant, where regions within Bangladesh are abbreviated: NW = north-west (Rajshahi and Rangpur Divisions); SW = south-west (Khulna Division); C = Dhaka and Mymensingh Divisions; NE = north-east (Sylhet Division); central coast = Barisal Division and the coast of Noakhali District; and SE = south-east (Chittagong Division). Where a species is restricted to the Chittagong Hill Tracts (the only significant hilly area, comprising three districts of Khagarachari, Rangamati and Bandarban) this is also noted. Note that the anglicised spelling of place names has in recent years been under a process of revision and the official spelling of what was long known as Chittagong is now Chottogram, but the former spelling has been retained here for ease of comparison to other references. Data on actual breeding records and non-breeding ranges are very few, so it has not been possible to give comprehensive details. In addition, space limitations have meant that the simple terms, 'breeds' and 'winters' are used to describe ranges and habitats for many species. The identification texts are based on Grimmett *et al.* (2011) with some updates. The vast majority of the illustrations have been taken from the same work and, wherever possible, the correct races for Bangladesh have been depicted. A small number of additional illustrations were executed for this book. Preparation of the text and plates for the original work included extensive reference to museum specimens combined with considerable work in the field. Habitats and habits (**HH**) are included to the extent that space allows and an alternative name (**AN**) and/or taxonomic note (**TN**) are given where relevant, mainly where they differ from Grimmett *et al.* (2011). Where a species is considered to be globally threatened with extinction in the BirdLife/IUCN Red List (HBW and BirdLife International 2019) or on the national Red List for Bangladesh (IUCN Bangladesh 2015) this is indicated using the following abbreviations: VU = Vulnerable, EN = Endangered and CR = Critically Endangered (see also Appendix

Key to maps

. Dhaka (capital city)

resident

former range (no recent records)

winter visitor

summer visitor

passage migrant (spring and/or autumn)

? Uncertain record or range (in the appropriate colour)

Note that isolated records are shown by a dot of the appropriate colour

2). Distribution maps accompany the text which, when using the key opposite, provide information on the extent and nature of each species' occurrence in the region. The maps and status notes have been prepared making use of a number of recent books (Halder 2010; Haque 2014; Khan 2008; Khan 2018; Siddiqui *et al.* 2008), papers summarising nationally notable records (Thompson *et al.* 1993; Thompson & Johnson 2003; Thompson *et al.* 2014), and reports and papers documenting the birds of particular sites and regions, such as Chowdhury (2020) and Round *et al.* (2014).

PLUMAGE TERMINOLOGY

The figures below illustrate the main plumage tracts and bare-part features, and are based on Grant and Mullarney (1988–89). This terminology for bird topography has been used in the text (see below). Other terms have been used and are defined in the Glossary. Juvenile plumage is the first plumage on fledging, and in many species it is looser, more fluffy, than subsequent plumages. In some families, juvenile plumage is retained only briefly after leaving the nest (e.g. pigeons), or hardly differs from adult plumage (e.g. many babblers), while in other groups it may be retained for the duration of long migrations or for many months (e.g. many waders). In some species (e.g. *Aquila* eagles), it may be several years before all juvenile feathers are finally moulted. The relevance of the juvenile plumage to field identification therefore varies considerably. Some species reach adult plumage after their first post-juvenile moult (e.g. larks), whereas others go through a series of immature plumages. The term 'immature' has been employed more generally to denote plumages other than adult and is used either where a more exact terminology has not been possible, or where more precision would give rise to unnecessary complexity. Terms such as 'first-winter' (resulting from a partial moult from juvenile plumage) or 'first-summer' (plumage acquired prior to the breeding season of the year after hatching) have, however, been used where it was felt that this would be useful.

Many species assume a more colourful breeding plumage, which is often more striking in the male compared with the female. This can be realised either through a partial (or in some species complete) body moult (e.g. waders) or results from the wearing-off of pale or dark feather fringes (e.g. redstarts and buntings).

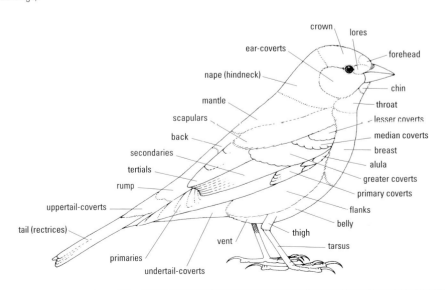

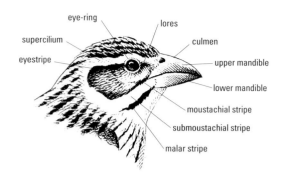

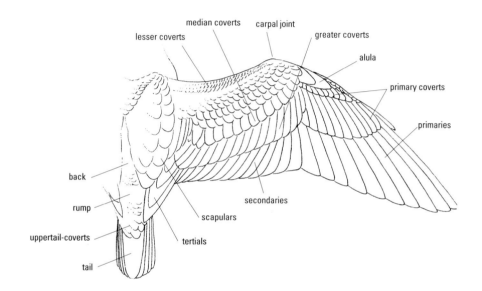

GEOGRAPHICAL SETTING

Located at the head of the Bay of Bengal, Bangladesh (20°34–26°38′ N and 88°01′–92°41′ E) has a land area roughly half the size of the United Kingdom, at 147,570km². With more than 160 million people, its population density of about 1,100/km² is more than twice that of the UK. The country is made up almost entirely of the Ganges–Brahmaputra–Meghna delta, the largest delta in the world, and much of the country is below 10m above sea level. It is surrounded by India but for a small south-east border with Myanmar. The Ganges (Padma), Brahmaputra (Jamuna) and Meghna rivers, laden with sediment that has travelled from the Himalayas and hills of north-east India, all feed into and create the delta. This sediment has settled across the low-lying land as the waters slow, and has created some of the most fertile land in the world. With a few exceptions in the far north-west and north-east, the south-east is the only part of Bangladesh that sits comfortably above sea level. Bordered by the Indian states of Tripura and Mizoram, and by Myanmar, this land – the Chittagong Hill Tracts – contrasts with the rest of the country, with steep narrow ridges running north–south, separated by fertile valleys and the Kaptai Reservoir; the highest peaks are just over 1,000m.

The country is ethnically homogeneous, and its name derives from Bangla, the language of the Bangali ethno-linguistic group, which comprises over 95% of the population. There are, however, 45 very diverse and distinctive ethnic minorities living in the country, predominantly in the forested and hill areas within the Chittagong Hill Tracts, other parts of Chittagong Division, Sylhet Division and Mymensingh Division.

Map 1. Map of Bangladesh showing regions and notable locations.

CLIMATE

Bangladesh has a climate that is broadly uniform across the country, with few exceptions. This tropical monsoon climate consists of a hot humid environment with seasonal heavy monsoon rains. The monsoon season begins towards the end of May and runs until mid-October. Following these sometimes seemingly relentless rains comes a transitional period when cyclonic storms periodically batter coastal areas. This is followed by a dry, cool winter season between December and February, with average peak temperatures around 25°C and average minimum temperatures around 12°C, but during cold snaps temperatures can drop to 2–4°C in Thakurgaon (far north-west) and Srimangal (north-east). March and April, before the rains, is the hot summer season, with average maximum temperatures of over 30°C, but they can reach 45°C; this is also the season of sudden violent 'nor'wester' storms, and cyclones are again a high risk in April and May. Rainfall data from 1950–2011 show that average annual precipitation over the country is 2,428mm, with 71% occurring during the monsoon. However, the far west is significantly drier (average annual precipitation around 1,400mm) and the east, especially the north-east, is significantly wetter (annual average exceeds 4,000mm). The monsoon causes severe issues for the country almost every year; the inundation of 25–33% of low-lying land is a normal part of ecology and human life, but in exceptional years flooding can affect two-thirds of the country.

The threat of climate change looms over Bangladesh more than most countries. Already, saline water is reaching further up the rivers of the delta during the dry season and, with sea-level rise, low-lying coastal land faces an increasing risk of permanent inundation. Combined with increasingly unpredictable rainfall and more frequent cyclones, these changes threaten agricultural productivity, livelihoods and habitats across the country, including the Sundarbans.

MAIN HABITATS, THREATS AND BIRD SPECIES

FORESTS

Bangladesh has three important forest types, although these are much reduced from their original extent. Forests continue to be under great pressure due to the high dependency of local people on wood for fuel, the high value of timber for furniture and the pressure to clear land for agriculture. This has been compounded in the far south-east by the arrival of Rohingya refugees from Myanmar, who have settled in and cleared forest lands.

Mangrove forest The Sundarbans is the largest block of mangrove forest in the world and is intersected by a complex network of tidal waterways, mudflats and small islands. The total area of the Sundarbans in both Bangladesh and India is estimated at 10,000km^2, and the Bangladesh area of reserved forest is about 6,000km^2. The forest covers the moribund delta of the Ganges River and receives fresh water in the wet season largely from distributaries of the Ganges–Padma River. The Sundarbans gets its name from the large number of Sundari *Heritiera fomes* trees. The area is generally free of permanent habitation and almost 70% is covered by mangrove forest, the remainder being waterways. In addition to being home to the world's largest population of the Endangered Masked Finfoot, characteristic and charismatic species resident here include a high diversity of kingfishers, notably Brown-winged and Ruddy, also Lesser Adjutant, Streak-breasted Woodpecker (its only location in South Asia), Mangrove Pitta and Mangrove Whistler. There are also extensive planted mangrove forests east of the Sundarbans on islands along the central coastline, and smaller residual and degraded patches of natural mangroves in the south-east; however, these lack most of the species characteristic of the Sundarbans.

A small tidal creek within the eastern Sundarbans.

Evergreen and semi-evergreen forest is found in patches in hilly areas of the eastern side of the country where rainfall is higher. These are mostly secondary or degraded forests, some with a mix of native trees and exotic plantations, and range from just a few metres above sea level to almost 1,000m altitude in the Chittagong Hill Tracts. In some areas former forest is now largely bamboo, grass-scrub, plantations, shifting cultivation (in the Chittagong Hill Tracts), or converted into tea estates (mainly in the north-east). These forests are among the most diverse for birds, with notable residents including White-cheeked Partridge, a range of woodpeckers and babblers, Greater and Lesser Necklaced Laughingthrushes, and Red-headed Trogon. Several species, such as Kalij Pheasant, occur at much lower altitudes in Bangladesh than in the rest of their ranges. In winter, birds are often concentrated in mixed feeding flocks that include migrants such as warblers and minivets from the hills of north-east India and the Himalayas, as well as resident birds; migrant chats and flycatchers can be sought in the undergrowth.

Early morning in winter along a trail through Lawacharra National Park in the north-east. This semi-evergreen forest is kept moist in the dry season by fog and dew, and is one of the richest forests in the country.

The remnant Sal forest in Bhawal National Park, just north of Dhaka, has been heavily coppiced in the past and is encroached by farmland.

Deciduous forest dominated by Sal *Shorea robusta* persists in small degraded patches in the north-west and in larger but previously coppiced areas in the central region, particularly in Modhupur National Park, but much former forest has been converted to agriculture and industry. The loss of mature trees has seen a decline in the owl diversity previously found in these forests, but in the spring (pre-monsoon) they still come alive with birds, including Indian Pitta and Indian Paradise-flycatcher when they arrive and establish territories.

COASTAL WETLANDS

The coastal ecosystems in Bangladesh (excluding the Sundarbans) are dominated by shallow silt-laden waters, extensive intertidal mudflats and deltoid islands known as chars, several of which have been planted with mangrove forests. In addition, mainly in the south-east there are sandy beaches, sand dune systems and on St. Martin's Island the only coral formations in the country. Some chars still feel wild and remote from the human bustle of much of the country, but fishing and grazing pressure is high, locally shorebirds are trapped, and tourism development is adversely affecting the beach areas and St. Martin's Island.

Nijhum Dwip National Park on the central coast includes extensive intertidal mudflats and the emerging silty island and grasses of Domar Char – a key habitat for shorebirds and Indian Skimmer.

Nevertheless, these extensive and ever-changing mudflats and islands are globally important for wintering waterbirds, particularly shorebirds. Several globally threatened species are present, including about 10% of the world population of the Critically Endangered Spoon-billed Sandpiper, about half of the world population of Vulnerable Indian Skimmer, and significant numbers of Endangered Spotted Greenshank and Great Knot. More abundant species include Common Shelduck, Eurasian Wigeon, sandplovers and stints.

FRESHWATER WETLANDS

With over half the country comprising floodplains, these could all be termed seasonal wetlands. However, most of these lands have over the centuries been converted to paddy-fields and intensively used to grow rice. Pressure on wetlands is immense, for fishing, grass-cutting, grazing and further conversion for agricultural use, while pollution increasingly affects wetlands. The many small and medium-sized depressions known as beels hold water for all or most of the year and are used by common waterbirds. Two types of freshwater wetland are of great ornithological interest.

One of the last remaining areas of mature freshwater swamp forest – accessible only by boat in the monsoon – is protected at Ratargul Special Biodiversity Conservation Area, about 25km north of Sylhet city in the north-east.

In the north-east, the haors are large saucer-shaped depressions bounded by natural river levees or low hills; in the monsoon they are deeply flooded and many coalesce to form inland seas, but in the dry season each contains large or small areas of residual permanent water and marshy vegetation in multiple beels. Until the early 20th century, the haors contained large areas of reed swamp and swamp forest. Swamp forest is flooded for half the year and dominated by Hijal *Barringtonia acutangula* and Koroch *Millettia pinnata* trees. These forests were largely converted to agriculture or intensively grazed in the mid- to late 20th century. Since around 2000, some significant areas of swamp forest have been regenerating through planting and local protection. In general, the haors with some level of protection and/or conservation effort are now the only ones of ornithological interest, and most of the others are dominated by intensive cultivation of a single rice crop in the dry season. Swamp-thickets and reed-swamp habitats are dense vegetation up to 2–3m tall comprising the reed *Phragmites karka* (though extensive reedbeds have largely been cleared for papermaking and cultivation), Dhol Kolmi *Ipomea fistulosa*, woody plants, and rarely Giant Cane *Arundo donax*. The haors are home to migratory winter-nesting Pallas's Fish-eagle, increasing flocks of Asian Openbill and large flocks of wintering waterfowl (particularly Common Teal, Northern Pintail, Ferruginous Duck and in some years much of the South Asian population of Fulvous Whistling-duck). A few Falcated Duck are regular, and formerly the haors were the main wintering area for Baer's Pochard, but now only small numbers are seen. Swamp-thickets and forests hide a special fauna of wintering passerines among the ubiquitous Dusky Warblers, including grasshopper-warblers such as Pallas's, two species of rubythroat and, for the very lucky, Firethroat.

A char close to Rajshahi city in the north-west. In the dry season, extensive areas of sand and silt, grasses and wet channels on these large riverine islands are important for specialised birds.

The main rivers – the Brahmaputra–Jamuna and Ganges–Padma (both rivers change their names downstream) – are broad and braided, stretching for 10km or more between their outer banks. In the dry season, they form complex networks of river channels and silty-sandy islands known as chars. These chars hold the last areas of tall wet grasses in the country (see Grasslands, below). In winter, sandbanks and backwaters hold modest numbers of waders and wildfowl, and are regularly the haunt of Black Stork and raptors such as Long-legged Buzzard and various eagles. However, nesting species such as River Lapwing, terns and Indian Skimmer have declined and are now rarely seen along the rivers.

GRASSLANDS

Tall wet grassland must once have covered much of Bangladesh but is now restricted to some of the chars, patches within the haors and some tea estates. These habitats are heavily used for grazing, grasses are cut for thatching and agriculture is spreading. Along the main rivers, a few specialised species nest, including Sand Lark, Graceful Prinia, Black-breasted Weaver and Blue-tailed Bee-eater. In riverine and haor grasslands, in the pre-monsoon period Bristled Grassbirds can be heard among the larks singing overhead. Along the Ganges in Rajshahi District in recent years species such as Sykes's Nightjar and Grey Francolin have been found, and there may be other discoveries to be made in the more remote remnants of this habitat.

Tall wet grassland in the late dry season along the Padma River in the north-west. Like most of this habitat, the grasses in the picture have been heavily exploited and the remaining tall grasses are being carried away for sale.

AGRICULTURAL LAND

Most land is used for two crops of rice a year (except for deeply flooded areas), although in winter mustard, pulses and a range of vegetables and other crops are also common, while maize is increasing and tall crops of jute are grown in some areas in the early monsoon. Birds are often few in number among growing crops, but the preparation of paddy-fields can attract Black Drongos, wagtails and Cattle Egrets. In winter, waders such as Grey-headed Lapwing and Wood Sandpiper may be found along the bunds separating irrigated paddy-fields, and ripening crops attract doves, Baya Weavers and Rose-ringed Parakeets. Agricultural intensification and the increasing use of agro-chemicals appear to be responsible for a decline in common farmland birds. Of particular concern in this landscape is the use of diclofenac and related anti-inflammatory drugs used to treat livestock ailments; these drugs have been identified as responsible for the now highly threatened status of Slender-billed and White-rumped Vultures in Bangladesh. Veterinary use of diclofenac has been banned in Bangladesh and while its illegal use has declined, the use of related drugs remains a conservation challenge.

VILLAGES AND TOWNS

In most of Bangladesh villages are spread out, with trees and small orchards around the dispersed houses. Village groves often hold a good diversity of common birds. Close to wetlands, tolerant homeowners may share their trees with nesting egrets and cormorants. More extensive orchards, such as those in Rajshahi (famous for mangos), are often home to Orange-headed Thrush and Indian Paradise-flycatcher, while in the south species such as Abbott's Babbler may be found in betel nut groves. In the north-east and a few parts of the south-east and far north-west are tea estates with extensive shade trees and small areas of scrub. Birds of open woodland and a few specialities such as Rufous-necked Laughingthrush can be found among the tea. Urban development has been rapid in recent decades, human populations are dense, and few urban parks have sufficient trees and bushes to support many birds. In general, older parks, botanical gardens, older university campuses and land reclamation areas prior to housing construction hold a typical avifauna of more widespread species and can be of interest for migrants.

A typical area of farmland in the dry season in Pabna District in the north-west. Once an extensive wetland, now birds are few and the small river and groundwater will be used to irrigate paddy-fields.

IMPORTANCE FOR BIRDS

SPECIES DIVERSITY

Up to the end of 2019 a total of 705 species had been recorded in Bangladesh, based on BirdLife International taxonomy as followed here. This comprises just over 6% of the word's birds and 51% of the 1,390 species recorded in South Asia (Praveen *et al.* 2019), a high percentage for a small country.

GLOBALLY THREATENED SPECIES AND ENDEMISM

Although Bangladesh has no endemic birds, 44 globally threatened species and a further 37 Near Threatened species have been recorded (Collar *et al.* 2001, HBW and BirdLife International 2019). Moreover, Bangladesh currently hosts, or in the past has hosted, a significant part of the global population of several of these species, notably Indian Skimmer, Spoon-billed Sandpiper and Great Knot on the coast in winter, Masked Finfoot resident in the Sundarbans, Bristled Grassbird breeding along the main rivers, and Baer's Pochard and Pallas's Fish-eagle in the haors of the north-east. BirdLife's Saving Asia's Threatened Birds (2003) provides a very useful perspective on the important species and habitats for threatened birds in Bangladesh in a continental context.

MIGRATION

Only 309 of the species recorded in Bangladesh are resident throughout the year, and for 11 of these (e.g. Purple Swamphen) small resident populations are considerably increased by wintering populations. A further 17 species are known only from 'old' records in the 19th or early 20th century; several of these were previously resident but are now nationally extirpated, such as Indian Peafowl.

A small number of species (11, including two that are more often passage migrants) are summer visitors to Bangladesh, which nest here. These include Indian Pitta, Indian Cuckoo and Indian Paradise-flycatcher, which migrate to spend the winter in southern India and Sri Lanka. A further 12 species are regular passage migrants, moving through Bangladesh from breeding grounds in the Himalayas and adjacent regions of China to winter in southern India (such as Brown-breasted Flycatcher and Forest Wagtail) and back again, or species that nest in north-east Asia and pass through on their way to and from wintering grounds in South-east Asia (such as Grey-tailed Tattler) or even in Africa in the cases of Amur Falcon and Common Cuckoo.

In total, 170 species (a quarter of the avifauna) are regular winter visitors. Another quarter of the avifauna (178 species) are currently considered to be vagrants (a few of these are shown in the main plates); most of these are migratory and many may prove to be regular but rare winter visitors or passage migrants with increased observer effort. This confirms the great importance for birds of Bangladesh's fertile farmland, village groves, forests and especially coastal and freshwater wetlands. Mild winters and an abundance of invertebrates, fish and seeds generated during the monsoon provide good conditions for a great diversity of species. Some of the most widespread and numerous winter visitors include Red-throated Flycatcher, Dusky Warbler and Brown Shrike. The origins of winter visitors are diverse and include species such as Grey-headed Canary-flycatcher and Lesser Racquet-tailed Drongo, which are altitudinal migrants presumably from the hills of north-east India; Black Kite, Bar-headed Goose and several duck species that have been satellite tracked to or from breeding grounds in Mongolia and China; and several species of wader and wildfowl that nest in the tundra and taiga of Siberia. One winter visitor with an unusual life cycle is Pallas's Fish-eagle, which nests in the winter in Bangladesh in isolated trees often close to villages in the haors, and leaves during the monsoon, when birds presumably spend the summer on the Tibetan plateau.

BIRDWATCHING AREAS

Details of some of the best and most accessible places for birdwatching are introduced here. Rural Bangladeshis in general are very hospitable, but remember that birdwatching is unfamiliar to most and may require some explanation to local people and officials. Also, where entry fees exist (usually in protected areas) they are higher (currently about US$5) for foreigners than for Bangladesh citizens.

In and close to the capital city **Dhaka** are a few well-watched sites but these also attract many people for recreation. The **National Botanical Gardens** in Mirpur, Dhaka, have extensive groves of planted trees, some ornamental lakes and an area of bamboo. Over the years 177 species have been recorded, but an early morning visit is best to avoid the crowds. Highlights include resident Orange-headed Thrush, Forest Wagtail on spring and autumn migration, and the chance of other winter visitors and migrants such as Indian Blue Robin. The adjacent zoo lake attracted a diversity of wildfowl in the past but is now too heavily disturbed. Elsewhere in the city nesting Red-headed Falcon and a small colony of Alexandrine Parakeet are present but elusive. **Jahangirnagar University** campus, an hour or more west of Dhaka in Savar, has several lakes and wooded areas and a bird list of over 200 species. It is worth at least a half-day morning visit, and notable regular birds include Grey-headed Fish-eagle, Brown Fish-owl and Stork-billed Kingfisher. The scrub, grasses and farmland in land reclamation areas to the north-east and south of Dhaka can be good for birding before housing estates are built there; for example, the Purbachal area was for several years a regular site for Yellow-wattled Lapwing and produced a few rarities. These sites require exploration and local knowledge to get the best from them. **Bhawal National Park** is only 40km north of the centre of Dhaka, but this may take two hours by road due to heavy traffic. Large numbers of visitors can be present on winter weekends, but the degraded Sal forest interspersed with paddy-fields can still be quite productive. This is probably the best site for the localised White-eyed Buzzard and has other notable regular species such as Indian Spotted Eagle, Orange-headed Thrush and, in the monsoon, Indian Pitta.

The lakes at Jahangirnagar University, just north-west of Dhaka city, hold large flocks of Lesser Whistling-duck and other common waterbirds in winter.

The low hills around Srimangal town in the north-east are the main tea-growing region of Bangladesh. The mixture of tea, scrub, shade trees and small lakes found in these tea estates can provide good birdwatching opportunities.

Elsewhere in the **central region** the most notable site is **Modhupur National Park** between the towns of Mymensingh and Tangail and at least three hours' drive from Dhaka. Although it deserves an evening and day of exploration, accommodation on site is difficult to arrange and staying in Mymensingh town necessitates an early morning start to reach the main track into the forest and central lake for the prime morning period. The Sal forest here was diverse and very productive for birds, particularly owls and winter visitors, and 211 species have been recorded. However, most of the large trees have now been felled and care may need to be taken due to robberies in recent years. In addition to the species found in Bhawal, wintering thrushes and flycatchers, Yellow-wattled Lapwing and resident owls are more likely to be seen here.

The **north-east region** has some of the most accessible and productive birdwatching sites in the country. The town of **Srimangal** is a tourist and tea-growing centre. About five hours from Dhaka, it is easily reached by train or road, has a wide range of guesthouses and hotels, and makes a good base. Within about 10km of the town 450 species have been recorded in three habitats or sites that are well known to local three-wheeler ('CNG') and hire-car drivers. **Lawacharra National Park** is the best-studied evergreen forest in the country and has a list of 266 species within only 1,250ha. The main trails north and south of the ticket counter in the centre of the forest are the most productive and, beyond the first 500m, are rarely visited by tourists. At times, the trails can be quiet for birds until a mixed-species flock is encountered in winter, but most resident and wintering species of evergreen forest birds found in Bangladesh have been recorded here, including such prizes as Red-headed Trogon, Blue-naped Pitta and in the monsoon Hooded Pitta. This is also an excellent forest for primates, with some of the world's most easily seen Western Hoolock Gibbons and Phayre's Leaf Monkeys. Immediately north and east of the town are easily accessible **tea estates**, which are worth an early morning or afternoon walk, particularly near small rivers, lakes and tall grassy patches. They are good for species of open woodland and specialists such as Rufous-necked Laughingthrush and Yellow-eyed Babbler. A short distance north-west of the town is the large wetland of Hail Haor and within the haor is the 170ha community managed sanctuary of **Baikka Beel**, with a bird list of 198 species. This is well worth a day's visit and takes almost an hour to reach due to the rough winding

Large numbers of many species of wintering waterbirds – in this case about 3,000 Fulvous Whistling-ducks – can be seen from the towers at Baikka Beel in the north-east, where they are protected by the local community.

village roads, but these also pass interesting fish ponds. With a visitor centre, two observation towers and a 1km nature trail through restored swamp forest, the site is well developed for birdwatching. A wide diversity of ducks and waders, often in substantial numbers (e.g. over 2,200 Ruff in January 2019) can be seen in winter in the open beel; Cotton Pygmy-goose and Pheasant-tailed Jacana are resident; a few non-breeding Pallas's Fish-eagles are usually present in winter; and wintering warblers and rubythroats are common in the scrub and swamp-thickets. Further afield but accessible as day trips from Srimangal are several good sites. About 60km west of Srimangal towards Dhaka, **Satchari National Park** only covers 243ha but 221 species have been seen in its excellent forest. From the ticket counter a short concrete track leads to steps and then a tall observation tower. In spring, there is a flowering tree adjacent to the top of the tower and this is a good place to see canopy species such as green-pigeons, Blossom-headed Parakeet, Vernal Hanging-parrot and Black Baza, and several rarities have been attracted to it. In addition, trails accessed from near the tower or the road to the east enter good forest. Also, west of Srimangal but more

The observation tower at Satchari National Park in the north-east gives close treetop views in the spring of birds visiting the flowers of the adjacent Palash or Flame-of-the-forest Tree (*Butea monosperma*). The rarities observed here include this male Purple-backed Starling, the fourth national record.

In Rema-Kalenga Wildlife Sanctuary in the north-east, a partnership between IUCN Bangladesh, the Bangladesh Forest Department and the local community, protects one of the last nesting colonies in Bangladesh of the Critically Endangered White-rumped Vulture. They also operate the vulture feeding station, shown here, and have established a 'vulture safe zone' in the wider area.

difficult to access, is **Rema-Kalenga Wildlife Sanctuary**. There is more extensive forest here within its 1,795ha and an 'eco-cottage' to stay in. In addition to a small colony of White-rumped Vultures, 220 species have been recorded, including a wide range of evergreen forest and forest-edge species. About two hours east of Srimangal along minor roads is **Hakaluki Haor**; at more than 18,000ha this Ecologically Critical Area contains a maze of beel wetlands and open grazed lands in the dry season. A long walk with a guide is needed to find the better beels, since these change between years according to which beels are rested from fishing or intensively fished; once located, large numbers of waterbirds can be found along with a few Falcated Duck and Baer's Pochard in most years.

Also in the north-east, the main complex of haor wetlands can be found to the west of Sunamganj town. The most notable of these is **Tanguar Haor**, covering 9,527ha and nestled just below the hills of Meghalaya (India); one of only two Ramsar sites in Bangladesh, it is afforded some protection and 179 species have been recorded. However, to reach this haor requires a rough drive of over two hours from Sunamganj, followed by a one-hour boat journey. A day trip is possible, but the tourist boats are not geared to entering

Large flocks of ducks – like this group of mainly Common Pochard, seen here – can be expected on a winter boat trip within Tanguar Haor in the north-east.

In the monsoon, the observation towers at Hakaluki Haor in the north-east give a bird's-eye view of the extensive wetland and, as shown here, the areas of regenerating swamp forest planted since the early 2000s.

the haor (which requires permission and a smaller boat), so keen birdwatchers need to be prepared to camp here and hire a boat to access the beels. However, some of the wildfowl and swamp-thickets can be accessed by foot from the usual entry point. In winter, typically more than 50,000 waterbirds are present and the large flocks of ducks usually include thousands of Red-crested Pochard, and the occasional Falcated Duck and Baer's Pochard occur on the beels, whilst the swamp-thickets hide a large diversity of elusive wintering warblers and other species including regular Firethroat. Pallas's Fish-eagle nests here.

Likri Hill, Sangu Wildlife Sanctuary, in the Chittagong Hill Tracts.

Lastly, two less studied but interesting sites are readily accessible by road to the north of Sylhet city in the north-east. **Khadimnagar National Park** is about 15km north of the city and has degraded evergreen forest and plantations where 168 species have been recorded. A similar distance further north is **Ratargul** swamp forest. This 'special biodiversity protection area' contains one of the few patches of mature swamp forest; accessible by boat and popular with tourists for its scenic value, birds have been little studied here but the forest and bordering thickets may reveal interesting species in winter.

The **south-east** has numerous protected areas and the Chittagong Hill Tracts. There are also some notable coastal sites. The Chittagong Hill Tracts have the only significant hills in Bangladesh, these support a somewhat different avifauna from the forests in the north-east or the low coastal hills of the south-east. In this region species absent from the rest of the country, such as Great Barbet, Streaked Spiderhunter, Mountain Imperial-pigeon and Nepal House Martin are relatively common, and many other birds absent from the rest of the country may be encountered. However, some of the best remaining forest in the hill tracts is difficult to access, requiring a trek of several days; in addition, foreigners require special permission to visit the three hill districts. The most accessible area here is **Kaptai National Park**, adjacent to Kaptai town and 57km east of Chittagong city. Here there is evergreen forest and plantations with a few trails, where 220 species have been seen, and there are chances of Great Slaty Woodpecker, Malay Night-heron and Grey Peacock-pheasant as well as many typical forest birds. About 65km north of Chittagong, close to the main road to Dhaka (and not within the hill tracts), are two adjoining forests, **Hazarikhil Wildlife Sanctuary** and **Baroyadhala National Park**. These have been less visited by birdwatchers but over 180 species have been recorded, including many of the typical eastern forest birds with potentially more to be discovered. South of Chittagong, next to the main road to the seaside resort of Cox's Bazar (and about 60km from the resort), are two easily accessed forests: **Medakacchapia National Park**, with tall stands of Garjan *Dipterocarpus turbinatus* trees but little undergrowth; and **Fashiakhali Wildlife Sanctuary**, with more diverse patches of forest. About 170 species have been recorded in this area, which is a regular site for Great Slaty Woodpecker. Just north of Cox's Bazar town (which has many hotels) and accessible by hiring a boat from the town jetty (and asking for Belekardia), is **Sonadia Island**, with mudflats, sand dunes and tidal creeks. Within this Ecologically Critical Area, 144 species have been recorded and it is the main wintering site for Spoon-billed Sandpiper, Spotted Greenshank and Great Knot in Bangladesh. There are several interesting sites south of Cox's Bazar town, including Himchari National Park, Teknaf Wildlife Sanctuary and St. Martin's Island, but the forests in this area are now heavily degraded and the area has been significantly impacted by the arrival of around a million refugees from Myanmar.

Kaptai National Park in the south-east offers the most accessible birdwatching in the Chittagong Hill Tracts, both within evergreen forest and by boat around the fringes of the Kaptai Reservoir where it borders the forest.

The chars, mudflats and shallow waters of the central coast – from Nijhum Dwip east to Sandwip – are a major wintering area for shorebirds and flocks of the endangered Indian Skimmer.

The **central coast** from the coast of Noakhali District westwards to south of the large island of Bhola has extensive intertidal mudflats and new island chars in a landscape that is constantly changing with the interactions of sediment from the Ganges, Brahmaputra and Meghna rivers, and tides and storms. Birdwatching in most of this area is best done on an expedition using a boat, for example by joining a survey with members of the Bangladesh Bird Club. Within this area **Nijhum Dwip National Park** and in particular Domar Char to the east of Nijhum Dwip Island are relatively more accessible by taking an overnight ferry from Dhaka to Hatiya (the island immediately to the north). Here, 172 species have been recorded, including some in the planted mangroves on Nijhum Dwip and many waterbirds on Domar Char. Until recently this was the favourite location of wintering flocks of Indian Skimmer but they now mainly use other more distant chars. Shorebird numbers are particularly high and diverse in early spring.

In the **south-west**, the **Sundarbans** is a large prime birdwatching area where more than 250 species have been recorded, including a distinctive set of mangrove specialists. All access involves travel by boat. Although the entire mangrove forest has some protected status, within it there are three main wildlife sanctuaries, of which Sundarbans East is the most visited and probably the best for birds. Khulna city is well connected by road and air with the rest of the country. Two main strategies are possible to visit this popular tourist destination. The visitor facilities at Karamjal in the northern part of Sundarbans, including a boardwalk, can work as a day trip by hiring a boat from Mongla port. In the pre-monsoon period, species such as Mangrove Pitta and Ruddy Kingfisher are possible here. However, the most productive and easiest way to visit the forest is to book on a tour of three or more days with one of several tour companies operating boats that provide an all-inclusive service. These usually go from Khulna or Mongla through the length of the forest along wide rivers to the Katka area of Sundarbans East, where birders can take a small boat into the side creeks to search for Masked Finfoot and Brown-winged Kingfisher, and can take a guided walk to an observation tower in an area where Streak-breasted Woodpecker is possible. Although there is a substantial population of Tigers, the chance of seeing one is very low.

Close encounters with Masked Finfoot are still possible along narrow mangrove-lined creeks in the eastern Sundarbans of Bangladesh – this is its global stronghold. A 2020 review by the photographer and colleagues estimated a world population of just 108–304 individuals (60–80% in the Sundarbans) and recommended that it should be considered Critically Endangered, needing urgent protection from threats such as hunting and fishing by-catch.

There are three widely separated locations with notable birding in the **north-west region**. Rajshahi city is on the north bank of the Ganges River, is easily reached from the rest of the country and has several hotels. From the riverside a boat may be hired for the day to explore the large char and nearby channels opposite the city. By walking several kilometres on this open sandy char, several very localised species may be seen. Between April and June Bristled Grassbird is widespread and can be seen song flighting over tall grasses, Blue-tailed Bee-eater is common and Sykes's Nightjar has recently been found. In winter, here and further upstream along the Ganges in Chapai Nawabganj District, large waterbirds such as Painted and Black Storks and Asian Woollyneck may be found along with a range of waterfowl. In the far north of this region are two more remote areas that have interesting birdwatching. The **Brahmaputra River** in Kurigram and Gaibandha Districts has birds typical of the main rivers along its many branch channels and islands, but exploring this area requires hiring a boat for a day from one of several small landing places. A rather different area around **Tetulia** in the far north-west is within sight of the Himalayas, which are separated from Bangladesh by a narrow strip of India. This area can be reached on a long overnight bus journey from Dhaka. The farmland and few tea estates here have a small population of Black Francolin and the potential for finding interesting wintering birds – for example, there are recent records of Crested Bunting.

CONSERVATION MEASURES

CULTURAL ATTITUDES

Although hunting is illegal and various conservation measures have been adopted, there are mixed cultural attitudes to nature and continued indirectly harmful practices as well as direct persecution of birds. Bangalis have a strong cultural appreciation of the beauty of nature and landscapes, but there is the pressure to tame the wild and protect livelihoods. Waterbirds are trapped for food or in some cases to protect fish farms. There is retributional killing of species that damage crops or (in the case of Tigers, for example) are perceived as threatening life. Colonies of resident waterbirds are probably much reduced from what they once were, but some families still tolerate and even protect egret colonies in their homestead trees. The various ethnic communities have to different extents a stronger identification with forests, based on their sustenance as well as considerable knowledge of and respect for nature. In these communities, subsistence hunting of birds and other wildlife is of cultural importance, and forest patches were also protected from shifting cultivation. However, large-scale encroachment by outsiders and forest loss have undermined sustainable exploitation, and despite over 20 years with a peace accord there remain tensions and a heavy military presence in the Chittagong Hill Tracts.

COMMITMENT TO INTERNATIONAL CONSERVATION AGREEMENTS

Bangladesh has ratified the United Nations Convention on Biological Diversity and thus accepted an international obligation to conserve the variety of animals and plants, and a commitment to ensure that the use of biological resources is sustainable. As a consequence, it has prepared a National Biodiversity Conservation Strategy and Action Plan (the current iteration covering the period 2016–21). The country has also ratified the Ramsar Convention on Wetlands, which requires the protection of wetlands of international importance and the fostering of their wise use. However, only two wetlands, Sundarbans Reserved Forest (601,700ha) and Tanguar Haor (9,500ha), have been designated as Ramsar Sites. International trade in wildlife is governed by Bangladesh being a party to the Convention on International Trade in Endangered Species of Wild Fauna and Flora (CITES). It is also a party to the United Nations (Bonn) Convention on Migratory Species (CMS), a global agreement for the conservation of migratory animals that lays the legal foundation for internationally coordinated conservation measures throughout migratory ranges. Key species in Bangladesh for which such measures are being advanced, e.g. through the development and implementation of action plans, include Spoon-billed Sandpiper and all the vulture species.

INSTITUTIONAL FRAMEWORK FOR CONSERVATION

The Ministry of Environment, Forest and Climate Change is the lead agency in central government for the planning, coordination and oversight of all environmental matters in Bangladesh. The Environment Conservation Act (1995) provides for conservation of the environment, improvement of environmental standards and pollution control. The Wildlife (Conservation and Security) Act (2012) more specifically provides for the conservation of biodiversity, wildlife and forests in the country. The Act (which replaced and updated a similar earlier law) enables the declaration of sanctuaries, national parks and community conservation areas, and sets out permissible activities in relation to these designations. The Bangladesh Biodiversity Act (2017) regulates access to biological resources and how these may be utilised and shared.

PROTECTED AND CONSERVATION AREAS

Bangladesh has two streams of conservation and protection designations for the environment: protected areas overseen by the Forest Department, and ecologically critical areas overseen by the Department of Environment.

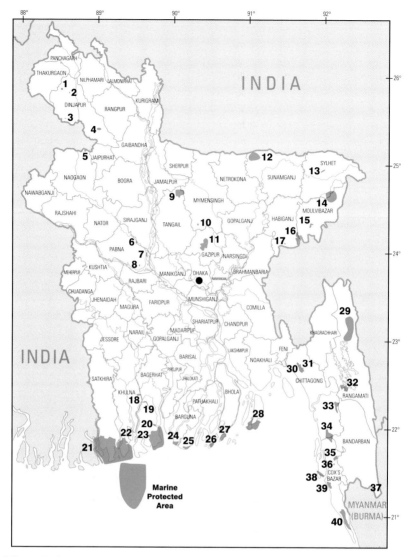

Map 2. Protected and conservation areas in Bangladesh (NP = National Park; WS = Wildlife Sanctuary; ECA = Ecologically Critical Area).

1   Singra NP
2   Birganj NP
3   Ramsagar NP
4   Nababgonj NP
5   Altadighi NP
6   Silonda-Nagdemra WS
7   Nagarbari-Mohonganj WS
8   Nazirganj WS
9   Modhupur NP
10 Kadigarh NP
11 Bhawal NP
12 Tanguar Haor ECA
13 Khadimnagar NP
14 Hakaluki Haor ECA

15 Lawachara NP
16 Rema Kalenga WS
17 Satchari NP
18 Dhangmari WS
19 Chandpai WS
20 Dudmukhi WS
21 Sundarban West WS
22 Sundarban South WS
23 Sundarban East WS
24 Tengragiri WS
25 Kuakata NP
26 Sonar Char WS
27 Char Kukri-Mukri WS
28 Nijhum Dwip NP

29 Pablakhali WS
30 Baraiyadhala NP
31 Hazarikhil WS
32 Kaptai NP
33 Dudpukuria Dhopachari WS
34 Chunati WS
35 Fashiakhali WS
36 Medakacchapia NP
37 Sangu WS
38 Sonadia ECA
39 Himchari NP
40 Teknaf WS

Protected areas

As of December 2019, the country has 39 protected areas, including 17 national parks, 21 wildlife sanctuaries and one 'special biodiversity conservation area', covering nearly 17.5% of the forest area and 1.8% of total land area (see Map 2), plus a further six smaller 'eco-parks'. All but four of these protected areas are governed and managed by the Forest Department. The designation of an area as a wildlife sanctuary means that the capturing, killing, shooting or trapping of wildlife is prohibited. National parks have similar restrictions on exploitation, are considered to be of outstanding scenic and natural beauty, and allow public access for education, research and recreation. In practice, there is little difference in management between the two categories, and basic visitor facilities have been developed in many protected areas of all types. Since the early 2000s collaborative management between the Forest Department and local communities has been adopted officially in many of these protected areas, and community groups are involved in protection and nature-related enterprises. The eco-parks differ in having recreational use as a major objective but do help to protect some smaller areas of accessible but usually degraded or planted forest.

Ecologically Critical Areas

As of December 2019, the country has 13 Ecologically Critical Areas. All of these are wetlands or coastal sites, and they range in area from 292,926ha (Sundarbans buffer area) to just 101ha (Gulshan–Baridhara Lake). All have ecological or human value and are threatened by over-exploitation, development pressure and/or pollution. These are multiple-use areas that include private as well as public lands. The designation is intended to limit changes in use and to prevent: the cutting of natural trees, hunting, damage to natural habitats and the operation of polluting industries. Some are waterways bordering Dhaka, but six are of high natural ecological importance and of interest for birds, and these are shown in Map 2.

In addition, in some other wetlands, and within some of the Ecologically Critical Areas, there are several smaller sanctuary areas protected by community groups and recognised/supported by local government.

Important Bird and Biodiversity Areas

There is a global effort by BirdLife International to identify, monitor and protect a network of sites for the conservation of the world's birds and other biodiversity. Important Bird and Biodiversity Areas (IBAs) are areas that (1) hold significant numbers of one or more globally threatened species; and/or (2) are one of a set of sites that together hold a suite of restricted-range species or biome-restricted species; and/or (3) have exceptionally large numbers of migratory or congregatory species. The first list of sites was published in 2004 (BirdLife International 2004) with minor amendments since. Twenty IBAs are currently recognised in Bangladesh and are shown on the map on page 32. Twelve of these IBAs are also either protected areas or Ecologically Critical Areas.

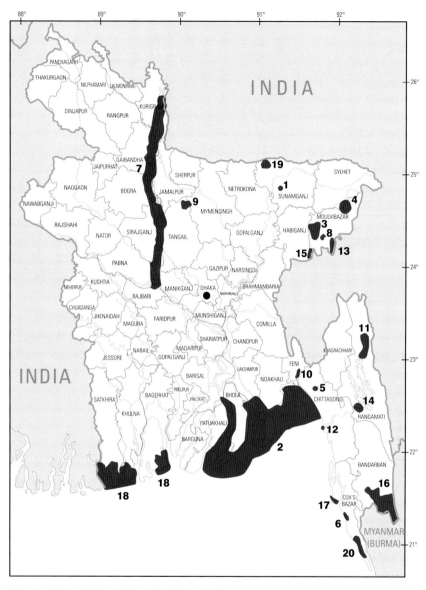

Map 3. Important Bird and Biodiversity Areas (IBAs) in Bangladesh.

1   Aila Beel
2   Ganges–Brahmaputra–Meghna Delta
3   Hail Haor
4   Hakaluki Haor
5   Hazarikhil Wildlife Sanctuary
6   Himchari National Park
7   Jamuna-Brahmaputra River
8   Lawachara/West Bhanugach Reserved Forest
9   Madhupur National Park
10 Muhuri Dam

11 Pablakhali Wildlife Sanctuary
12 Patenga Beach
13 Rajkandi Reserved Forest
14 Kaptai National Park
15 Rema Kalenga Wildlife Sanctuary
16 Sangu Matamuhari Reserved Forest
17 Sonadia Island
18 Sunderbans (East, South and West Wildlife Sanctuaries)
19 Tanguar Haor and Panabeel
20 Teknaf Wildlife Sanctuary

NATIONAL ORGANISATIONS

BANGLADESH BIRD CLUB

Rehena Manzil (5th floor), Cha 63/6 Progoti Swarani North Badda, Dhaka 1212. Email: enamuh@gmail.com
The Bangladesh Bird Club (BBC) promotes the appreciation and study of birds and their conservation. It holds monthly meetings in Dhaka, produces a quarterly magazine, organises exhibitions and other events, and conducts annual waterbird counts and other surveys. Since 2010 it has coordinated, with authorisation from the Bangladesh Forest Department, a national bird ringing programme and organised the training of bird-ringers. It played an important role in the banning of the veterinary drug diclofenac.

REGIONAL BIRD CLUBS

Several more informal local or regional bird clubs have started up in the 21st century, including ones in Chittagong, Rajshahi and Sherpur. These are mainly focused on bird photography and birding within those districts.

JAHANGIRNAGAR UNIVERSITY

Many of the few active bird researchers in Bangladesh studied zoology here. Each January, it organises a bird fair on its campus, which is a good birding site, and awards a 'big bird' prize for a notable photographed rarity documented in the previous year.

INTERNATIONAL ORGANISATIONS

BIRDLIFE INTERNATIONAL

birdlife.org
The David Attenborough Building, Pembroke St, Cambridge CB2 3QZ, UK. Email: birdlife@birdlife.org
A global partnership of conservation organisations (NGOs) that strives to conserve birds, their habitats and global biodiversity, working with people towards sustainability in the use of natural resources. There are 115 BirdLife Partners worldwide – typically one per country or territory – and this number is growing. BirdLife is the world's leading authority on the status of the world's birds, their habitats and the urgent problems facing them. The BirdLife Partnership has made a commitment to: prevent extinction in the wild; maintain and where possible improve the conservation status of all bird species; conserve the sites and habitats important for birds and other biodiversity; sustain the vital ecological systems that underpin human livelihoods, enrich the quality of people's lives, and in the process empower people, contribute to the alleviation of poverty and strive to ensure sustainability in the use of natural resources.

ORIENTAL BIRD CLUB (OBC)

orientalbirdclub.org
PO Box 324, Bedford MK42 0WG, UK. Email: mail@orientalbirdclub.org
OBC's aims are to: encourage an interest in wild birds of the Oriental region and their conservation; promote the work of regional bird and nature societies; and collate and publish information on Oriental birds.

IUCN BANGLADESH

The International Union for Conservation of Nature (IUCN) is a global partnership with its headquarters in Switzerland. The Bangladesh office and national programme were established in the early 1990s. A major activity was developing an updated national Red Data Book, which was completed in 2015. Its main bird-related initiatives include vulture conservation (supporting two 'Vulture Safe Zones', controlling lethal veterinary drugs and rehabilitating vultures captured by villagers) and research on waterfowl, including satellite tracking.

GLOSSARY

See also the section on plumage terminology (pages 11–12), which covers bird topography.

Arboreal: tree-dwelling.

Axillaries: the feathers in the 'armpit' at the base of the underwing.

Beel: small and medium-sized depression that holds water for all or much of the year.

Cap: a well-defined patch of colour or bare skin on the top of the head.

Carpal: the bend of the wing, or carpal joint.

Carpal patch: a well-defined patch of colour on the underwing in the vicinity of the carpal joint.

Casque: an enlargement on the upper surface of the bill, in front of the head, as on hornbills.

Char: a sandy-silty island in major river or coastal delta.

Cere: a fleshy (often brightly coloured) structure at the base of the bill containing the nostrils.

Culmen: the ridge of the upper mandible.

Eclipse plumage: a female-like plumage acquired by males of some species (e.g. ducks and some sunbirds) during or after breeding.

Edgings or edges: outer feather margins, which can frequently result in distinct paler or darker panels of colour on the wings or tail.

Flight feathers: the primaries, secondaries and tail feathers (although frequently used to denote the primaries and secondaries alone).

Fringes: complete feather margins, which can frequently result in a scaly appearance to body feathers or wing-coverts.

Gape: the mouth and fleshy corner of the bill, which can extend back below the eye.

Graduated tail: a tail in which the longest feathers are the central pair and the shortest the outermost, with those in between intermediate in length.

Gregarious: living in flocks or communities.

Gular pouch: a loose and pronounced area of skin extending from the throat (e.g. as in hornbills).

Gular stripe: a usually very narrow (and often dark) stripe running down the centre of the throat.

Hackles: long and pointed neck feathers which can extend across the mantle and wing-coverts (e.g. on junglefowl).

Hand: the outer part of the wing, from the carpal joint to the tip.

Haor: large saucer-shaped depression bounded by natural river levees or low hills in the north-east; completely inundated during the monsoon.

Hepatic: used with reference to the rufous-brown morph of some (female) cuckoos.

Iris (plural, **irides**): the coloured membrane surrounding the pupil of the eye, which can be brightly coloured.

Lappet: a wattle, particularly one at the gape.

Leading edge: the front edge of the forewing.

Local: occurring or common within a small or restricted area.

Mandible: the lower or upper half of the bill.

Mask: a dark area of plumage surrounding the eye and often covering the ear-coverts.

Morph: a distinct plumage type which occurs alongside one or more other distinct plumage types exhibited by the same species.

Nomenclature: the scientific naming of species and subspecies, and of the genera, families and other categories in which species may be classified.

Nominate: the first-named race of a species, which has its scientific racial name the same as the specific name.

Nuchal: relating to the hindneck, used with reference to a patch or collar.

Ocelli: eye-like spots of iridescent colour; a distinctive feature in the plumage of peafowls.

Orbital ring: a narrow circular ring of feathering or bare skin surrounding the eye.

Primary projection: the extension of the primaries beyond the longest tertial on a closed wing; this can be of critical importance in identification (e.g. of larks or *Acrocephalus* warblers).

Race (subspecies): a geographical population whose members all show constant differences (e.g. in plumage or size) from those of other populations of the same species.

Rectrices (singular, **rectrix**): the tail feathers.

Remiges (singular, **remex**): the primaries and secondaries.

Rictal bristles: bristles, often prominent, at the base of the bill.

Shaft streak: a fine line of pale or dark colour in the plumage produced by the feather shaft.

Speculum: the often glossy panel across the secondaries of (especially) dabbling ducks, frequently bordered by pale tips to these feathers and a greater-covert wingbar.

Subspecies: see race.

Subterminal band: a dark or pale band, usually broad, situated inside the outer part of a feather or feather tract (used particularly in reference to the tail).

Taxonomy: the science of classification of species, subspecies, genera, families and other categories in which species may be classified.

Terminal band: a dark or pale band, usually broad, at the tip of a feather or feather tract (especially the tail); cf. subterminal band.

Terrestrial: living or occurring mainly on the ground.

Trailing edge: the rear edge of the wing, often darker or paler than the rest of the wing; cf. leading edge.

Vent: the area around the cloaca (anal opening), just behind the legs (should not be confused with the undertail-coverts).

Vermiculated: marked with narrow wavy lines, usually visible only at close range.

Wattle: a lobe of bare, often brightly coloured, skin attached to the head (frequently at the bill-base), as on mynas and wattled lapwings.

Wingbar: generally a narrow, well-defined dark or pale bar across the upperwing, and often referring to a band formed by pale tips to the greater or median coverts (or both, as in 'double wingbar').

Wing-linings: the entire underwing-coverts.

Wing panel: a pale or dark band across the upperwing (often formed by pale edges to the remiges or coverts), broader and generally more diffuse than a wingbar.

FAMILY SUMMARIES

ORDER: GALLIFORMES

PARTRIDGES, PHEASANTS AND ALLIES Phasianidae

Stout-bodied birds with short, stout bill and short, rounded wings. They nest on the ground, but many species roost in trees at night. They are good runners, often preferring to escape on foot rather than taking to the air. Their flight is powerful and fast but, except in the case of the migratory quail, it cannot be sustained for long periods. Typically, they forage by scratching the ground with strong feet to expose food hidden among dead leaves or in the soil. They mainly eat seeds, fruit, buds, roots and leaves, complemented by invertebrates.

ORDER: ANSERIFORMES

WHISTLING-DUCKS, GEESE AND DUCKS Anatidae

Stocky with short legs and webbed feet. Bill is short, flattened and rounded at tip. Aquatic and highly gregarious, typically migrating, feeding, roosting and resting together, often in mixed flocks. Most species are chiefly vegetarian when adult, feeding on seeds, algae, plants and roots, often supplemented by aquatic invertebrates. Their main foraging methods are diving, surface-feeding or dabbling, and grazing. They also upend, wade, filter and sieve water and debris for food, and probe with the bill. They have a direct flight with sustained fast wingbeats, and characteristically often fly in V-formation.

ORDER: PODICIPEDIFORMES

GREBES Podicipedidae

Aquatic birds adapted for diving from the surface and swimming underwater to catch fish and aquatic invertebrates. Their strong legs are placed near the rear of their almost tail-less body, and their feet are lobed. In flight, grebes have an elongated appearance, with the neck extended and the feet hanging lower than the humped back. They usually feed singly but may form loose congregations outside the breeding season.

ORDER: PHOENICOPTERIFORMES

FLAMINGOS Phoenicopteridae

Large wading birds with a long neck, very long legs, webbed feet and pink plumage. The angled bill is highly specialised for filter-feeding. Flamingos often occur in huge numbers and are found mainly on salt lakes and lagoons. Greater Flamingo is the only member of the family recorded as a vagrant in Bangladesh.

ORDER: PHAETHONTIFORMES

TROPICBIRDS Phaethontidae

Aerial seabirds with long wings and elongated central tail feathers. They range over tropical and subtropical waters, and nest mainly on oceanic and offshore islands. Graceful and pigeon-like flight with flapping and circling alternating with long glides. Usually solitary but may congregate with flocks of feeding terns. They feed by first hovering to locate prey (mainly fish and squid) and then plunge-diving on half-closed wings. Red-billed Tropicbird is the only member of the family recorded in Bangladesh offshore waters.

ORDER: COLUMBIFORMES
PIGEONS AND DOVES Columbidae

Pigeons and doves have a stout, compact body, a rather short neck, and a small head and bill. Their flight is swift and direct, with fast wingbeats. Most species are gregarious outside the breeding season. Seeds, fruits, buds and leaves form their main diet, but many species also eat small invertebrates. They have soft plaintive cooing or booming voices that are often monotonously repeated.

ORDER: CAPRIMULGIFORMES
FROGMOUTHS Podargidae

Frogmouths have the same cryptic colouring, soft plumage, wide gape and nocturnal habits as nightjars, but differ in some of their habitats. They are more arboreal than nightjars, nesting and roosting in trees, and hunting from them at night by pouncing on prey. Hodgson's Frogmouth is the only member of the family recorded in Bangladesh but its status is a mystery.

NIGHTJARS Caprimulgidae

Small to medium-sized birds with long, pointed wings, and gaping mouths with long bristles that help them catch insects in flight. Nightjars are crepuscular and nocturnal in habit, with soft owl-like, cryptically patterned plumage. By day they perch on the ground or lengthwise on a branch and are difficult to detect. They eat flying insects that are caught on the wing. Typically, they fly erratically to and fro over and among vegetation occasionally wheeling, gliding and hovering to pick insects from foliage. Most are easily located by their calls.

TREESWIFTS Hemiprocnidae

Treeswifts have long wings and forked tails. Unlike 'true' swifts, they will perch on exposed branches. Like 'true' swifts, they have short bills with a wide gape and very short legs. They spend much time on the wing, catching flying insects. Crested Treeswift is the only member of the family recorded as a vagrant in Bangladesh.

SWIFTS Apodidae

Swifts have long pointed wings, compact bodies, short bills with a wide gape and very short legs. They spend most of the day swooping and wheeling in the sky with great agility and grace. Typical swift flight is a series of rapid, shallow wingbeats interspersed with short glides. They feed entirely in the air, drink and bathe while swooping low over the water, and regularly pass the night in the air. Swifts eat mostly tiny insects, caught by flying back and forth among aerial concentrations of these with their large mouths open; they also pursue individual insects.

ORDER: CUCULIFORMES
CUCKOOS, MALKOHAS AND COUCALS Cuculidae

Cuckoos have elongated bodies with a fairly long neck, a tail varying from medium length to long and graduated, and a quite long, downcurved bill. Almost all cuckoos

are arboreal. They eat hairy caterpillars. Male cuckoos of most species are very noisy in the breeding season, calling frequently during the day, especially if cloudy, and often into the night. When not breeding, they are silent and unobtrusive, and as a result their status and distribution at this season are very poorly known. Cuckoos are notorious for their nest parasitism.

Malkohas are larger than other cuckoos, plumper-bodied with stouter bills and very long, graduated tails. They are usually seen singly or in pairs in the middle storey of forest. Malkohas raise their own young. Green-billed Malkoha is the only species found in Bangladesh.

Coucals are large, skulking birds with long graduated tails and weak flight. They are terrestrial, frequenting dense undergrowth, bamboo, tall grassland or scrub jungle. Coucals eat small animals and invertebrates.

ORDER: GRUIFORMES

FINFOOTS Heliornithidae

Medium-sized aquatic birds with a comparatively long, thick neck, the bill is thick and tapering, the toes have wide lobes and the tail is relatively long and stiff. They swim, dive and run well, but rarely fly. There is only one species in Bangladesh, Masked Finfoot.

RAILS, CRAKES, GALLINULES AND COOTS Rallidae

Small to medium-sized birds, with moderate to long legs for wading and short, rounded wings. With the exception of Common Moorhen and Eurasian Coot, which spend much time swimming in the open, rails are mainly terrestrial. Many occur in marshes. They fly reluctantly and feebly, with legs dangling, for a short distance and then drop into cover again. However, some species, such as Eurasian Coot, are long-distance migrants. Most are heard more often than seen and are generally noisy at dusk and at night. Their calls consist of strident or raucous repeated notes. They eat insects, crustaceans, amphibians, fish and vegetable matter.

CRANES Gruidae

Stately, long-necked and long-legged, cranes have a tapering body and long inner secondaries which hang over the tail. Their flight is powerful, with the head and neck extended forwards and the feet stretched out behind. Flocks of cranes often fly in V-formation and sometimes soar at considerable heights. Most cranes are gregarious outside the breeding season, and flocks are often very noisy. Cranes have a characteristic resonant and far-reaching musical, trumpet-like call. They take a wide variety of plant and animal food. The bill is used to probe and dig for plant roots and to graze and glean vegetable material above the ground. Both sexes have a spectacular and beautiful dance that takes place throughout the year.

ORDER: OTIDIFORMES

BUSTARDS Otididae

Medium-sized to large terrestrial birds of extensive grasslands. They have fairly long legs, stout bodies, long necks and crests, and neck plumes, which are exhibited in display. The wings are broad and long, and in flight the neck is outstretched. Their flight is powerful and can be very fast. When feeding, bustards have a steady, deliberate gait. They are more or less omnivorous, and feed opportunistically on large insects, such as grasshoppers and locusts, young birds, shoots, leaves, seeds and fruits. Males perform elaborate and spectacular displays in the breeding season. Sadly, the two species that occurred in the past were extirpated in Bangladesh, probably by the early 20th century.

ORDER: PROCELLARIIFORMES

SHEARWATERS AND PETRELS Procellariidae

Long-winged, marine species that come to shore only to breed. They feed on zooplankton, squid, fish and offal, seized on or below the water surface. Gregarious, often gathering in flocks at food concentrations. Typically, they fly by a combination of rapid, rather stiff wingbeats interspersed with long glides (gliding or 'shearing' being more pronounced in strong winds). Have 'tubenoses' through which they are able to expel salt. Short-tailed Shearwater is the only member of the family recorded in Bangladesh (as a vagrant).

ORDER: CICONIIFORMES

STORKS Ciconiidae

Large or very large birds with long bills, necks and legs, long and broad wings and short tails. In flight, the legs are extended and the neck is outstretched. They have powerful, slow-flapping flight and frequently soar for long periods, often at great heights. They capture fish, frogs, snakes, lizards, large insects, crustaceans and molluscs while walking slowly in marshes, at the edge of lakes and rivers, and in grassland.

ORDER: PELECANIFORMES

IBISES AND SPOONBILL Threskiornithidae

Large birds with long necks and legs, partly webbed feet, and long, broad wings. Ibises have long, downcurved bills and forage by probing in shallow water, mud and grass. Spoonbills have long spatulate bills and catch floating prey in shallow water. Only one species occurs in the region, Eurasian Spoonbill.

HERONS AND BITTERNS Ardeidae

Medium-sized to large birds with long legs for wading. The diurnal herons (including egrets) have slender bodies and long heads and necks; the night herons are squatter, with shorter necks and legs. They fly with leisurely flaps, with the legs outstretched and projecting beyond the tail, and nearly always with the neck and head drawn back. They frequent marshes and the shores of lakes and rivers. Typically, herons feed by standing motionless at the water's edge, waiting for prey to swim within reach, or by slow stalking in shallow water or on land.

Bitterns usually skulk in reed swamps, although occasionally one may forage in the open, and they can clamber around in tangled stems with surprising agility. Normally they are solitary and crepuscular, and they are most often seen flying low over tall swamp-thickets with slow wingbeats, soon dropping into cover again. When in danger, bitterns freeze, pointing the head and neck upward and compressing their feathers so that the whole body appears elongated. Bitterns are characterised by their booming territorial calls. Herons and bitterns feed on a wide variety of aquatic prey.

PELICANS Pelecanidae

Large aquatic, gregarious, fish-eating birds. The wings are long and broad, and the tail is short and rounded. They have characteristic long, straight, flattened bills, hooked at the tip, and with a large expandable pouch suspended from the lower mandible. Pelicans often fish cooperatively by swimming forward in a semicircular formation, driving fish into shallow water; each bird then scoops fish into its pouch before swallowing the food. Pelicans fly together in V-formation or in lines, and often soar for considerable periods in thermals. They are powerful fliers, proceeding by steady flaps and with the head drawn back between the shoulders. When swimming, the closed wings are typically held above the back. Formerly probably resident, these impressive birds are now only vagrants to Bangladesh.

ORDER: SULIFORMES

FRIGATEBIRDS Fregatidae

Large aerial seabirds that rarely land on water, and roost and nest in trees and bushes. They are agile in the air and can soar for long periods. Noted for chasing other seabirds, especially boobies, until they drop or disgorge their food, frigatebirds also capture their own prey by diving to the water surface. They are chiefly storm-driven visitors to the coast, typically during cyclones. Lesser Frigatebird is the only member of the family recorded in Bangladesh.

BOOBIES Sulidae

Large seabirds. They forage on the wing, scanning the sea, and on sighting fish or squid they plunge-dive at an angle. Flight is direct, with alternating periods of flapping and gliding. Masked Booby is the only member of the family recorded as a vagrant in Bangladesh marine waters.

CORMORANTS Phalacrocoracidae

Medium-sized to large aquatic birds. They are long-necked, with hook-tipped bills of moderate length and long, stiff tails. Cormorants swim with the body low in the water, the neck straight and the head and bill pointing a little upwards. They eat mainly fish, which are caught by underwater pursuit. In flight, the neck is extended and the head is held slightly above the horizontal. Typically they perch for long periods on trees, posts or rocks in an upright posture with spread wings and tail.

DARTERS Anhingidae

Large aquatic birds adapted for hunting fish underwater. Darters have long, slender necks and heads, long wings and very long tails. Only one species in the family has been recorded in the region, Oriental Darter.

ORDER: CHARADRIIFORMES

THICK-KNEES Burhinidae

Medium-sized to large waders with long legs and short, stout bills. Mainly crepuscular or nocturnal, with cryptically patterned plumage. They eat invertebrates and small animals.

OYSTERCATCHERS Haematopodidae

Oystercatchers are waders that usually inhabit the seashore and are only vagrants inland. They have all-black or black-and-white plumage. The bill is long, stout, orange-red and adapted for opening the shells of bivalve molluscs. Eurasian Oystercatcher is the only family member recorded in the region.

STILTS AND AVOCETS Recurvirostridae

Stilts and avocets are waders that have characteristic long bills, and longer legs in proportion to their body than any other birds except flamingos. They inhabit marshes, lakes and pools. Black-winged Stilt is the only stilt and Pied Avocet is the only avocet recorded in Bangladesh.

PLOVERS AND LAPWINGS Charadriidae

Plovers and lapwings are small to medium-sized waders with rounded heads, short necks and short bills. Typically, they forage by running in short spurts, pausing and standing erect, then stooping to pick up invertebrate prey. Their flight is swift and direct.

PAINTED-SNIPES Rostratulidae

Waders that frequent marshes and superficially resemble snipes but have spectacular plumages. The bill is long, and the wings are short and broad. The female is more brightly coloured than the male as the male takes care of the nest and young. Greater Painted-snipe is the only species recorded in Bangladesh.

JACANAS Jacanidae

Jacanas characteristically have very long toes, which enable them to walk over floating vegetation. They inhabit freshwater lakes, ponds and marshes. Like painted-snipes, the females leave parenting duties to the males after laying eggs.

SNIPES, CURLEWS, SANDPIPERS AND STINTS Scolopacidae

Woodcocks and snipes are small to medium-sized waders with very long bills, fairly long legs and cryptically patterned plumage. If approached, they usually crouch on the ground and 'freeze', preferring to rely on their plumage pattern to escape detection. They generally inhabit marshy ground. Godwits and curlews are wading birds with quite long to very long legs and a long bill. Sandpipers and stints are small to medium-sized, rather plump waders, with short, medium or longish bills and short (stints) or medium-long legs. All species feed mainly by probing with their bills in soft substrates and also by picking from the surface. Their diet consists mostly of small invertebrates.

BUTTONQUAILS Turnicidae

These small, plump terrestrial birds are now placed in the order Charadriiformes having previously been regarded as a family in the order Gruiformes. They are found in a wide variety of habitats with a dry, often sandy substrate and low ground cover under which they can readily run or walk. Buttonquails are very secretive and fly with great reluctance, with weak whirring beats low over the ground, dropping quickly into cover. They feed on grass and weed seeds, grain, greenery and small insects, picking food from the ground or scratching with the feet.

CRAB-PLOVER Dromadidae

A mainly white wader. The distinctively shaped thick black bill is adapted for preying on crabs and other crustaceans which it hunts chiefly on coastal mudflats and reefs. Usually found singly, in pairs or in small parties, hundreds may occur at traditional roost sites. Mainly crepuscular. There is only one species in the family.

PRATINCOLES Glareolidae

Pratincoles have arched and pointed bills, wide gapes and long, pointed wings. Most pratincoles are short-legged. They catch most of their prey in the air, although they also feed on the ground. All pratincoles live near water.

GULLS, TERNS AND SKIMMERS Laridae

Gulls are medium-sized to large birds with relatively long, narrow wings, usually a stout bill, moderately long legs and webbed feet. Immatures are brownish and cryptically patterned. In flight, gulls are graceful and soar easily on updraughts. All species swim buoyantly and well. They are highly adaptable, and most species are opportunistic feeders with a varied diet, including invertebrates. Most are gregarious.

Terns are small to medium-sized aerial birds with gull-like bodies, but they are generally more delicately built. The wings are long and pointed, typically narrower than those of gulls, and their flight is buoyant and graceful. Terns are highly vocal, and most species are gregarious. *Sterna* terns generally have deeply forked tails. They mainly eat small fish and crabs caught by hovering and then plunge-diving from the air, often submerging completely, but they also pick prey from the water surface.

Skimmers are distinguished by their long, strong scissor-like bills with elongated lower mandibles. They feed by skimming the water surface with the bill open and lower mandible partly immersed to snap up fish. Indian Skimmer is the only species in the region.

JAEGERS Stercorariidae

Aerial seabirds, with a strong, hooked bill, long, pointed wings, short legs and webbed feet. Jaegers feed by chasing other seabirds, especially terns, until they drop or disgorge their food. They are usually found in marine waters, some distance from land, but are occasionally found inshore and may occur inland after cyclonic storms.

ORDER: STRIGIFORMES

BARN-OWLS Tytonidae

Large-headed with a heart-shaped facial disc, long legs and strong feet. Habits are similar to the typical owls. Common Barn-owl is the only member of the family recorded in Bangladesh.

TYPICAL OWLS Strigidae

Owls have a large, rounded head, big forward-facing eyes surrounded by a broad facial disc, and a short tail. Most are nocturnal and cryptically coloured and patterned, making them inconspicuous when resting during the day. When hunting, owls either quarter the ground or scan and listen for prey from a perch. Their diet consists of small animals and invertebrates. Owls are usually located by their distinctive and often weird calls, which are diagnostic of the species and advertise their presence and territories.

ORDER: ACCIPITRIDAE

OSPREY Pandionidae

A specialised raptor similar to other hawks. They are adapted to predate fish, having a reversible outer toe and spiny toe pads to help grasp prey. They hover over open water and plunge-dive, feet first, to catch fish. The wings are long and narrow, and the tail is short. The female is larger than the male but otherwise the sexes are similar in appearance. There is only one species in the family.

HAWKS, EAGLES, HARRIERS AND VULTURES ETC. Accipitridae

A large and varied family of raptors, ranging from the Besra to the huge Himalayan Griffon in size. In most species, the vultures being an exception, the female is larger than the male and is often duller and brownish. Members of the family feed on mammals, birds, reptiles, amphibians, fish, crabs, molluscs and insects – dead or alive. All have a hooked, sharp-tipped bill and very acute sight, and all except the vultures have powerful feet with long, curved claws. They frequent all habitat types, ranging from dense forests, grasslands and mountains to fresh waters.

ORDER: TROGONIFORMES

TROGONS Trogonidae

Brightly coloured, short-necked, medium-sized birds with a long tail, short rounded wings and a rather short, broad bill. Usually found alone or in widely separated pairs. Characteristically, they perch almost motionless in an upright posture for long periods in the middle or lower storey of dense forests. Trogons are insectivorous but also eat leaves and berries. They capture flying insects on the wing when moving from one vantage point to another, twisting with the agility of a flycatcher. Red-headed Trogon is the only member of the family recorded in Bangladesh.

ORDER: BUCEROTIFORMES

HORNBILLS Bucerotidae

Medium-sized to large birds with a massive bill and a variable-sized casque. Mainly arboreal, feeding chiefly on wild figs (*Ficus*), berries and drupes, supplemented by small animals and insects. Flight is powerful and slow, and for most species consists of a few wingbeats followed by a sailing glide with the wingtips upturned. In all but the smaller species, the wingbeats make a distinctive loud puffing sound, audible over some distance. Hornbills often fly one after another in a follow-my-leader fashion. Usually found in pairs or small parties, sometimes in flocks of up to 30 or more where food is abundant.

HOOPOES Upupidae

Hoopoes have a distinctive appearance with long downcurved bills, short legs and rounded wings. They are insectivorous and forage by pecking and probing the ground. Flight is undulating, slow and butterfly-like. Common Hoopoe is the only species occurring in the region.

ORDER: CORACIIFORMES

BEE-EATERS Meropidae

Brightly coloured birds with a long downcurved bill, pointed wings and very short legs. They catch large flying insects on the wing by making short, swift sallies like a flycatcher from an exposed perch such as a treetop, branch, post or telegraph wire; insects are pursued in a lively chase with swift, agile flight. Some species also hawk insects in flight like swallows. Most species are sociable. Their flight is graceful and undulating, a few rapid wingbeats followed by a glide.

ROLLERS Coraciidae

Stoutly built, medium-sized birds with a large head and short neck. They mainly eat large insects. Typically, they occur alone or in widely spaced pairs. Flight is buoyant, with rather rapid deliberate wingbeats.

KINGFISHERS Alcedinidae

Small to medium-sized birds, with a large head, long strong bill and short legs. Most kingfishers spend long periods perched singly or in well-separated pairs, watching intently before plunging swiftly downwards to seize prey with the bill; they usually return to the same perch. They eat mainly fish, tadpoles and invertebrates; larger species also consume frogs, snakes, crabs, lizards and rodents. Their flight is direct and strong, with rapid wingbeats, often close to the water surface.

ORDER: PICIFORMES

ASIAN BARBETS Megalaimidae

Stocky, short-tailed birds with a stout bill. Arboreal, and usually found in treetops. Despite their bright coloration, they can be very difficult to see, especially when silent, their plumage blending remarkably well with the foliage. They often sit motionless for long periods. Barbets call persistently and monotonously in the breeding season, sometimes throughout the day; outside the breeding season, they are usually silent. They are chiefly frugivorous, many species favouring figs (*Ficus*). Their flight is strong and direct, with deep woodpecker-like undulations.

WRYNECK, PICULETS AND WOODPECKERS Picidae

Chiefly arboreal, and usually seen clinging to, or climbing up, vertical trunks and lateral branches. Typically, they work up trunks and along branches in jerky spurts, directly or in spirals. Some species feed regularly on the ground, searching mainly for termites and ants. Most have powerful bills, for boring into wood to extract insects and for excavating nest-holes. Woodpeckers feed chiefly on ants, termites, grubs and the pupae of wood-boring beetles. Most woodpeckers also hammer rapidly against tree trunks with their bill, producing a loud rattle, known as 'drumming', which is used to advertise and defend their territories. Their flight is strong and direct, with marked undulations. Many species can be located by their characteristic loud calls. Piculets are tiny woodpeckers with similar habits. Eurasian Wryneck is unusual as it behaves more like a passerine, but often twists its neck to look over its back.

ORDER: FALCONIFORMES

FALCONS Falconidae

Small to medium-sized birds of prey which resemble members of the family Accipitridae in having a hooked bill, sharp, curved talons, and remarkable powers of sight and flight. Like other raptors, they are mainly diurnal, although a few are crepuscular. Some falcons kill flying birds in a surprise attack, often by stooping at great speed (e.g. Peregrine); others hover and then swoop on prey on the ground (e.g. Common Kestrel) and several species hawk insects in flight (e.g. Amur Falcon).

ORDER: PSITTACIFORMES

PARROTS AND PARAKEETS Psittacidae

Parrots have a short neck and short, stout, hooked bill with the upper mandible strongly curved and overlapping the lower mandible. Most parrots are noisy and highly gregarious. They associate in family parties and small flocks and gather in large numbers at concentrations of food, such as paddy-fields. Their diet is almost entirely vegetarian: fruit, seeds, buds, nectar and pollen. The flight of *Psittacula* parakeets is swift, powerful and direct. The hanging-parrots are much smaller with short tails lacking streamers. They habitually sleep upside-down.

ORDER: PASSERIFORMES

PITTAS Pittidae

Brilliantly coloured, terrestrial, forest passerines. They are of medium size, stocky and long-legged, with a short, square tail, stout bill and an erect carriage. Most of their time is spent foraging for invertebrates on the forest floor, flicking leaves and other vegetation, and probing with their strong bill into leaf litter and damp earth. Pittas usually progress on the ground by long hopping bounds. Typically, they are skulking and are often most easily located by their high-pitched whistling calls or songs. They sing in trees or bushes.

TYPICAL BROADBILLS Eurylaimidae

Small to medium-sized plump birds with rounded wings and short legs, most species having a distinctively broad bill. Typically, they inhabit the middle storey of forest and feed mainly on invertebrates gleaned from leaves and branches. Broadbills are active when foraging but are often unobtrusive and lethargic at other times.

OLD WORLD ORIOLES Oriolidae

Medium-sized colourful passerines with a short, stout bill. Arboreal and usually keep hidden in the leafy canopy. Orioles have beautiful, fluty, whistling songs and harsh grating calls. They are usually seen singly, in pairs or in family parties. Their flight is powerful and undulating, with fast wingbeats. They feed mainly on insects and fruit.

WHISTLERS Pachycephalidae

Whistlers are reminiscent of chats and flycatchers but are more strongly built with a thick rounded head, short thick neck and short heavy bill. They pick insects from branches and foliage but also flycatch. Mangrove Whistler is the only family member recorded in the region.

VIREOS AND ALLIES Vireonidae

Most vireo species are found in the Americas, but recent genetic studies have shown that White-bellied Erpornis (previously regarded as a yuhina but now in its own monospecific genus) and the *Pteruthius* shrike-babblers are best treated as members of this family. The only species found in Bangladesh is White-bellied Erpornis, a crested warbler-like forest dweller of the mid-storey. It often hangs tit-like upside-down.

CUCKOOSHRIKES, MINIVETS AND ALLIES Campephagidae

Cuckooshrikes are arboreal, insectivorous birds that usually remain high in trees. They are of medium size, with long pointed wings, moderately long, rounded tails and an upright carriage when perched.

Minivets are small to medium-sized, mostly brightly coloured passerines with moderately long tails and an upright stance when perched. They are arboreal, and feed on insects by flitting about in the foliage to glean prey from leaves, buds and bark, sometimes hovering in front of a sprig or making short aerial sallies. They usually remain in pairs in the breeding season and in small parties when not breeding. When feeding and in flight, they continually utter contact calls.

WOODSWALLOWS Artamidae

Plump birds with long pointed wings, a short tail and legs, and a wide gape. They feed on insects, usually captured in flight, and spend prolonged periods on the wing. They are sociable and perch close together on a bare branch or wire, and often waggle their tail from side to side. Ashy Woodswallow is the only member of the family recorded in Bangladesh.

VANGAS AND ALLIES Vangidae

Mainly a Malagasy and African family, it includes the flycatcher-shrikes *Hemipus* and woodshrikes *Tephrodomis*. Flycatcher-shrikes are small pied birds with arboreal flycatching habits and an upright stance when perched. Bar-winged Flycatcher-shrike is the only species of this genus recorded in Bangladesh. Woodshrikes are medium-sized, arboreal, insectivorous passerines. The bill is stout and hooked, the wings are rounded and the tail is short.

IORAS Aegithinidae

Ioras are a small group of lively passerines that feed in trees, mainly on insects and especially on caterpillars. Common Iora is the only species recorded in Bangladesh.

FANTAILS Rhipiduridae

Small, confiding, arboreal birds, perpetually on the move in search of insects. Characteristically, they erect and spread their tails like fans, and droop their wings, while pirouetting and turning from side to side with jerky, restless movements. When foraging, they flit from branch to branch, making frequent aerial sallies after winged insects. They call continually. Fantails are usually found singly or in pairs, and often join mixed foraging parties with other insectivorous birds. White-throated Fantail is the only species found in Bangladesh.

DRONGOS Dicruridae

Medium-sized passerines with characteristic black and often glossy plumage, long, often deeply forked tails, and a very upright stance when perched. They are arboreal and insectivorous, catching larger-winged insects by aerial sallies from a perch. Usually found singly or in pairs. Their direct flight is swift, strong and undulating. Drongos are rather noisy and have a varied repertoire of harsh calls and pleasant whistles; some species are good mimics.

MONARCHS Monarchidae

Most species are small to medium-sized birds, with a medium-length or long tail. They feed mainly on insects. Male paradise-flycatchers are notable for their very elongated central tail feathers. Black-naped Monarch and three species that were once treated as subspecies of Asian Paradise-flycatcher are the representatives of this family in Bangladesh.

SHRIKES Laniidae

Medium-sized, predatory passerines with a strong stout bill, hooked at the tip of the upper mandible, strong legs and feet, a large head and a long tail with graduated tips. Shrikes search for prey from a vantage point, such as the top of a bush or small tree or post. They swoop down to catch invertebrates or small animals on the ground or in flight. Over long distances their flight is typically undulating. Their calls are harsh, but most have quite musical songs and are good mimics. Shrikes typically inhabit open country with scattered bushes or light scrub.

CROWS, MAGPIES AND JAYS Corvidae

These are all robust perching birds which differ considerably from each other in appearance, but which have a number of features in common: a fairly long straight bill, very strong feet and legs, and a tuft of nasal bristles extending over the base of the upper mandible. The sexes are alike or almost alike in plumage. They are strong fliers. Most are gregarious, especially when feeding or roosting. Typically, they are noisy birds, uttering loud, discordant squawks, croaks or screeches. Crows are highly inquisitive and adaptable.

FAIRY FLYCATCHERS AND ALLIES Stenostiridae

Recent genetic studies have brought together a number of species, including Grey-headed Canary-flycatcher (the only species of the family found in Bangladesh) as a separate family. Grey-headed Canary-flycatcher is similar to a typical flycatcher in shape and behaviour, perching upright with frequent aerial sallies to catch winged insects.

TITS Paridae

Tits are small, active, highly acrobatic passerines with short bills and strong feet. Their flight over long distances is undulating. They are mainly insectivorous, although many species also depend on seeds, particularly from trees in winter, and some also eat fruit. They probe bark crevices, search branches and leaves, and frequently hang upside-down from twigs. Tits are chiefly arboreal, but also descend to the ground to feed, hopping about and flicking aside leaves and other debris. In the non-breeding season, most species join roving flocks of other insectivorous birds.

LARKS Alaudidae

Terrestrial, cryptically coloured passerines, which are generally small and often have a very elongated hind claw. They usually walk and run on the ground, and their flight is strong and undulating. Larks take a wide variety of food, including insects, molluscs, arthropods, seeds, flowers, buds and leaves. Many species have a melodious song, which is often delivered in a distinctive, steeply climbing or circling aerial display, but also from a conspicuous low perch. They live in a wide range of open habitats, including grassland and cultivation.

CISTICOLAS AND ALLIES Cisticolidae

Includes the cisticolas *Cisticola*, prinias *Prina* and tailorbirds *Orthotomus*.

Cisticolas are a group of tiny, short-tailed, insectivorous passerines. The tail is longer in winter than in summer. They are found in grassy habitats, and many have aerial displays.

Prinias have long, graduated tails, longer in winter than in summer. Most inhabit grassland, marsh vegetation or scrub. They forage by gleaning insects and spiders from vegetation, and some species also feed on the ground. When perched, the tail is often held cocked and slightly fanned. Their flight is weak and jerky.

Tailorbirds have a long, downcurved bill, short wings and a graduated tail, the latter held characteristically cocked. They sew leaves together to make a nest, hence their name.

REED-WARBLERS Acrocephalidae

Recent genetic work has established this as a separate family, including *Acrocephalus* and *Iduna* warblers as well as Thick-billed Warbler.

Acrocephalus warblers are medium-sized to large warblers with a prominent bill and a rounded tail. They usually occur singly. Many species are skulking, typically keeping low in dense vegetation. Most frequently found in marshy habitats, they clamber about readily on the vertical stems of reeds and other marsh plants. Their songs are harsh and often monotonous.

Iduna warblers are medium-sized warblers, with a large bill, a square-ended tail and a distinctive domed head shape with a rather sloping forehead and peaked crown. Their songs are harsh and varied. They clamber about vegetation rather clumsily.

CUPWINGS Pnoepygidae

Previously treated as a genus in a wider group of wren-babblers within a very large family of babblers Timaliidae, recent genetic studies have separated out this family. They are small, rotund, tail-less and ground-dwelling, with a proportionately large bill, legs and feet. Only one species, Pygmy Cupwing, has occurred in Bangladesh, presumably as a rare altitudinal migrant.

GRASSHOPPER-WARBLERS AND GRASSBIRDS Locustellidae

The family includes the *Locustella* warblers and grassbirds. Now included in *Locustella*, based on recent genetic studies, are several species previously regarded as *Bradypterus* bush-warblers.

Locustella warblers are very skulking, medium-sized warblers with rounded wings, usually found singly. Characteristically, they keep low down or on the ground among dense wetland vegetation, walking furtively and scurrying off when startled. They fly at low level, flitting between plants, or rather jerkily over longer distances, ending in a sudden dive into cover.

Grassbirds are brownish warblers with a longish tail. They inhabit damp tall grassland. Males perform song flights in the breeding season.

MARTINS AND SWALLOWS Hirundinidae

Gregarious, rather small passerines with a distinctive slender, streamlined body, long pointed wings and a small bill. The long-tailed species are often called swallows, and the shorter-tailed species termed martins. All hawk day-flying insects in swift, agile, sustained flight, sometimes high in the air. Many species have a deeply forked tail, which affords greater manoeuvrability. Hirundines catch most of their food while flying in the open. They readily perch on exposed branches and wires.

BULBULS Pycnonotidae

Medium-sized passerines with soft, fluffy plumage, rather short and rounded wings, a medium-long to long tail, slender bill and short, weak legs. Bulbuls feed on berries and other fruits, often supplemented by insects, and sometimes also nectar and the buds of trees and shrubs. Many species are noisy, especially when feeding. Typically, bulbuls have a variety of cheerful, loud, chattering, babbling and whistling calls. Most species are gregarious outside of the breeding season.

LEAF-WARBLERS Phylloscopidae

Now included in *Phylloscopus*, based on recent genetic studies, are several species previously regarded as being in a separate genus *Seicercus*.

Leaf-warblers are rather small, slim and short-billed warblers. Useful identification features are voice, strength of supercilium, colour of underparts, rump, bill and legs, and presence or absence of wingbars, of coronal bands or of white on the tail. The coloration of the upperparts and underparts and the presence or prominence of wing-bars are affected by wear. Leaf-warblers are fast-moving and restless, hopping and creeping about actively and often flicking their wings. They mostly glean small insects and spiders from foliage, twigs and branches, often first disturbing prey by hovering and fluttering; they also make short flycatching sallies.

BUSH-WARBLERS Scotocercidae

Recent genetic studies have separated out a number of disparate taxa and grouped them as bush-warblers, including the *Cettia* and *Horornis* bush-warblers, tesias, *Abroscopus* warblers and Mountain Tailorbird.

The bush-warblers are medium-sized warblers with rounded wings and tail that inhabit marshes, grassland and forest undergrowth. They are usually found singly. Bush-warblers call frequently and are usually heard more often than seen. *Cettia* species have surprisingly loud voices, and some can be identified by their distinctive melodious songs. *Tesia* species are almost tail-less and largely terrestrial. When excited, these birds flick their wings and tail.

OLD WORLD WARBLERS AND PARROTBILLS Sylviidae

A much-reduced family (previously including the Acrocephalidae, Locustellidae and Phylloscopidae warblers for example), now including only the *Sylvia* warblers, and

incorporating Yellow-eyed Babbler and the parrotbills (previously included in a much larger Timaliidae family or in their own family Paradoxornithidae).

The *Sylvia* warblers are small to medium-sized passerines with a fine bill. Typically, they inhabit bushes and scrub, and feed chiefly by gleaning insects from foliage and twigs; they sometimes also consume berries in autumn and winter.

The parrotbills have a stout bill, strong legs and feet, and a long tail. They frequent stands of bamboo and tall grasses, including within evergreen forest.

WHITE-EYES AND YUHINAS Zosteropidae

Recent taxonomic studies have grouped the yuhinas (previously included in the Timaliidae) with the white-eyes.

Yuhinas are crested passerines with a fine, pointed bill. Restless, they are often found in flocks and at times exhibit tit-like feeding behaviour. One species, Striated Yuhina, is recorded in Bangladesh.

White-eyes are small or very small insectivorous passerines with a slightly down-curved, pointed bill, brush-tipped tongue, and a white ring around each eye. White-eyes frequent forest edge and bushes in gardens. Indian White-eye is the only species recorded in Bangladesh.

SCIMITAR-BABBLERS AND ALLIES Timaliidae

Previously this family comprised a much larger group of babblers, but recent taxomomic work has separated out the Pnoepygidae (cupwings), Pellorneidae (ground babblers) and Leiotrichidae (laughingthrushes and allies). Of relevance to Bangladesh, the family currently includes the scimitar-babblers, as well as the *Stachyris* and *Cyanoderma* babblers, Pin-striped Tit-babbler and Chestnut-capped Babbler. They are small to medium-sized passerines, with soft, loose plumage, short or fairly short wings, and strong legs and feet. The sexes are alike. The scimitar-babblers have longish, downcurved bills.

GROUND BABBLERS Pellorneidae

A new family based on taxonomic studies, it comprises some species previously included in Timaliidae (e.g. the *Pellorneum* babblers) and also Swamp Grass-babbler (previously considered a prinia) and Indian Grass-babbler (previously placed in Cisticolidae). Varied morphologically and in habits and habitat, most species are skulking and favour undergrowth or dense, tall grasses.

LAUGHINGTHRUSHES AND ALLIES Leiothrichidae

Another new family split from the Timaliidae, this still comprises a large number of genera, including the *Alcippe* fulvettas, Jungle Babbler and *Garrulax* laughingthrushes. Laughingthrushes are medium-sized, long-tailed babblers that are gregarious even in the breeding season. They often feed on the ground and their flight is short and clumsy.

TREECREEPERS Certhiidae

Small, quiet, arboreal passerines with a slender, downcurved bill and a stiff tail which they use as a prop when climbing, like that of the woodpeckers. Treecreepers forage by creeping up vertical trunks and along the underside of branches, spiralling upwards in a series of jerks in search of insects and spiders; on reaching the top of a tree, they fly to the base of the next one. Their flight is undulating and weak, and is usually only over short distances. Treecreepers are not gregarious within their own species, but outside the breeding season they often join mixed parties of other insectivorous birds. One species, Bar-tailed Treecreeper, has a surprising record in Bangladesh.

NUTHATCHES Sittidae

Nuthatches are small, energetic, compact passerines with a short tail, large strong feet and a long bill. They are agile tree climbers and can move with ease upwards, downwards, sideways and upside-down over trunks or branches, progressing by a series of jerky hops; they do not use the tail as a prop. Their flight is direct over short distances and undulating over longer ones. Nuthatches eat insects, spiders, seeds and nuts. They are often found singly or in pairs; outside the breeding season, they often join foraging flocks of other insectivorous birds.

STARLINGS AND MYNAS Sturnidae

Robust, medium-sized passerines with a strong bill and strong legs, moderately long wings and a square tail. The flight is direct, strong and fast in the more pointed-wing species (*Sturnus*) and rather slower with more deliberate flapping in the more rounded-winged ones. Most species walk with an upright stance in a characteristic, purposeful, jaunty fashion, broken by occasional short runs and hops. Their calls are often loud, harsh and grating, and the song of many species is a variety of whistles; mimicry is common. Most are highly gregarious at times. Some starlings are mainly arboreal and feed on fruits and insects; others are chiefly ground-feeders and are omnivorous. Many are closely associated with human cultivation and habitation.

THRUSHES Turdidae

Including the genera *Turdus* and *Zoothera*, these are medium-sized passerines with rather long, strong legs, a slender bill and fairly long wings. On the ground they progress by hopping. All eat insects, and some eat fruit as well. Several species are chiefly terrestrial and others arboreal. Most thrushes have loud and varied songs, which are used to proclaim and defend their territories when breeding. Many species gather in flocks outside the breeding season. The family also includes the cochoas, which are fairly large, robust, colourful birds with a fairly broad bill. Shy, unobtrusive, arboreal and frugivorous. Recent taxonomic work has assigned some genera to Muscicapidae.

CHATS AND OLD WORLD FLYCATCHERS Muscicapidae

Chats are a diverse group of small or medium-sized passerines that includes the chats, blue robins, magpie-robins, redstarts, forktails, wheatears and rock-thrushes. Most are terrestrial or partly terrestrial, some are arboreal, and some are closely associated with water. Their main diet is insects, and they also consume fruits, especially berries. They forage mainly by hopping about on the ground in search of prey, or by perching on a low vantage point before dropping to the ground onto insects or making short sallies to catch them in the air. Found singly or in pairs.

Flycatchers are small insectivorous birds with a small, flattened bill, and bristles at the gape that help in the capture of flying insects. They normally have a very upright

stance when perched. Many species frequently flick the tail and hold the wings slightly drooped. Generally, flycatchers frequent trees and bushes. Some species regularly perch on a vantage point, from which they catch insects in mid-air via short aerial sallies or by dropping to the ground, often returning to the same perch. Other species capture insects while flitting among branches or by picking them from foliage. Flycatchers are usually found singly or in pairs; a few join mixed parties of other insectivorous birds.

FAIRY-BLUEBIRDS Irenidae

Medium-sized passerines with a fairly long, slender bill, the upper mandible of which is downcurved at the tip; arboreal, typically frequenting thick foliage in the canopy. They search leaves for insects and also feed on berries and nectar. Their flight is swift, usually over a short distance. Represented by only one species in Bangladesh, Asian Fairy-bluebird.

LEAFBIRDS Chloropseidae

Medium-sized green-and-yellow birds with a slender downcurved bill. They are arboreal and feed on nectar from flowering trees, fruit and invertebrates.

FLOWERPECKERS Dicaeidae

Flowerpeckers are very small passerines with a short bill and tail, and a tongue adapted for nectar-feeding. They usually frequent the tree canopy and feed mainly on soft fruits, berries and nectar; also on small insects and spiders. Flowerpeckers are very active, constantly flying about. Normally they live singly or in pairs; some species form small parties outside the breeding season.

SUNBIRDS AND SPIDERHUNTERS Nectariniidae

Sunbirds have a bill and tongue adapted to feed on nectar; they also eat small insects and spiders. The bill is long, thin and curved for probing the corollas of flowers. The tongue is very long, tubular and extensible far beyond the bill, being used to draw out nectar. Sunbirds feed mainly at the blossoms of flowering trees and shrubs. They flit and dart actively from flower to flower, clambering over the blossoms, often hovering momentarily in front of them and clinging acrobatically to twigs. Sunbirds usually occur singly or in pairs, although several may congregate in flowering trees and some species join mixed foraging flocks. They have sharp, metallic calls and high-pitched trilling and twittering songs.

Spiderhunters are small, robust arboreal forest birds with a very long downcurved bill. Very active with fast, dashing flight. Usually found singly or in pairs. They feed on nectar and small invertebrates.

WEAVERS Ploceidae

Small, rather plump, passerines with a large conical bill. Adults feed chiefly on seeds and grain, supplemented by invertebrates. Weavers inhabit grassland, marshes, cultivation and very open woodland. They are highly gregarious, roosting and nesting communally, and are noted for weaving their elaborate roofed nests from grasses.

MUNIAS Estrildidae

Small, slim passerines with a short, stout conical bill. They feed chiefly on small seeds which they pick from the ground or gather by clinging to stems and pulling the seeds directly from the heads. Outside the breeding season all species are gregarious. Their flight is fast and undulating.

OLD WORLD SPARROWS Passeridae

Small passerines with a thick, conical bill. This family includes *Passer*, the true sparrows, some of which are closely associated with human habitation. Most species feed on seeds taken on or near the ground. The *Passer* sparrows are rather noisy, using a variety of harsh, chirping notes.

PIPITS AND WAGTAILS Motacillidae

Small, slender, terrestrial birds with long legs, relatively long toes and a thin, pointed bill. Some wagtails exhibit wide geographical plumage variation. All walk with a deliberate gait and run rapidly. The flight is undulating and strong. Most wagtails wag their long tail up and down, as do some pipits. They feed mainly by picking insects from the ground as they walk, or by making short, rapid runs to capture insects they have flushed; they also catch prey in mid-air. They occur in scattered flocks in autumn and winter.

FINCHES Fringillidae

Small to medium-sized passerines with a strong, conical bill. They forage on the ground, some species feeding on the seed-heads of tall herbs, and the blossoms and berries of bushes and trees, as well as on insects. Finches are highly gregarious outside the breeding season. Their flight is fast and undulating. Only Common Rosefinch is recorded in Bangladesh.

BUNTINGS Emberizidae

Small to medium-sized, terrestrial passerines with a strong, conical bill adapted for shelling seeds, usually of grasses. They forage by hopping or creeping on the ground. Their flight is undulating. Buntings are usually gregarious outside the breeding season, feeding and roosting in flocks. They occur in a wide variety of open habitats.

Rufous-throated Partridge *Arborophila rufogularis* 27cm

Rare, presumed resident in NE and SE but only one recent record from NE. **ID** Has rufous sides to neck and upper breast, with sides of neck boldly splashed with black. Race *intermedia* found in Bangladesh has black throat. Also has diffuse greyish-white supercilium, diffuse white moustachial stripe, unbarred mantle, grey breast, bold pattern on wing-coverts, and bold white streaking on flanks. Sexes are similar. **Voice** A mournful double whistle, *wheea-whu*, repeated constantly and on ascending scale. **HH** Secretive, keeps to ground in dense understorey of evergreen forests.

White-cheeked Partridge *Arborophila atrogularis* 28cm

Rare, resident in NE and SE. **ID** White supercilium and cheeks, black mask and throat, black streaking on upper breast, black-streaked orange-yellow hindneck, barred upperparts, and absence of rufous streaking on flanks. **Voice** Accelerating and ascending series of 12–18 far-carrying, throaty *whew* notes, ending abruptly. **HH** Difficult to see, keeps to ground on slopes of small hills in dense bamboo thickets and undergrowth in evergreen and semi-evergreen forest.

Black Francolin *Francolinus francolinus* 34cm

Rare resident, confined to extreme NW in Thakurgaon District, but with past records in C, SE and NE. **ID** Male has black face with white ear-covert patch, rufous collar, black upper mantle spotted with white, and black underparts with white spotting on flanks. Female has rufous patch on hindneck, buffish supercilium and cheeks divided by dark stripe behind eye, and blackish barring to white underparts. Shows blackish tail in flight. **Voice** A loud, penetrating, repeated, harsh *kar-kar, kee, ke-kee*. **HH** Found singly, in pairs or in groups of up to five birds. Rests and roosts in thick ground cover. If much disturbed, escapes by running away swiftly, but otherwise flushes easily, flying off strongly and at great speed. Active in early mornings and late afternoons. Calls at any time of day during the breeding season, often from a tree or stump. Requires good ground cover and water close by. Tall grass and scrub with cultivation such as sugar cane. **Threatened** Nationally (EN).

Grey Francolin *Francolinus pondicerianus* 33cm

Rare, local resident rediscovered in 2018 in Ganges riverine chars in Rajshahi District of NW, formerly in C. **ID** Rather plain buffish face, and buffish-white throat with fine necklace of dark spotting. Upperparts are finely barred with buff, chestnut and brown. Underparts are buffish and finely barred with dark brown. Shows rufous tail in flight. **Voice** A rapidly repeated *khateeja-khateeja-khateeja*; also softer, more whistling *kila-kila-kila*, and a high, whirring *khirr-khirr*. **HH** Typically in pairs or in groups of up to eight birds which roost together in small thorny trees or shrubs. Digs and scratches in the ground with bill and feet. Very fast on their legs and usually escape by running; seldom flies. When pressed, rises with a loud whirr of wings, scatters in different directions and alights again after only 50–100m. Male usually calls from the ground. Riverine grassy and scrubby areas.

Swamp Francolin *Francolinus gularis* 37cm

Former resident in SW, C and NE, presumed to be extirpated. **ID** Rufous-orange throat, buff supercilium and cheek-stripe (separated by dark eyestripe), finely barred upperparts, and bold white streaking on underparts. Sexes similar. Shows rufous primaries and tail in flight. **Voice** A loud *kew-care* when alarmed, occasional *qua, qua, qua*, ascending in tone, and a harsh *chukeroo, chukeroo, chukeroo* preceded by several chuckles and croaks. May sound similar to Grey Francolin but louder. **HH** Found in pairs or in groups of up to six birds. In marshes often wades through shallow water or mud and climbs up onto reeds in deep water. Reluctant to fly, but if flushed it rises clumsily and noisily with loud chuckling and whirring of wings. Roosts in thorny trees and on broken reeds in swamps. Tall wet grassland and swamps. **Threatened** Globally (VU).

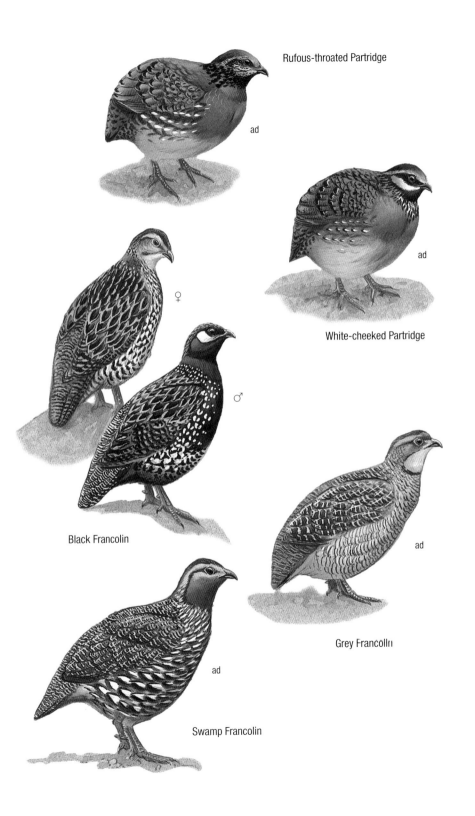

Rufous-throated Partridge

ad

White-cheeked Partridge

ad

♀

♂

Black Francolin

Grey Francolin

ad

Swamp Francolin

ad

Indian Peafowl *Pavo cristatus*

♂ 180–230cm, ♀ 90–100cm

Former rare resident, last recorded in Madhupur NP area of C in early 1980s. **ID** Male has blue neck and breast, and spectacular glossy green train of elongated uppertail-covert feathers with numerous ocelli. Female lacks train; has whitish face and throat, bronze-green neck, brown upperparts and white belly. Primaries of female are brown (chestnut in male). First-year male lacks train and is similar to female, but head and neck are usually blue, and primaries are chestnut with dark brown mottling. Second-year male more closely resembles adult male but has a short train, which lacks ocelli and is barred with green and brown. Length of train increases until fifth or sixth year. **Voice** Trumpeting, far-carrying and mournful *kee-ow, kee-ow, kee-ow*. Also, a series of short, gasping screams, *ka-an... ka-an... ka-an*, repeated 6–8 times, and *kok-kok* and *cain-kok* when alarmed. **HH** Gregarious, keeping in small flocks of usually one cock and three to five hens when breeding, and often in separate parties of adult males and of females with immatures in the non-breeding season. Roosts in tall trees. Emerges from dense thickets in early mornings and afternoons to feed.

Green Peafowl *Pavo muticus*

♂ 180–300cm, ♀ 100–110cm

Former resident? Records from the 19th and early 20th century in the Chittagong Hill Tracts of SE, although the validity of these records is questionable. **ID** Male has erect, tufted crest, and is mainly green, with long green train of elongated uppertail-covert feathers with numerous ocelli. Female lacks long train, otherwise similar to male but upperparts are browner. **Voice** Male has very loud, far-carrying *ki-wao* or *yee-ow*, often repeated. Female gives loud *aow-aa* with emphasis on first syllable, often repeated at short intervals. **HH** Habits similar to those of Indian Peafowl, but extremely shy and generally does not emerge into forest clearings and edges. Dense forest near streams and clearings. **Threatened** Globally (EN).

Grey Peacock-pheasant *Polyplectron bicalcaratum*

♂ 64cm, ♀ 48cm

Rare resident in SE, one record in NE. **ID** Greyish with white throat and long, broad tail. Male has prominent purple and green ocelli, particularly on wing-coverts and tail; tail is long and broad, and has short tufted crest. Female and immature male are smaller and browner, with shorter tail and smaller and duller ocelli. **Voice** Makes a deep guttural *hoo*, rapidly repeated about seven times, and soft chuckling notes; also, *ok-kok-kok-kok*. **HH** Singly, in pairs or family groups. Rarely seen and almost impossible to flush. Dense undergrowth in evergreen and semi-evergreen forest. **Threatened** Nationally (VU).

Red Junglefowl *Gallus gallus*

♂ 65–75cm, ♀ 42–46cm

Locally fairly common resident, mainly found in protected areas, absent in NW. **ID** Male has rufous-orange hackles, blackish-brown underparts, rufous wing-panel, white tail-base, and long greenish-black, sickle-shaped tail. There is an eclipse plumage, after the summer moult, when the hackles are replaced by short, dark brown feathers, and the central tail feathers are lacking. Female has 'shawl' of elongated feathers (edged golden-buff, black-centred), rufous head, and naked reddish face. Immature male much duller than adult male; hackles less developed (with black centres); lacks elongated central tail feathers. **Voice** Male's loud *cock-a-doodle-doo* is very similar to a crowing domestic cockerel; both sexes make cackling and clucking notes. **HH** In small groups. Often wary and secretive. If flushed, rises, cackling, with a clatter of wings. In early mornings and late afternoons, comes into the open to forage. Roosts in trees and bamboo clumps. Inhabits forest undergrowth and forest edges, including mangroves.

Kalij Pheasant *Lophura leucomelanos*

♂ 65–73cm, ♀ 50–60cm

Local and uncommon resident in protected areas in E. **ID** Both sexes have red facial skin and downcurved tail. Male of race *lathami*, the form recorded in the country, is mainly blue-black, with white barring on back and rump. Female is reddish-brown, with greyish-buff fringes producing scaly appearance. **Voice** A loud, whistling chuckle or *chirrup*. When flushed gives guinea-pig-like squeaks and chuckles, sharply repeated *koorchi, koorchi, koorchi* or whistling *psee-psee-psee-psee*. **HH** Found in pairs or family parties. Spends much time digging and scratching for food. Emerges into the open to forage in early mornings and late afternoons. Roosts in trees. Evergreen and semi-evergreen forest with dense undergrowth. **Threatened** Nationally (VU).

Indian Peafowl

♀

♂

♂

♀

Green Peafowl

♀

Grey Peacock-pheasant

♂

♂

♀

Red Junglefowl

♀

♂

Kalij Pheasant

Common Quail *Coturnix coturnix* 20cm

Rare winter visitor and possible passage migrant. **ID** Male has black 'anchor' mark on throat and buff gorget, although head pattern is variable and black anchor lacking in some. Some males have rufous face and throat, with or without black anchor. Female has less striking head pattern and lacks black anchor. **Voice** Song is a far-carrying *whit, whit-tit* repeated in quick succession. **HH** Usually alone or in pairs. Very secretive and keeps out of sight. If flushed rises with rapid, whirring wingbeats and brief glides on downturned wings. Tall grasses such as in riverine chars and wetlands.

Rain Quail *Coturnix coromandelica* 18cm

Rare resident in W and SE. **ID** Male similar in appearance to male Common, but has more strongly marked head pattern, variable black breast patch and streaking on flanks, and cinnamon sides to neck and breast. Female smaller than female Common, with unbarred primaries. **Voice** Utters a loud, metallic and high-pitched *whit-whit* repeated in runs of 3–5 calls. **HH** Habits similar to Common Quail's. Tall grasses in riverine chars and wetlands, also nearby crops.

Asian Blue Quail *Synoicus chinensis* 14cm

Rare presumed resident in C and NE. **ID** Small size. Male has black-and-white patterned throat, slaty-blue flanks, and chestnut belly. Female similar to Common Quail but noticeably smaller, with rufous-buff forehead and supercilium, barred breast and flanks, and more uniform upperparts. **Voice** Typical call is a high-pitched series of two or three descending piping notes, *ti-yu* or *quee-kee-kew*; repeated *tir* notes may be uttered when flushed. **HH** Usually flushes singly, dropping into grass after a short flight and very reluctant to fly again. Habits are similar to Common's. Favours wetter habitats than other quails: edges of marshes, tall grassland and fields. **AN** King Quail. **TN** Formerly in the genus *Coturnix*.

Common Buttonquail *Turnix sylvaticus* 13cm

Former resident, last recorded in 19th century in C. **ID** Very small with pointed tail. Bill greyish, and legs pinkish to greyish. More heavily marked above than Yellow-legged Buttonquail, with buff edges to scapulars and tertials forming prominent lines. Has rufous hindneck and mantle fringed with buff. Buff-fringed, dark-centred coverts result in spotted appearance, but less prominently so than in Yellow-legged. Underparts are similar to many Yellow-legged, with orange-buff lower throat and breast, and bold black spotting on sides of breast (becoming chestnut spotting on flanks). Female has brighter and more extensive rufous on neck than male. **Voice** Female has a booming call which lasts for *c*. 1 second and is repeated every 1–2 seconds for 30 seconds or more. **HH** Habits similar to Barred's. Tall grassland. **AN** Small Buttonquail.

Yellow-legged Buttonquail *Turnix tanki* 15–16cm

19th century record and six records since 1980s in NE, SW and C. **ID** Yellow legs and bill (with variable dark culmen and tip). Bold black spotting to buff coverts and upper flanks. Upperparts more uniform than Common Buttonquail, varying from grey and finely dotted, to being more heavily marked with black and diffusely with rufous (with buff edges to some feathers, although not as prominent as on Common). Some (breeding?) females distinctive, with unmarked rufous nape and upper mantle, rufous-orange throat, sides of neck and breast, and blackish crown. Other females less striking with buff crown-stripe, indistinct rufous collar, and orange-buff breast. Rufous collar lacking in male. **Voice** Low-pitched hoot, repeated with increasing strength to resemble a human-like moan. **HH** Habits similar to Barred's.

Barred Buttonquail *Turnix suscitator* 15cm

Local uncommon resident. **ID** Grey legs, and bold black barring on sides of neck, breast and wing-coverts. Orange-buff flanks and belly clearly demarcated from barred breast. Most show buff crown-stripe, and speckled supercilium is often apparent. Female usually has black throat and centre of breast. Male usually has greyish- or buffish-white throat. **Voice** Advertising call of female is *drr-r-r-r-r*, similar to the sound of a distant motorcycle, lasting for 15 seconds or more, and usually prefaced by three or four long, deep *groo* notes, as well as a far-carrying booming call, *hoon-hoon-hoon-hoon*. **HH** Usually found singly or occasionally in pairs. Very secretive and prefers to escape by walking away quickly. Flies with great reluctance, with weak, whirring beats low over the ground, dropping quickly into cover. Scrub, grassland, tea estates, open areas bordering forest.

Common Quail ♀ ♂

Rain Quail ♀ ♂

Asian Blue Quail ♀ ♂

Common Buttonquail ad

Yellow-legged Buttonquail ♂ ♀

Barred Buttonquail ♂ ♀

Fulvous Whistling-duck *Dendrocygna bicolor* 51cm

Winter visitor, locally common in NE, scarce elsewhere. **ID** Larger than Lesser Whistling-duck, with bigger, squarer head and larger bill. Adult from adult Lesser by warmer rufous-orange head and neck, dark blackish line down hindneck, dark striations on neck, more prominent pale streaking on flanks, indistinct chestnut-brown patch on forewing, and white band across uppertail-coverts. Often associates with Lesser. **Voice** Very noisy in flight and at rest; a repeated whistle *k-weeoo*. **HH** Shallow freshwater wetlands, especially haors, with emergent vegetation.

Lesser Whistling-duck *Dendrocygna javanica* 42cm

Widespread resident and winter visitor; resident birds and visitors congregate in winter. **ID** Smaller and more neatly proportioned than Fulvous Like Fulvous, has rather weak, deep-flapping flight, appears very dark on the upperwing and underwing, and is very noisy with repeated whistling. From Fulvous by greyish-buff head and neck, dark brown crown, lack of well-defined dark line down hindneck, bright chestnut patch on forewing, and chestnut uppertail-coverts. **Voice** Incessant wittering call in flight; at rest, a clear, whistled *whi-whee*, also a subdued quacking. **HH** Flooded grassland and paddy-fields, shallow freshwater wetlands, ponds, beels and haors, preferring those with emergent vegetation; in monsoon, wooded areas close to wetlands.

Bar-headed Goose *Anser indicus* 71–76cm

Scarce winter visitor to coast, rare passage migrant elsewhere. **ID** Yellowish legs and black-tipped yellow bill. Adult has white head with black banding across crown, and white line down grey neck. Juvenile has white face and dark grey crown and hindneck. Plumage paler steel-grey, with more uniform pale grey forewing compared with Greylag. **Voice** Honking flight call but notes more nasal and more slowly uttered compared with Greylag. **HH** Feeds mainly at night in cultivation or grassland on coastal islands; roosts by day on sandbanks. Winters mainly on coastal chars with sandbanks and short grassland.

Greylag Goose *Anser anser* 75–90cm

Scarce winter visitor. **ID** Large, grey goose, with stout pink bill and pink legs and feet. Juvenile is similar to adult, but has less prominent pale fringes to upperparts, flanks and belly. Shows pale grey forewing in flight. See Appendix 1 to compare with vagrant Greater White-fronted Goose. Feeds chiefly by grazing, mainly at night; rests on large lakes, rivers and open grassland by day. **Voice** Utters loud cackling and honking, deeper than in other 'grey' geese with repeated deep *aahng-ahng-ung*. **HH** Forages mainly at night and early morning; spends day swimming on large wetlands or rivers, or loafing on sparse grass and sandbanks in chars. Large wetlands, large rivers and coastal islands.

Fulvous Whistling-duck

ad

ad

juv

Lesser Whistling-duck

ad

ad

juv

ad

ad

Bar-headed Goose

juv

ad

ad

ad

Greylag Goose

juv

Common Shelduck *Tadorna tadorna* 58–67cm

Winter visitor, locally common on coast, rare inland. **ID** Adult has greenish-black head and neck, and largely white body with chestnut breast-band and black scapular stripe. White upperwing- and underwing-coverts contrast with black remiges in flight. Female slightly smaller than male, has narrower chestnut breast-band and lacks knob on bill. Adult eclipse duller and greyer, with less distinct breast-band. Juvenile lacks breast-band and has sooty-brown crown, hindneck and upperparts, and white forehead, cheeks, foreneck and underparts. Flight pattern similar to adult (though less contrasting) but shows white trailing edge to secondaries. **Voice** Low whistling call (male); rapid *gag-ag-ag-ag-ag* (female). **HH** Dabbles on mud, also wades or upends in shallows. Shallow coastal waters, also large wetlands and large rivers.

Ruddy Shelduck *Tadorna ferruginea* 61–67cm

Common winter visitor. **ID** Rusty-orange, with buff to orange head; white upperwing- and underwing-coverts contrast with black remiges in flight. Breeding male has black neck-collar, which is less distinct or absent in non-breeding plumage. Female very similar to male but lacks neck-collar and often has diffuse whitish face patch. Juvenile as female, but with browner and duller upperparts and underparts, and greyish tone to head. **Voice** Honking *aakh* and trumpeted *pok-pok-pok-pok* when taking off. **HH** Grazes on banks of rivers, wetlands and coastal chars, also wades, dabbles and upends in shallows. Large rivers, also large wetlands and coastal islands, especially those with sandbanks and sandy islets.

African Comb Duck *Sarkidiornis melanotos* 56–76cm

Rare winter visitor to NE and C, former resident. **ID** Whitish head, speckled with black, and whitish underparts with incomplete narrow breast-band. Upperwing and underwing blackish. Male has blackish upperparts glossed with bronze, blue and green, with fleshy 'comb' at base of bill and yellowish-buff wash to sides of head and neck in summer; comb much reduced in winter. Female much smaller with duller upperparts and no comb. Juvenile has pale supercilium contrasting with dark crown and eyestripe, buff scaling on upperparts, and rufous-buff underparts with dark scaling on sides of breast. **Voice** Generally silent but utters low croak when flushed. **HH** Grazes in marshes and wet grassland, also wades and dabbles in shallows. Pools with plentiful aquatic vegetation in well-wooded areas. **AN** Knob-billed Duck.

Cotton Pygmy-goose *Nettapus coromandelianus* 30–37cm

Uncommon resident. **ID** Male has broad white band across wing, and female has white trailing edge to wing. Male has white head and neck, black cap, greenish-black upperparts, and black breast-band. Eclipse male, female and juvenile are duller and have dark stripe through eye. **Voice** Male utters sharp staccato cackle *car-car-carawak* or *quack-quacky duck* at rest and in flight. Female makes weak *quack*. **HH** Generally in pairs in breeding season and in small flocks at other times. Forages by dabbling and grazing among floating vegetation; picks food from the surface and dips head and neck under water. Usually escapes predators by flying off strongly and rapidly. Perches readily in trees. Wetlands partly covered with vegetation and flooded paddy-fields.

White-winged Duck *Asarcornis scutulata* 66–81cm

Rare former resident, last reliably seen in 1980s, in Chittagong Hill Tracts of SE. **ID** Large size. White upperwing- and underwing-coverts, and white head variably speckled with black. Head, neck and breast can be mainly white on some. Sexes similar, although female duller with more heavily speckled head. Juvenile similar to female, but duller and browner, with brownish head. **HH** Flies to feeding grounds at dawn; roosts in forest trees at night. Small stagnant and slow-flowing freshwater wetlands, often with dead trees, in tropical forest. **TN** Formerly in genus *Cairina*. **Threatened** Globally (EN).

juv

juv

♂

Common Shelduck

Ruddy Shelduck

♂

♂

♀

♂

♀

♂

juv

♂
eclipse

♂

♀

Cotton Pygmy-goose

♂

♀

African Comb Duck

♂

♂
variant

juv

♂

White-winged Duck

Red-crested Pochard *Netta rufina*　　　　　　　　　53–57cm

Uncommon winter visitor, locally numerous in Tanguar Haor in NE. **ID** Large, with square-shaped head. Shape at rest and in flight more like dabbling duck. Male has red bill, rusty-orange head, and white flanks which contrast with black breast and ventral region. Female has pale cheeks contrasting with brown cap, and brown bill with pink towards tip. Both sexes have largely white flight feathers on upperwing, and whitish underwing. Eclipse male very similar to female, but with reddish iris and bill. **Voice** Silent away from breeding grounds. **HH** Feeds chiefly by diving; occasionally by upending and head-dipping. Large wetlands (haors) with deep open water and plentiful submerged and fringing vegetation; occasionally rivers.

Common Pochard *Aythya ferina*　　　　　　　　　42–49cm

Common winter visitor. **ID** Large, with domed head. Pale grey flight feathers and grey forewing result in upperwing pattern different from other *Aythya* ducks. Male has chestnut head, black breast, and grey upperparts and flanks. Female has brownish head and breast contrasting with paler brownish-grey upperparts and flanks; usually shows indistinct pale patch on lores, and pale throat and streak behind eye. Eye of female is dark and bill has grey central band. Does not show white undertail-coverts of Ferruginous Duck. Eclipse male and immature male recall breeding male but are duller with browner breast. **Voice** Silent away from breeding grounds. **HH** Highly gregarious. Feeds chiefly by diving in open water. Where there is disturbance, it is mainly a nocturnal feeder. Large wetlands with open water and large rivers. **Threatened** Globally (VU).

Baer's Pochard *Aythya baeri*　　　　　　　　　41–46cm

Rare winter visitor to NE and previously C. **ID** Greenish cast to dark head and neck, which contrast with chestnut-brown breast. White patch on fore flanks visible above water, and white undertail-coverts. Male has white iris. Female and immature male have duller head and breast than adult male. Female has dark iris and pale and diffuse chestnut-brown loral spot. **Voice** Silent away from breeding areas. **HH** A shy duck that has declined greatly, usually found singly, previously in small flocks. Feeds mainly by diving in shallower waters, often associates with Ferruginous Duck. Large wetlands; mainly recorded in Tanguar and Hakaluki haors, which were formerly two of the main wintering sites globally. **Threatened** Globally (CR) and nationally (CR).

Ferruginous Duck *Aythya nyroca*　　　　　　　　　38–42cm

Locally common winter visitor, mainly in NE. **ID** Smallest *Aythya* duck, with dome-shaped head. Breeding male is unmistakable, with rich chestnut head, neck and breast, and white iris. Female is chestnut-brown on head, neck, breast and flanks, and has dark iris. Eclipse male resembles female but is brighter on head and breast and has white iris. In flight, shows extensive white wingbar extending farther onto outer primaries than other *Aythya* species; striking white belly (less pronounced in female). **Voice** Silent away from breeding grounds. **HH** Shy and feeds mainly at night. Not easily flushed, preferring to hide among aquatic vegetation. Social, usually in dispersed flocks. Freshwater wetlands with extensive submerged vegetation.

Tufted Duck *Aythya fuligula*　　　　　　　　　40–47cm

Frequent winter visitor. **ID** Breeding male is glossy black, with prominent crest and white flanks. Eclipse/immature males are duller, with greyish flanks and less pronounced crest. Female is dusky brown, with paler flanks; some females may show scaup-like white face patch, but they usually also show tufted nape and squarer head. Female has yellow iris; dark in female Common and Baer's Pochards, and Ferruginous Duck. See Appendix 1 to compare with vagrant Greater Scaup. **Voice** Silent away from breeding grounds. **HH** Gregarious; sometimes in flocks of several hundred. Feeds in the day mainly by diving. Large wetlands, especially haors, also principal rivers with large areas of open water deep enough to allow diving.

Pink-headed Duck *Rhodonessa caryophyllacea*　　　　　　　　　60cm

May be extinct. Former presumed rare resident. **ID** Long neck and body and triangular head. Male has combination of pink head and hindneck, and dark brown foreneck and body; bill pink. Female similar, but with paler, dull brown body, greyish-pink head, and brownish crown and hindneck. In flight, shows pale fawn secondaries and contrasting dark forewing, and pale pink underwing with dark body. **HH** Shy and secretive. Feeds by dabbling on water surface but can also dive; occasionally perches in trees. Secluded pools and marshes in elephant-grass jungle. **Threatened** Globally (CR).

Red-crested Pochard

Common Pochard

♂ imm

♀

Baer's Pochard

Ferruginous Duck

♀ with scaup-like head

Tufted Duck

Pink-headed Duck

Garganey *Spatula querquedula*
37–41cm

Common winter visitor and spring passage migrant. **ID** Male has white stripe behind eye, and brown breast contrasting with grey flanks; shows blue-grey forewing in flight. Female has more patterned head than female Common Teal, with pale supercilium, whitish loral spot, pale line below dark eyestripe, dark cheek-bar, and whiter throat; in flight, shows prominent white belly, pale grey forewing and broad white trailing edge to wing. Eclipse male is similar to female but has upperwing pattern of breeding male. See Appendix 1 to compare with vagrant Baikal Teal. **Voice** Male has dry crackling call when alarmed; female has Common Teal-like quack. **HH** Feeds mainly by dabbling, preferring to keep among emergent vegetation. All types of freshwater wetlands with abundant vegetation, rare in coastal wetlands. **TN** Formerly in genus *Anas*.

Common Teal *Anas crecca*
34–38cm

Common winter visitor. **ID** Male has chestnut head with green band behind eye, white stripe along scapulars, and yellowish patch on undertail-coverts. Female has rather uniform head, lacking pale loral spot and dark cheek-bar of female Garganey, with less prominent supercilium; also, bill often shows orange at base, and undertail-coverts have prominent white streak at sides. Eclipse male and juvenile much as female. In flight, both sexes have broad white band along greater coverts, and green speculum with narrow white trailing edge; forewing brown. See Appendix 1 to compare with vagrant Baikal Teal. **Voice** Male has distinctive, soft, throaty whistle, *preep-preep*. Female is usually silent but utters a sharp *quack* if flushed. **HH** Feeds by grazing, dabbling, head-dipping and upending, also foraging in fields by night. Shallow freshwater and coastal wetlands.

Northern Shoveler *Spatula clypeata*
44–52cm

Common winter visitor. **ID** Long spatulate bill and bluish forewing. Male has dark green head, white breast, chestnut flanks, and blue forewing. Female recalls female Mallard in plumage but has greyish-blue forewing and lacks white trailing edge. Eclipse male recalls female, but is more rufous-brown, especially on flanks and belly, and has upperwing pattern of breeding male. In sub-eclipse resembles breeding male but has black scaling on breast and flanks, and whitish facial crescent between bill and eye. **Voice** Usually silent; female has descending series of quacks. **HH** Usually in pairs or small parties. Often feeds by sweeping bill from side to side while swimming, but also upends or fully immerses its head. All types of freshwater, rare along coast. **TN** Formerly in genus *Anas*.

Falcated Duck *Mareca falcata*
48–54cm

Rare winter visitor mainly to NE. **ID** Male has bottle-green head with maned hindneck, elongated black-and-grey tertials, and black-bordered yellow patch at sides of vent; pale grey forewing in flight. Female has rather plain greyish head (with maned appearance), dark bill, and greyish-white fringes to exposed tertials; greyish forewing and white greater-covert bar in flight, but does not show striking white belly. Eclipse male similar to female, but has dark crown, hindneck and upperparts, and pale grey forewing. **Voice** Distinctive, loud, piercing whistle in flight. **HH** Usually singly or in small flocks. Feeds mainly by dabbling and upending; usually keeps close to emergent vegetation. Large freshwater wetlands, rare in large rivers. **TN** Formerly in genus *Anas*.

Gadwall *Mareca strepera*
46–56cm

Locally fairly common winter visitor. **ID** White patch on inner secondaries in all plumages (can be indistinct in female); lacks metallic speculum shown by Mallard. Male mainly grey, with white belly and black rear end; bill dark grey. Female similar to female Mallard; orange sides to dark bill, clear-cut white belly and white inner secondaries are best features. Eclipse male similar to female but has more uniform grey upperparts and pale grey (rather than blackish-grey) tertials, and upperwing pattern of breeding male. **Voice** Generally silent outside breeding season. **HH** Feeds mainly by dipping head into shallow water; sometimes by upending. Freshwater wetlands and rivers with extensive aquatic and emergent vegetation, scarce at coast. **TN** Formerly in genus *Anas*.

Eurasian Wigeon *Mareca penelope*
45–51cm

Uncommon to locally common winter visitor. **ID** Male has yellow forehead and forecrown, chestnut head, and pinkish breast; white forewing in flight. Female has rather uniform brownish head, breast and flanks. In all plumages, white belly and rather pointed tail in flight. Eclipse male similar to female, but more rufous on head and breast, and has white forewing. **Voice** Male has distinctive whistled *wheeooo* call. **HH** Highly gregarious. Feeds chiefly by grazing on waterside grasslands and in wet paddy-fields. Mainly coastal chars, also large rivers and large wetlands (haors). **TN** Formerly in genus *Anas*.

Garganey

Common Teal

Northern Shoveler

Falcated Duck

Gadwall

Eurasian Wigeon

Indian Spot-billed Duck *Anas poecilorhyncha* 58–63cm

Scarce resident, rare outside NE. **ID** From other *Anas* species by yellow-tipped black bill, greyish-white head and neck with black crown and eyestripe, blackish spotting on breast, white scalloping on flanks, and largely white tertials. In flight, wings appear dark except for white on tertials and white underwing-coverts. Male has prominent red loral spot and is more strongly marked than female and juvenile (the red loral spot is less conspicuous on female and lacking on juvenile). **Voice** As Mallard. **HH** Habits very similar to Mallard, but is slower to take off from water. Feeds by dabbling, head-dipping, upending, and walking among marsh vegetation. Wetlands (beels, haors and backwaters in large rivers) with extensive emergent vegetation.

Mallard *Anas platyrhynchos* 50–65cm

Scarce winter visitor, mainly to Tanguar Haor, rare elsewhere in NE, NW and C. **ID** In all plumages, has white-bordered purplish speculum. Male has yellow bill, dark green head and purplish-chestnut breast, mainly grey body, and black rear end. Female is pale brown and boldly patterned with dark brown. Bill is variable, patterned mainly in dull orange and dark brown. Eclipse male is similar to female, but with (less heavily marked) rusty-brown breast, blackish (glossed green) crown and eyestripe, and uniform olive-yellow bill. **Voice** Male has soft rasping *kreep* and female distinctive laughing *quack-quack-quack-quack*. **HH** Sociable; associates with other ducks or in small flocks. Feeds in flooded paddy-fields (at night) and marshes by dabbling, head-dipping, grazing or upending. Large shallow wetlands.

Northern Pintail *Anas acuta* 51–56cm

Common winter visitor. **ID** Long neck and pointed tail. Male has chocolate-brown head, with white stripe down sides of neck. Female has comparatively uniform buffish head, slender grey bill, and (as male) shows white trailing edge to secondaries and greyish underwing in flight. Eclipse male resembles female, but has grey tertials, and bill pattern and upperwing pattern as breeding male. **Voice** Male utters mellow *prop-proop*, recalling male Common Teal; female has descending series of weak quacks, recalling Mallard, and low croak when flushed. **HH** Highly gregarious. Feeds mainly by upending, dabbling and head-dipping in shallow water; also grazes on land. Forages at night, in early morning and evening in marshes and flooded paddy-fields; roosts by day on open waters with aquatic vegetation; freshwater and coastal wetlands.

Little Grebe *Tachybaptus ruficollis* 25–29cm

Uncommon resident, more numerous as winter visitor. **ID** Small size, often with puffed-up rear end. Shows whitish secondaries in flight. In breeding plumage, has rufous cheeks and neck-sides and yellow patch at base of bill. In non-breeding plumage, has buff cheeks, foreneck and flanks. Juvenile is similar to non-breeding but has brown stripes across cheeks. See Appendix 1 to compare with vagrant Black-necked Grebe. **Voice** Drawn-out whinnying trill in breeding season, and sharp *wit* in alarm. **HH** Often singly or in pairs among aquatic vegetation when breeding; in non-breeding season more easily seen on open water, often in small loose groups. Swims buoyantly. Can often be detected by its bubbling call. Beels, ponds, reservoirs, ditches and slow-moving rivers.

Great Crested Grebe *Podiceps cristatus* 46–51cm

Uncommon and local winter visitor, mainly to NE. **ID** Large and slender-necked, with pinkish bill. Black crown does not extend to eye; has white cheeks and foreneck in non-breeding plumage. Rufous-orange ear-tufts and white cheeks and foreneck in breeding plumage. **Voice** Silent in non-breeding season. See Appendix 1 to compare with vagrant Red-necked Grebe. **HH** An expert diver. Usually feeds singly. Rises clumsily from the water, pattering over the surface, and flying with rapid wingbeats. Swims with body low in the water and neck held erect. In flight has an elongated appearance, with the neck extended. Favours large areas of deep open water.

Indian Spot-billed Duck

♂

♂

♀

Mallard

♀

♂

♀

♂

♀

Northern Pintail

non-br

non-br

br

Little Grebe

non-br

non-br

br

Great Crested Grebe

Rock Dove *Columba livia*   33cm

Common and widespread resident. **ID** Grey with metallic green and purple on neck. Grey tail with blackish terminal band, and broad black bars across greater coverts and tertials/secondaries. Feral populations found in Bangladesh vary greatly in coloration and pattern although some resemble wild birds. **Voice** Deep, repeated *gootr-goo, gootr-goo.* **HH** Lives in colonies all year. Wild birds roost on cliffs. Nests in hole or crevice. Feeds chiefly in cultivation, mainly on seeds, also eats green shoots. Feral birds live in villages and towns; wild birds not recorded. **AN** Common Pigeon, Feral Pigeon.

Oriental Turtle-dove *Streptopelia orientalis*   33cm

Locally common resident. **ID** Vinaceous-pink head, neck and underparts, rufous-scaled scapulars and wing-coverts, dusky underparts, and barring on neck. In flight, has dusky-grey underwing and grey sides and tip to tail. Juvenile lacks neck-barring, has buffish-grey head and underparts, and pale buff fringes to dark-centred feathers of upperparts. **Voice** A hoarse, mournful repeated *goor... gur-grugroo.* **HH** Found singly or in pairs when breeding, and in small parties in winter. Ground feeder and often forages on dusty tracks. Flight is fast and direct, with fast wingbeats, and on take-off and alighting the tail is widely fanned. Open evergreen, deciduous and mangrove forest.

Eurasian Collared-dove *Streptopelia decaocto*   32cm

Common and widespread resident. **ID** Sandy-brown with black half-collar, white sides to tail, and white underwing-coverts. Juvenile lacks neck-collar, and feathers of upperparts are fringed with buff. **Voice** A repeated cooing *kukkoo... kook.* **HH** Habits are similar to those of Oriental Turtle-dove. Breeds all year, varying locally. Nests low down in a bush or small tree. Open country with cultivation and groves.

Red Turtle-dove *Streptopelia tranquebarica*   23cm

Fairly common and quite widespread resident. **ID** Male has blue-grey head with black half-collar, pinkish-maroon upperparts, and pink underparts. Compared with Eurasian Collared, female has darker buffish-grey underparts, darker fawn-brown upperparts, greyer underwing-coverts, white (rather than grey) vent, and is smaller with shorter tail. Juvenile lacks neck-collar, and feathers of upperparts and breast are fringed with buff. **Voice** A harsh, rolling, repeated *groo-gurr-goo.* **HH** Habits are similar to those of Oriental Turtle-dove. Breeds all year, varying locally. Nests in trees 3–8m above ground. Light woodland, tea estates, village groves and trees in open country. **AN** Red Collared-dove.

Eastern Spotted Dove *Spilopelia chinensis*   30cm

Scarce resident, mainly in SE. **ID** From Western Spotted by duller upperparts narrowly fringed buff (with upperparts appearing dark- rather than pale-spotted), and yellow to orange iris and greyish eye-ring (iris and eye-ring reddish in Western). **Voice** Song described as typically less protracted than Western since it lacks repetition of main stressed note. **HH** Habits same as Western Spotted. **TN** Race *tigrina* found in Bangladesh is treated here as within *chinensis* rather than *suratensis*.

Western Spotted Dove *Spilopelia suratensis*   30cm

Common and widespread resident. **ID** Upperparts are broadly spotted with pinkish-buff. Has extensive black-and-white chequered patches on sides of neck, vinaceous-pink-tinged neck and breast, and dark grey-brown rump and tail with blackish base to outer tail feathers. Juvenile is paler and browner, lacks chequered patch on sides of neck, and has faintly barred mantle and scapulars and narrow rufous fringes to wing-coverts. **Voice** A soft, mournful *krookruk-krukroo... kroo-kroo-kroo.* **HH** Habits are similar to those of Oriental Turtle-dove. When disturbed, bursts upwards with a noisy clatter of wings, then glides down to settle nearby. Breeds almost all year, varying locally. Nests fairly low down in a tree, thorn bush or bamboo clump. Cultivation, habitation and open woodland. **TN** Formerly placed in *Streptopelia* or *Stigmatopelia*. Western and Eastern Spotted Doves previously treated as conspecific as Spotted Dove *Streptopelia chinensis*.

Rock Dove

ad

ad

ad

Oriental Turtle-dove

juv

Eurasian Collared-dove

ad

♀

♂

Red Turtle-dove

ad

Eastern Spotted Dove

ad

Western Spotted Dove

Grey-capped Emerald Dove *Chalcophaps indica* 27cm

Uncommon local resident. **ID** Stocky, broad-winged, short-tailed pigeon with emerald-green upperparts. Typically very rapid flight, and allows only brief views before being flushed from forest floor, when shows black-and-white banding on back. Male has grey crown, white forehead and supercilium, and deep vinaceous-pink head-sides and underparts; has white shoulder patch. Female is warm brown on crown, neck and underparts, has forehead and supercilium suffused with grey, and shoulder patch is generally warm brown. Juvenile resembles female but has dark grey barring on buffy-white forehead, narrow buff fringes (and some dark subterminal bars) to neck and underparts, dark brown tertials with chestnut tips, brown primaries with chestnut edges and tips, rufous-brown rump, and brownish bill. **Voice** Mournful booming *tk-hoon...tk-hoon*. **HH** Singly or in pairs. Feeds on ground, often on forest tracks. Frequently seen flying away rapidly and directly through forest, only a few metres above ground. Evergreen and semi-evergreen forest, rarer in deciduous forest and mangroves. **AN** Emerald Dove.

Orange-breasted Green-pigeon *Treron bicinctus* 29cm

Scarce resident in E and Sundarbans. **ID** Has grey central tail feathers in both sexes (at rest, tail appears grey rather than green). Male from other green-pigeons by orange breast bordered above by lilac band and yellowish-green forehead merging into pale blue-grey hindcrown and nape. Mantle uniformly green. Female has yellow cast to breast and belly, and grey hindcrown and nape. **Voice** Very similar to Thick-billed. **HH** Normally in small flocks, gathering in large flocks where food is plentiful. Often in mixed feeding parties with other fruit-eating birds. Usually at tops of tall trees. Clambers about twigs with great agility to reach fruit and berries, sometimes hanging upside-down. Keeps well-concealed in foliage; when approached 'freezes'. When flushed, wings make a loud clatter as birds burst out of tree. Evergreen, semi-evergreen and mangrove forest.

Ashy-headed Green-pigeon *Treron phayrei* 27cm

Scarce resident in E and C. **ID** Both sexes from Thick-billed by thin blue-grey bill (without prominent red base) and lack of prominent greenish orbital skin. Male has maroon mantle; further differences from male Thick-billed are diffuse orange patch on breast, greenish-yellow throat, and uniform dark chestnut undertail-coverts. Maroon mantle is best feature from Orange-breasted. Female lacks maroon mantle. Green central tail feathers, greyish cap, yellowish throat and white undertail-coverts help separate from female Orange-breasted. Tail shape and paler green coloration help separate from female Wedge-tailed. **Voice** Whistling song is similar to, but stronger and higher-pitched than that of Thick-billed. **HH** Habits like those of Orange-breasted. Mainly evergreen and semi-evergreen forest, rare in deciduous forest. **TN** Treated by some authorities as part of Pompadour Green-pigeon *T. pompadora* complex.

Thick-billed Green-pigeon *Treron curvirostra* 27cm

Scarce resident in E and C. **ID** Both sexes from Ashy-headed by thick bill with red base, prominent greenish orbital skin, pronounced whitish scaling on vent. Male has maroon mantle, and green breast without orange wash. **Voice** Calls and song typical of green-pigeons: pleasant wandering whistles and quiet, warbling song comprises series of whistling and cooing notes, rising and falling in pitch. **HH** Habits like Orange-breasted. Evergreen and semi-evergreen forests, occasional in deciduous forest.

Yellow-footed Green-pigeon *Treron phoenicopterus* 33cm

Widespread and locally common resident. **ID** Large size, grey cap and greenish-yellow forehead and throat, broad olive-yellow collar, pale greyish-green upperparts, mauve shoulder patch, yellowish band at base of tail, and yellow legs and feet. Sexes are similar, although female is duller. **Voice** Similar to Thick-billed. **HH** Habits like Orange-breasted. Deciduous and semi-evergreen forest, tea estates and fruiting trees around villages and cultivation.

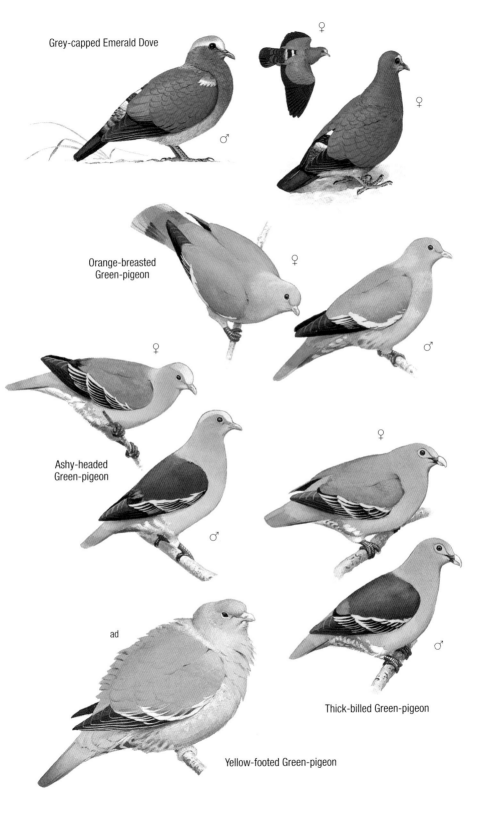

Grey-capped Emerald Dove

♀

♂

♀

Orange-breasted
Green-pigeon

♀

♂

♀

Ashy-headed
Green-pigeon

♂

♀

♂

ad

Thick-billed Green-pigeon

Yellow-footed Green-pigeon

Pin-tailed Green-pigeon *Treron apicauda* 42cm

Rare resident or visitor in E. **ID** Large green-pigeon with extended and pointed central tail feathers. Grey tail (with greenish tip to central feathers), contrasting with lime-green rump and uppertail-coverts, are additional features from female Wedge-tailed. Green crown, wing-coverts and back are additional differences from male Wedge-tailed. Has blue cere and bill-base and naked blue lores (lores are feathered on Wedge-tailed). Male has longer central tail feathers, pale orange wash to breast, and more pronounced grey cast to upper mantle compared with female. **Voice** Distinctive, deep, tuneful short melody: *oou...ou-ruu...oo-ru...ou-rooou.* **HH** Habits like Orange-breasted. Evergreen forest.

Wedge-tailed Green-pigeon *Treron sphenurus* 33cm

Rare resident in NE. **ID** Male from male Ashy-headed by larger size, long and wedge-shaped tail, less extensive maroon patch on upperparts (confined to inner wing-coverts, scapulars and lower mantle), pale cinnamon undertail-coverts, darker olive-green back and rump, and only indistinct, fine yellow edges to greater coverts and tertials. In flight, shows more uniform tail, lacking pale grey terminal band of Ashy-headed. Orange wash to crown, and very long pale cinnamon undertail-coverts are further differences. Female is mainly green, lacking orange coloration to crown and breast and maroon on upperparts of male. Undertail-coverts are yellowish-white with grey-green centres. From female Ashy-headed by differences in tail shape and colour, lack of prominent yellow in wing, uniform green head (lacking grey crown). Tail shape, maroon on upperparts (male), and dull olive-green rump, uppertail-coverts and central tail feathers are best distinctions from Pin-tailed. **Voice** Series of mellow whistles like other green-pigeons, but deeper with more hooting than Thick-billed. **HH** Habits like Orange-breasted, but less gregarious than other *Treron.* Evergreen and semi-evergreen forest, including one nesting record.

Green Imperial-pigeon *Ducula aenea* 43–47cm

Rare resident in E forests. **ID** From Mountain Imperial by dark metallic bronze-and-green upperparts, uniform dark green tail (distinctly banded on Mountain Imperial), and maroon undertail-coverts, contrasting with, and appearing darker than, grey breast and belly. **Voice** A lengthy, deep, hollow and resonant *currr-whoo,* with an emphasis on the second syllable. **HH** Keeps singly or in pairs in canopy of forest. Evergreen forest.

Mountain Imperial-pigeon *Ducula badia* 43–51cm

Rare resident in Chittagong Hill Tracts, vagrant elsewhere. **ID** Very large pigeon, with pinkish-grey head and underparts, brownish upperparts, and pale terminal band to tail. Juvenile has rufous fringes to mantle and wing-coverts, with chestnut leading edge to wing, and tail pattern is less well defined. **Voice** Deep, resonant, booming double note preceded by a click heard at close range. **HH** Singly or in pairs; gathers in small flocks when feeding on fruiting trees. Evergreen forest.

Pin-tailed
Green-pigeon

Wedge-tailed
Green-pigeon

ad

Green Imperial-pigeon

ad

Mountain Imperial-pigeon

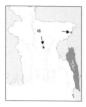

Great Eared-nightjar *Lyncornis macrotis* 40–41cm

Rare resident in E and C. **ID** A very large, richly coloured nightjar. At rest, shows prominent ear-tufts, and is generally more richly marked with golden-buff and rufous than other nightjars. Has fine rufous barring on black ear-coverts and throat, buff underparts boldly barred with dark brown, and tail broadly banded with golden-buff and dark brown. In flight, appears large, with slow and buoyant flight action (often feeding high in the air), and lacks white or buff spots on wings or tail. **Voice** Song is a clear, wailing *pee-wheeeu*. **HH** Evergreen and semi-evergreen forest and second growth, vagrant to Dhaka. **TN** Previously included in the genus *Eurostopodus*.

Grey Nightjar *Caprimulgus jotaka* 32cm

Rare and localised resident. **ID** Dark grey-brown and heavily marked with black. More cold-coloured and less strongly patterned than Large-tailed Nightjar, with greyer upperparts and lacking diffuse warm rufous-brown nuchal collar. Breast is dark grey-brown, lacking warm buff or brown tones. Further, has bold, irregular black markings on scapulars, usually lacking prominent pale edges, variable but rather poorly defined rufous-buff spotting on coverts, and broader dark banding on tail (with less white at tip than Large-tailed). **Voice** Song is a rapid series of loud, knocking *tuck* or *SCHurk* notes. **HH** Habits like Large-tailed. Forest edges, wooded areas. **TN** Treated here as a separate species from Jungle Nightjar *C. indicus*.

Sykes's Nightjar *Caprimulgus mahrattensis* 23cm

Rare presumed resident restricted to NW. **ID** Small, grey nightjar. Has finely streaked crown, black marks on scapulars, large white patches on sides of throat, and irregular buff spotting on nape forming indistinct collar. Compared with Indian Nightjar, which is similarly proportioned, crown is much less heavily marked, lacks well-defined rufous-buff nuchal collar, scapulars are relatively unmarked, and central tail feathers are more strongly barred. **Voice** Has continuous churring song; low, soft *chuck-chuck* in flight. **HH** By day rests on ground among clumps of tall grasses. Sandy islands in Padma (Ganges) River.

Large-tailed Nightjar *Caprimulgus macrurus* 33cm

Common widespread resident. **ID** Larger, longer-tailed and more warmly coloured and strongly patterned than Grey, with pale rufous-brown nuchal collar, complete white throat, well-defined buff edges and bold wedge-shaped black centres to scapulars, broad buff tips to coverts forming wingbars, and more extensive white or buff in outer tail feathers. **Voice** Series of loud, resonant calls: *chaunk-chaunk-chaunk*, repeated at rate of *c.* 100 per minute. **HH** A typical nightjar. Crepuscular and nocturnal. By day, perches on ground or lengthwise on branch and is difficult to detect. Feeds on insects caught in flight. Typically flies erratically to and fro over and among vegetation, occasionally wheeling, gliding and hovering to pick insects from foliage. Most easily located by calls given between sunset and dark and in the earliest dawn. Forest edges and clearings, well-wooded areas.

Indian Nightjar *Caprimulgus asiaticus* 24cm

Rare presumed resident, status uncertain. **ID** Grey, sandy-grey to brownish-grey in coloration. Best told by combination of small size and relatively short wings and tail, boldly streaked crown, rufous-buff markings on nape forming distinct collar, bold black centres and broad buff edges to scapulars, prominent buff or rufous-buff spotting on wing-coverts, and pale, relatively unmarked central tail feathers. Similar in appearance to Large-tailed, but much smaller, with shorter tail; note also broken patches of white on sides of throat and more uniform tail. **Voice** Far-carrying *chuk-chuk-chuk-chuk-tukaroo*, likened to a ping-pong ball bouncing to rest on a hard surface; male gives short sharp *qwit-qwit* in flight. **HH** Habits like Large-tailed. Dry, open areas including grassy areas bordering mangrove forest and large rivers.

Savanna Nightjar *Caprimulgus affinis* 23cm

Rare resident in NE and NW. **ID** A medium-sized, dark brownish-grey nightjar. Less strikingly marked than other nightjars; crown and mantle are finely vermiculated, and lack bold dark streaking; has more uniform coverts with fine dark vermiculations and irregular rufous-buff markings, and scapulars are usually edged with rufous-buff. Male has largely white outer tail feathers. **Voice** Strident and shrill *dheet*. **HH** Habits like Large-tailed. Tea estates, open wooded areas.

Great Eared-nightjar

ad

Grey Nightjar ♀ ♂

♂

Sykes's Nightjar ♂ ♂

♀

Large-tailed Nightjar ♂ ♂

♀

Indian Nightjar ad

♂ ♀

Savanna Nightjar ♂

Hodgson's Frogmouth *Batrachostomus hodgsoni* 27cm

Rare presumed resident, only two records. **ID** Male is rufous-brown. Upperparts are heavily marked with black, especially on head, with irregular bold whitish markings particularly on scapulars and upper mantle (forming 'collar' on some birds). Underparts are heavily and irregularly marked with black, white and rufous. Female is more uniformly rufous, with irregular black-tipped white spots on upper mantle, scapulars and underparts. **Voice** Usual call is a series of soft, burring *gwaa* notes with rising inflection. **HH** Evergreen and deciduous forest.

Brown-backed Needletail *Hirundapus giganteus* 23cm

Local resident in Chittagong Hill Tracts. **ID** Largest and most powerful of the needletails; 'needles' of tail larger and longer than on other species and often easy to see in the field. Compared with Silver-backed (see Appendix 1) has white lores, brown throat concolorous with rest of underparts, and less contrasting pale 'saddle' (uniformly pale brown, lacking silvery-white centre), with longer and rounded or point-ended tail. Juvenile has dark fringes to white undertail-coverts. **Voice** Makes a soft rippling *trp-trp-trp-trp-trp-trp*. **HH** Roosts communally in old tree hollows. Hawks over evergreen forest and forest clearings.

Himalayan Swiftlet *Aerodramus brevirostris* 14cm

Uncommon and local winter visitor. **ID** A stocky brown swiftlet with slight gloss to upperparts and pronounced indentation to tail. Has paler grey-brown underparts than upperparts, and distinct pale grey rump-band. **Voice** Low, rattling call and twittering *chit-chit* at roost. **HH** Typically singles or small groups seen. Wanders erratically over large distances to feed. Feeds over open fields, wetlands and open wooded areas. **TN** Often placed in the genus *Collocalia*.

Asian Palm-swift *Cypsiurus balasiensis* 13cm

Common, widespread resident. **ID** Small and very slim with fine scythe-shaped wings and deeply forked tail (usually held closed). Rapid fluttering wingbeats are interspersed with short glides. Throat is slightly paler than rest of underparts; may also show a slightly paler rump. Crested Treeswift is much larger (see Appendix 1), with weaker, more fluttering flight, and is browner in coloration (and does not show whiter belly and undertail-coverts). **Voice** A trilling *te-he-he-he-he*. **HH** Usually hawks insects around palms (in which it nests). Particularly active foraging in the evening and readily joins mixed flocks with other swifts and hirundines. Twists and turns in the air with great agility. Open country; closely associated with palms.

Pacific Swift *Apus pacificus* 15–18cm

Scarce passage migrant and winter visitor. **ID** Dark swift with prominent white rump and deeply forked tail. Best told from House Swift by longer, deeply forked tail, and slimmer-bodied and longer-winged appearance. **Voice** Has a *sreee* call. **HH** Favours hawking over open ridges or hilltops, rarely large flocks seen on passage. **AN** Fork-tailed Swift. **TN** Subspecies occurring in Bangladesh uncertain. Himalayan breeding *leuconyx* ('Blyth's Swift'), sometimes considered a separate species, most likely (and depicted in plate). Long-distance migrant nominate form could occur; this is larger, with pale scaling on underparts.

House Swift *Apus nipalensis* 15cm

Common and widespread resident. **ID** A small, stocky swift with prominent white throat and rump-band. From Pacific Swift by smaller size, shorter and broader wings, stout body and rather big head, and much shallower tail fork. Flight is weaker. **Voice** Rapid and shrill *sik-sik-sik-sik... sik-sik-sik-sik-sik-sik*. **HH** Usually associates in large scattered flocks and keeps within a wide vicinity of its nesting areas when breeding. Breeds and roosts communally. Depends entirely on human habitations for nesting; only nests in larger settlements. Builds a typical swift nest, each placed haphazardly one upon another, usually under eaves or in verandas of buildings, or under bridges or arches. Inhabits cities, towns and larger villages. **TN** Often treated as conspecific with Little Swift *A. affinis*.

Hodgson's Frogmouth (not to scale)

♂

♀

ad

Brown-backed Needletail

ad

Himalayan Swiftlet

ad

Asian Palm-swift

leuconyx

ad

ad

Pacific Swift

House Swift

PLATE 14: CUCKOOS I

Greater Coucal *Centropus sinensis* — 48cm

Common and widespread resident. **ID** Adult from adult breeding Lesser Coucal by much larger size, black underwing-coverts (difficult to see in the field), and brighter and more uniform chestnut wings. Juvenile has brownish-black head and body, with chestnut spotting on crown and nape (becoming barred on mantle), and diffuse whitish barring on entire underparts; chestnut-brown coverts and flight feathers are barred with dark brown, and tail is narrowly barred with buff or greyish-white. Immature resembles adult, although head and body are duller black and has barred (juvenile) flight feathers and tail. **Voice** Deep, resonant and primate-like *hoop-hoop-hoop-hoop-hoop-hoop*, descending and then rising towards end of the series. **HH** Walks sedately with tail held horizontal, or skulks in dense vegetation. Tall grasses and thickets near cultivation.

Lesser Coucal *Centropus bengalensis* — 33cm

Uncommon and local resident, mainly in E. **ID** Smaller than Greater, with stouter bill, duller chestnut mantle and wings (including browner tertials and primary tips), and chestnut underwing-coverts. Eyes are dark (red in Greater). Often shows buff streaking on some scapulars and wing-coverts (unlike Greater). Adult non-breeding has dark brown head and mantle with prominent buff shaft streaks, and dark brown and rufous barring on rump and very long uppertail-coverts. Wings and tail are as adult breeding. Juvenile is similar to adult non-breeding, but has less distinct pale shaft streaking on upperparts, dark barring on crown, mantle and back, dark brown barring on wings, and narrow rufous barring on tail. Immature has head and body as adult non-breeding, but wings and tail are barred as juvenile. **Voice** Series of deep resonant *pwoop-pwoop-pwoop* notes, very similar to Greater, but usually slightly faster and more interrogative, initially ascending, then descending and decelerating. **HH** Habits similar to Greater. Tall wetland vegetation and grasses, tea estates and shrubberies.

Green-billed Malkoha *Phaenicophaeus tristis* — 38cm

Locally fairly common resident except in NW. **ID** Large and very long-tailed. Greyish-green in coloration, with lime-green bill, red eye patch, white-streaked supercilium, and broad white tips to tail feathers. **Voice** Low croaking *ko... ko... ko*, and a chuckle when flushed. **HH** Very skulking, creeps and clambers unobtrusively through branches in mid-canopy and lower down in thick vegetation. Evergreen, semi-evergreen, deciduous and mangrove forest and thickets. **TN** Placed by some authorities in genus *Rhopodytes*.

Jacobin Cuckoo *Clamator jacobinus* — 33cm

Uncommon but widespread summer/monsoon visitor. **ID** Black and white with crest. Has white patch at base of primaries, and prominent white tips to tail feathers. Juvenile has browner upperparts, grey wash to throat and upper breast, and buffish wash to rest of underparts. Has smaller crest than adult, with smaller white wing patch, and paler bill. **Voice** Metallic *piu... piu... pee-pee piu, pee-pee piu* or *piu...piu...piu* uttered frequently day and night. **HH** Conspicuous, often perching in open. Chiefly arboreal but may forage in low bushes and on ground. Open woodland, well-wooded areas and wetlands with tall vegetation. **AN** Pied Cuckoo.

Chestnut-winged Cuckoo *Clamator coromandus* — 47cm

Scarce and local summer/monsoon visitor mainly to E. **ID** Prominent crest, whitish collar, chestnut wings, and orange wash to throat and breast. Long black tail has narrow greyish-white feather tips. Lacks white in wing. Juvenile has shorter crest, rufous fringes to upperparts, buff collar, whitish throat and breast, broad buff tips to rectrices, and paler bill. Immature is similar to adult, but retains some buff tips to scapulars, coverts and tail feathers. **Voice** Series of double metallic whistles, *breep breep*, and harsh, grating scream. **HH** Arboreal and rather retiring. Evergreen and semi-evergreen forest, rare in deciduous forest.

Western Koel *Eudynamys scolopaceus* — 43cm

Common and widespread resident. **ID** Large, with long and broad tail. Male is glossy black (with green iridescence), with a dull lime-green bill and brilliant red eye. Female is brown above (with faint green gloss), spotted and barred with white and buff, and white below, strongly barred with dark brown. Also has striking red eye. Juvenile is blackish, with white or buff tips to wing-coverts and tertials, and variable white barring on underparts; tail is black, but shows pronounced rufous barring in some. **Voice** Loud, rising and increasingly anxious repeated *ko-el...ko-el...ko-el* and a bubbling, more rapid repeated *koel... koel*. **HH** Typically keeps concealed in dense foliage when not feeding. Open woodland, groves, gardens and cultivation. **AN** Asian Koel.

Greater Coucal

juv

ad

Green-billed
Malkoha

ad

imm

br

non-br

Lesser Coucal

juv

ad

Jacobin Cuckoo

♂

♀

ad

juv

♂ imm

♀ imm

Chestnut-winged
Cuckoo

Western Koel

Asian Emerald Cuckoo *Chrysococcyx maculatus* 18cm

Rare summer/monsoon visitor to E. **ID** Male has emerald-green upperparts. Female has rufous-orange crown and nape and unbarred bronze-green mantle and wings. Has yellow bill with dark tip. Juvenile is similar to female but with rufous fringes to mantle and wing-coverts, and rufous-orange wash to barred throat and breast. **Voice** Clear, loud three-note whistle, sharp *chweek* flight call and loud, descending *kee-kee-kee-kee*. **HH** Usually keeps to leafy canopy of tall trees, but on arrival in spring often flies about conspicuously. Very active, moving rapidly from branch to branch and making sallies to capture flying insects. Has habit of perching along a branch, rather than across it. Flight is fast and direct. Evergreen and semi-evergreen forest.

Violet Cuckoo *Chrysococcyx xanthorhynchus* 17cm

Rare resident or summer visitor to E. **ID** Male has purple upperparts. Female has uniform bronze-brown upperparts, with only slight greenish tinge, and white underparts with brownish-green barring. Male has orange bill; orangish with darker tip in female. Juvenile similar to juvenile Asian Emerald, but upperparts (especially wings) more heavily barred with rufous and lacks rufous-orange wash to throat. **Voice** A disyllabic and repeated *che-wick*, particularly in flight, and an accelerating trill. **HH** Habits are very poorly known, presumably similar to those of Asian Emerald. Evergreen and semi-evergreen forest.

Banded Bay Cuckoo *Cacomantis sonneratii* 24cm

Scarce local resident in E. **ID** White supercilium (finely barred with black and encircles brown ear-coverts), finely barred white underparts, and fine and regular dark barring on upperparts. Juvenile has broader (and more diffuse) barring on underparts, and crown and nape have some buff barring. **Voice** A shrill, whistled *pi-pi-pew-pew*, first two notes at same pitch, third and fourth descending. **HH** Calls from bare branches of treetops, usually holding tail depressed, wings drooping and rump feathers fluffed out. Dense evergreen and semi-evergreen forest.

Plaintive Cuckoo *Cacomantis merulinus* 23cm

Common and widespread resident. **ID** Adult has orange underparts. On hepatic female, base colour of underparts is pale rufous, upperparts are dull rufous-brown with regular dark barring, and tail is strongly barred. Juvenile has bold streaking on rufous-orange head and breast, and is distinct from hepatic and juvenile Grey-bellied. **Voice** Mournful whistle *tay... ta... tee*, second note lower and third note higher than first; also *tay... ta... ta... tay* repeated with increasing speed and ascending pitch. **HH** Keeps mainly to leafy tops of trees and bushes, except in dry season when some birds move into open wetlands among short and taller vegetation. Very active and restless. When calling, holds wings loosely, tail depressed and rump feathers fluffed out. Open forest, groves, well-wooded villages.

Grey-bellied Cuckoo *Cacomantis passerinus* 23cm

Rare and localised resident or summer visitor, mainly to W. **ID** Grey morph adult is grey with white vent and undertail-coverts. On hepatic female, compared with Plaintive, base colour of underparts is mainly white, upperparts are brighter rufous with crown and nape only sparsely barred, and tail is unbarred. Juvenile varies. Some have brownish-black upperparts (without distinct barring), dusky-grey underparts with indistinct buffish-grey barring mainly on belly and flanks (some are more heavily barred on underparts), and tail grey-brown with whitish notching. Others are barred rufous on upperparts, and underparts are similar to hepatic female; tail is dark brown, barred with rufous (and is similarly patterned to tail of Plaintive). Intermediates occur, e.g. with uniform grey upperparts and strong rufous barring on tail. **Voice** Clear *pee-pipee-pee... pipee-pee*, ascending in scale and higher-pitched with each repetition. **HH** Habits like Plaintive. Open forest and groves.

♂

♀

Asian Emerald
Cuckoo

juv

♀

Violet Cuckoo

♂

juv

ad

juv

Banded Bay Cuckoo

ad

juv

Plaintive Cuckoo

ad

♀
hepatic

ad

♀
hepatic

rufous
juv

grey
juv

Grey-bellied Cuckoo

Square-tailed Drongo-cuckoo *Surniculus lugubris* 25cm

Locally fairly common resident in E. **ID** Adult is glossy black, except for fine white barring on very long undertail-coverts, white thighs, and tiny white patch on nape (difficult to see in field). Best told from a drongo by square-ended tail, fine, downcurved black bill and white-barred undertail-coverts. Juvenile is similar, but dull black, spotted with white. **Voice** Distinctive series of ascending whistles *pee-pee-pee-pee-pee-pee*, broken off and quickly repeated. **HH** Resembles a drongo, especially when perched upright, but is less active and flies like a cuckoo. Perches on a bare branch when calling, but otherwise usually keeps in canopy foliage of trees and bushes. Evergreen and semi-evergreen forest. **TN** Fork-tailed Drongo-cuckoo *S. dicruroides* (which could occur as a vagrant) is often considered conspecific with this species.

Large Hawk-cuckoo *Hierococcyx sparverioides* 38cm

Rare winter visitor to E. **ID** Larger than Common Hawk-cuckoo, with browner mantle (contrasting with slate-grey head), blackish chin, grey streaking on throat and breast, irregular rufous breast-band, broad dark brown barring on underparts, and broader and stronger dark tail-banding. Underwing-coverts are white, barred with dark brown. Juvenile has strongly barred rather than spotted underparts, and has broader tail-banding. Immature has darker slate-grey head than immature Common Hawk-cuckoo, with blackish chin and grey throat streaking. **Voice** (Not heard in Bangladesh) Similar to Common, but less shrill. **HH** Habits similar to Common, but more secretive.

Common Hawk-cuckoo *Hierococcyx varius* 34cm

Common and widespread resident. **ID** Smaller than Large Hawk-cuckoo, with whitish or greyish chin and throat, uniform grey upperparts, more rufous on underparts, indistinct barring on belly and flanks, and narrower tail-banding. Underwing-coverts are rufous and only faintly barred. In juvenile, flanks typically less heavily marked than in Large, with spots or chevron marks rather than barring (although some very similar), while rufous tail-banding and tail-tip are typically brighter and more clearly defined. **Voice** Shrill and manic repeated *pee-pee-ah... pee-pee-ah*, rising in pitch to hysterical crescendo, becoming more vocal during hot weather; often calls throughout night. **HH** Usually keeps well hidden among foliage of trees, even when calling. *Accipiter*-like in flight; flies low with a few fast wingbeats followed by a glide. Village groves, lightly wooded areas, open forest including mangroves.

Indian Cuckoo *Cuculus micropterus* 33cm

Common summer visitor and presumed resident (few records outside breeding season when not calling). **ID** From Eurasian and Oriental (see Appendix 1) by browner mantle and broader, more widely spaced black barring on underparts. Tail has broader (diffuse) dark subterminal band, broader white barring on outer tail feathers and larger white spots on central feathers. Eyes brown or reddish-brown (yellow in Eurasian; yellow or brown in Oriental). Female has rufous-buff wash to base of grey breast, and rufous suffusion to whitish barring of lower breast. There is no hepatic female morph. Juvenile distinctive with broad and irregular white tips to feathers of crown, nape, scapulars and wing-coverts; throat and breast are creamy-white with irregular brown markings, and barring on rest of underparts is broader and more irregular than on Eurasian and Oriental. **Voice** Descending four-note whistle, *kwer-kwah... kwah-kurh*. **HH** Frequents treetops and canopy; sometimes flies hawk-like above forest. Often calls at night. Forests and well-wooded country.

Eurasian Cuckoo *Cuculus canorus* 32–34cm

Rare, probably under-recorded passage migrant. **ID** For differences from Indian, see that species. Best distinguished from Oriental (see Appendix 1) by song; see that species, but rarely heard in Bangladesh. Non-hepatic female has rufous wash to lower border of grey breast. Hepatic female is rufous-brown above and whitish below, and strongly barred dark brown all over. Juvenile is very variable, some superficially resembling grey adult, others hepatic female, with whitish fringes to upperparts and white nape patch. **Voice** Two-note whistle, *Cuck-oo... cuck-oo*. **HH** Habits similar to Indian, less vocal at night. Often perches in open when calling. Open woodland and bushy country. **AN** Common Cuckoo.

juv

ad

Square-tailed
Drongo-cuckoo

Large Hawk-cuckoo

imm

juv

ad

ad

imm

Common
Hawk-cuckoo

juv

♀

♂

Indian Cuckoo

juv

♂

♀
hepatic

Eurasian Cuckoo

Slaty-legged Crake *Rallina eurizonoides* 25cm

Rare presumed resident or summer visitor, with one breeding record from Bagerhat in SW. **ID** Greenish or grey legs and feet and more extensive black-and-white barring on underparts are best features from adult Ruddy-breasted. Also, olive-brown mantle contrasts with rufous neck and breast. Juvenile has dark olive-brown head, breast and upperparts. More prominent and extensive black-and-white barring on underparts are best features from juvenile Ruddy-breasted. **Voice** A *kek-kek, kek-kek*, a loud drumming croak, a subdued *kok*, and a *krrrr* alarm call. **HH** A typical rail though partly nocturnal. Very skulking, emerging from thick cover in early morning and at dusk like other rails. Nests among dense vegetation, sometimes away from water. Marshes in forest and well-wooded country.

Eastern Water Rail *Rallus indicus* 23–28cm

Rare and local winter visitor, mainly to NE. **ID** From Slaty-breasted by finer bill (which is more noticeably downcurved), brown crown, hindneck and upperparts (which are prominently streaked), diffuse pale supercilium and darker eyestripe, brownish wash across breast, and pinkish legs. **Voice** A metallic, strident *skrink, skrink*, beginning explosively and repeated after a few seconds. **HH** A typical rail. In common with other rails it is secretive, keeping within thick cover, except for early morning and dusk; typically bolts into cover with head and tail lowered at the least sign of danger. Marshes, wet fallow fields bordered by rank vegetation, and reedbeds. **AN** Brown-cheeked Rail. **TN** Formerly treated as conspecific with Western Water Rail *R. aquaticus*.

Slaty-breasted Rail *Lewinia striata* 27cm

Scarce resident with most records from NE. **ID** Longish bill with red at base (stouter and straighter than in Eastern Water Rail). Legs are olive-grey. Adult has chestnut crown and nape, slate-grey foreneck and breast, white barring and spotting on upperparts, and white barring on dark grey belly, flanks and undertail-coverts. Juvenile is duller, with crown and nape being dark-streaked olive-brown (some with rufous tinge), upperparts are paler olive-brown and more sparsely marked with white, and underparts are browner with less pronounced barring on flanks. **Voice** A sharp *cerrk*. **HH** Dense vegetation in marshes, nearby paddy-fields and mangroves. **TN** Formerly in the genus *Gallirallus*.

Ruddy-breasted Crake *Zapornia fusca* 22cm

Uncommon resident, the most widespread crake. **ID** From other crakes by combination of dull chestnut underparts, unmarked dark olive-brown upperparts, indistinct dark brown-and-white barring on rear flanks and undertail-coverts (much more restricted than in Slaty-legged Crake) and red legs. Juvenile dark olive-brown, with white-barred undertail-coverts and fine greyish-white mottling/barring on rest of underparts. Legs duller and iris brown (rather than red as in adult). **Voice** Single soft *crake* at considerable intervals; a loud metallic *twek* repeated at 2–3-second intervals, often followed by squeaky trilling. **HH** A typical rail. Skulking. Like other rails walks with rhythmic movements of head and neck, and jerks tail. Nests on marshy ground among grass, reeds or rice plants; stalks sometimes bent over to form a canopy. Well-vegetated edges of shallow wetlands and marshes including small waterways and wet fields. **TN** Formerly in the genus *Porzana*.

Brown Crake *Zapornia akool* 28cm

Rare presumed resident (not known to migrate); recently rediscovered in NW, previously recorded in 19th century. **ID** Olive-brown upperparts, grey face and breast, and olive-brown flanks and undertail-coverts; underparts lack barring. Has red iris, greenish bill and pinkish-brown to purple legs. Juvenile is similar to adult but has dull iris and paler grey underparts. **Voice** Calls include a shrill rattle, a long, drawn-out vibrating whistle and a short plaintive note. **HH** A typical rail though less secretive than most species. Like other rails flies reluctantly and feebly, with legs dangling, for a short distance before dropping into cover again. Nests in or near marshes. Inhabits reedy marshes and vegetation bordering watercourses. **TN** Formerly in the genus *Porzana*.

Baillon's Crake *Zapornia pusilla* 17–19cm

Scarce to rare winter visitor. **ID** Adult has rufous-brown upperparts (extensively marked with white), grey underparts, and black-and-white barring on flanks. Legs and bill are green. Juvenile has buff underparts. **Voice** Song may be heard on migration: a rattling rasp, *trrrrr-trrrrr*, also a *tyiuk* alarm call. **HH** A typical rail. Secretive, though not particularly shy, mainly crepuscular. Reeds and well-vegetated edges of large wetlands, particularly haors, marshes with dense emergent vegetation, and overgrown fallow, wet paddy-fields. **TN** Formerly in the genus *Porzana*.

ad

juv

Slaty-legged Crake

ad

Eastern Water Rail

ad

juv

Slaty-breasted Rail

juv

Ruddy-breasted
Crake

ad

Brown Crake

ad

juv

ad

Baillon's Crake

White-breasted Waterhen *Amaurornis phoenicurus*   32cm

Fairly common resident. **ID** Adult has grey upperparts and white face, foreneck and breast; undertail-coverts rufous-cinnamon. Bill and legs are greenish or yellowish, with swollen reddish base to upper mandible. Juvenile has greyish face, foreneck and breast, and olive-brown upperparts; bill and legs are darker. **Voice** Highly vocal when breeding; calls include a metallic *krr-kwaak-kwaak* and a *kook... kook... kook*, often following loud roars, croaks and chuckles. **HH** Less shy than others in family; often feeds in open and on quite dry land. Clambers among bushes with ease. Inhabits thick cover close to all wetland types.

Watercock *Gallicrex cinerea*   ♂43cm, ♀36cm

Locally frequent monsoon visitor, scarce in winter/dry season. **ID** Breeding male is mainly greyish-black, with yellow-tipped red bill and red shield and horn. Upperparts fringed with grey and buff. Legs bright red. First-summer male has broad rufous-buff fringes to plumage. Non-breeding male and female have buff underparts with fine barring, and buff fringes to dark brown upperparts. Legs greenish. Juvenile has uniform rufous-buff underparts, and rufous-buff fringes to upperparts. Male is much larger than female, with heavier bill. **Voice** Series of 10–12 *kok-kok-kok* notes followed by a deep, hollow *utumb-utumb-utumb* repeated 10–12 times, and then by 5–6 *kluck-kluck-kluck* notes. Males call persistently in the breeding season, in the monsoon. **HH** Mainly crepuscular, also emerging from cover in cloudy weather. More often seen flying in breeding season than other rails. In breeding season, in dense grassland on islands in main rivers, also large wetlands including flooded fields; in dry season, in smaller wetlands bordered by dense vegetation.

Purple Swamphen *Porphyrio porphyrio*   45–50cm

Chiefly a winter visitor, but some presumed to breed locally. **ID** Large size, purplish-blue coloration with variable greyish head, and huge red bill and frontal shield. Female smaller than male. Juvenile is duller than adult, with greyer neck and underparts, more olive-brown above, duller red bill (which is blackish at first), and duller legs and feet. **Voice** An explosive, nasal, rising *quinquinkrrkrr* in alarm; also a soft *chuck-chuck*. **HH** Typically in small parties but can be found in quite large dispersed flocks in large wetlands. Diurnal, and where undisturbed is not shy. Forages mainly within tall wet grasses and reedbeds, also in dense patches of water hyacinth; readily clambers about in vegetation but also swims out to forage in lotus beds. **TN** The subspecies of the Indian Subcontinent is sometimes treated as a separate species, Grey-headed Swamphen *P. poliocephalus*.

Common Moorhen *Gallinula chloropus*   32–35cm

Locally common winter visitor and scarce resident. **ID** White lateral undertail-coverts with black central stripe, and usually shows white line along flanks. Breeding adult has blackish head and neck, slate-grey underparts, and dark olive-brown upperparts; red bill with yellow tip and red frontal shield. Non-breeding adult has duller bill and legs. Juvenile has dull green bill, and is mainly brown with whitish throat, grey wash to breast and flanks, and variable whitish patch on belly. Immature resembles non-breeding adult, but has duller body plumage, with grey of underparts washed brown and buff, whitish throat, and less prominent pale border to the flanks. **Voice** A *cuk cuk cuk* and *kekuk* in alarm; also a loud explosive *kurr-ik* and *kark*; and soft muttering *kook... kook*. **HH** Usually forages in open; spends most of day swimming among aquatic plants; also feeds on land. Lakes, pools and marshes with emergent vegetation.

Eurasian Coot *Fulica atra*   36–38cm

Locally common winter visitor. **ID** Blackish, with white bill and frontal shield. Shows paler trailing edge to secondaries in flight. Immature duller than adult with whitish throat. Juvenile grey-brown with whitish throat and breast. **Voice** Calls include high-pitched *pyee* and series of long, soft *dp... dp* notes. **HH** Gregarious, diurnal and not shy. Forages chiefly for aquatic vegetation in open water, mainly by diving; also by sieving plant material from the surface. When disturbed, prefers to skitter away along the water surface rather than fly. Takes to the air with difficulty, after prolonged pattering across the water. Large dispersed flocks found in open water with emergent vegetation in larger beels in NE. **AN** Common Coot.

juv

♂ br

♀

ad

White-breasted
Waterhen

Watercock

♀ juv

♂ non-br
transitional

Purple Swamphen

juv

juv

ad

Common Moorhen

ad

juv

ad

Eurasian Coot

Masked Finfoot *Heliopais personatus* 56cm

Rare resident in E half of Sundarbans. **ID** A large, grebe-like bird with huge yellow bill, green legs and feet, and long pointed tail. Male has black forehead, throat and foreneck, white stripe extending down neck-sides, and (in breeding condition) small yellow horn at base of bill. Female has white throat and foreneck. Immature is similar to female but has grey forehead. **Voice** A high-pitched bubbling sound and a loud grunting quack. **HH** Prefers narrow tidal creeks in dense mangrove forest, possibly avoiding areas with cyclone damage and high salinity. Nests in branches of mangrove trees close to creeks. **Threatened** Globally (EN) and nationally (EN).

Bengal Florican *Houbaropsis bengalensis* 66cm

Former resident; extirpated, with no records since 19th century. **ID** Larger and stockier than Lesser Florican, with broader head and thicker neck. Breeding male has black head, neck and underparts, and white wing-coverts. In flight, wings are entirely white except for black tips. Non-breeding male is similar to female but with white wing-coverts, and retains some black on belly. Female is larger than male and has buffish neck and underparts, and pale buff wing-coverts with dark flight feathers. Immature is similar to female but has banding on flight feathers, and wing-coverts are more heavily marked. **Voice** Shrill, metallic *chik-chik-chik* when disturbed. **HH** Very shy. Forages in short-grass areas in early morning and late afternoon. Most easily seen in breeding season when males perform striking displays, leaping above grassland in early morning and evening. Grassland with scattered bushes. **Threatened** Globally (CR).

Lesser Florican *Sypheotides indicus* 46–51cm

Former summer visitor; extirpated, with no records since 19th century. **ID** Smaller and slimmer than Bengal, with smaller head, finer neck and more rapid wingbeats. Breeding male has black head, neck and underparts, and white wing-coverts. Differs from Bengal by white throat, spatulate-tipped head plumes, and white 'hind-collar'; also, white in wing is more restricted, and there is rufous banding across dark flight feathers. Non-breeding male similar to female but has whiter wing-coverts and black underwing-coverts. Female and immature differ from Bengal by dark crescent below eye, dark stripes on foreneck, and rufous-barred primaries. **Voice** Short whistle when disturbed. **HH** Less shy than Bengal. When flushed, flies off with fast wingbeats, covering some distance before landing and running into cover. Dry grassland. **Threatened** Globally (EN).

Sarus Crane *Antigone antigone* 156cm

Former rare resident; extirpated, with no records since 1989–91. **ID** A huge, mainly pale grey crane with reddish legs and very large bill. Adult is grey, with bare red head and upper neck and bare ashy-green crown. In flight, black primaries contrast with rest of wing. Immature has rusty-buff feathering to head and neck, and upperparts are marked with brown; older immatures are similar to adult but have dull red head and upper neck and lack greenish crown of adult. **Voice** Very loud trumpeting, usually a duet by pairs at rest or in flight. **HH** Like other cranes has powerful flight with head and neck extended, and legs and feet outstretched behind. Cultivation in well-watered country. **TN** Formerly in the genus *Grus*. **Threatened** Globally (VU).

♀

♂

Masked Finfoot

Bengal Florican

♂

imm

♀

♂

♀ br

♀

♀

♂ br

♂

♂ non-br

Lesser Florican

juv

ad

Sarus Crane

Lesser Adjutant *Leptoptilos javanicus* 110–120cm

Local resident in Sundarbans in SW, local visitor and former resident in NE, rare visitor elsewhere. **ID** Only confusion species is Greater Adjutant which is a vagrant to Bangladesh (see Appendix 1). Flies with neck retracted, as Greater Adjutant, giving rise to different profile compared with other storks. Smaller than Greater, with slimmer bill that has straighter ridge to culmen. From adult breeding Greater Adjutant by smaller size, glossy black mantle and wings (lacking paler panel across greater coverts – although this is much less distinct in non-breeding and immature Greater), and white undertail-coverts; neck ruff is largely black (appearing as black patch on sides of breast in flight). Also, has pale frontal plate, denser hair-like feathering on back of head (forming small crest) and down hindneck, and lacks neck pouch. Breeding adult has red tinge to face and neck, copper spots at tips of median coverts, and narrow white fringes to scapulars and inner greater coverts. Juvenile is similar to adult, but upperparts are dull black, and head and neck are duller and more densely feathered. **Voice** Largely silent away from nest. **HH** Usually found singly. Forages by walking slowly on dry ground or in shallow water, and grabs prey with its bill. Semi-colonial or colonial when nesting in large trees. Frequents tidal mud among mangrove forest, also wet paddy-fields, marshes and pools. **Threatened** Globally (VU) and nationally (VU).

Painted Stork *Mycteria leucocephala* 93–100cm

Scarce non-breeding visitor mainly to W and coast. **ID** Adult has downcurved yellow bill, bare orange head (redder in breeding season) and pinkish legs; white barring on mainly black upperwing-coverts, pinkish tertials, and black barring across breast. In flight, underwing appears mainly dark, with whitish barring on coverts. Juvenile dirty greyish-white, with grey-brown (feathered) head and neck and brown lesser coverts; bill and legs duller than adult's. Has distinctive appearance in flight, with extended drooping neck and downcurved bill, long wings with rather deep flapping beats, and long trailing legs. **Voice** Largely silent away from nesting colonies. **HH** Like other storks has a powerfully slow-flapping flight and frequently soars for long periods, often at great heights. Found singly and in small parties. Large rivers, coasts and large freshwater wetlands. Forages by wading slowly in shallow water with bill open and partly submerged, feeling for prey. **Threatened** Nationally (CR).

Asian Openbill *Anastomus oscitans* 68cm

Widespread resident. **ID** Stout, dull-coloured 'open bill'. Largely white (breeding) or greyish-white (non-breeding), with black flight feathers and tail; legs usually dull pink, brighter in breeding condition. Juvenile has brownish-grey head, neck and breast, and brownish mantle and scapulars slightly paler than blackish flight feathers. Often forages in medium-sized to large flocks. See Appendix 1 to compare with vagrant White Stork. **HH** Usually feeds, mainly on molluscs, by submerging its head and opened bill into shallow water and probing bottom mud. Moves locally depending on water conditions. Like other storks it regularly soars on thermals. Nests in trees, often over water, forming large colonies, sometimes mixed with other waterbirds. Inhabits freshwater wetlands with shallow water; also forages in wet paddy-fields.

Asian Woollyneck *Ciconia episcopus* 75–92cm

Rare visitor mainly to NW where it is now resident, with two recent nesting sites. **ID** Stocky, largely blackish stork with 'woolly' white neck, black 'skullcap', and white vent and undertail-coverts. Adult has black of body and wings glossed with greenish-blue, purple and copper. Bill is black, with variable amounts of red, and legs and feet are dull red. Juvenile similarly patterned to adult, but has duller brown body and wings, and feathered forehead. In flight, upperwing and underwing entirely dark. **Voice** Largely silent away from nest. **HH** Usually found singly or in pairs. Habits are similar to those of other storks. Hunts on dry or marshy ground and in wet grasslands; rarely wades. Nest site in Bangladesh is in a telecommunications tower. Mainly found along main rivers and associated wet paddy-fields. **AN** Woolly-necked Stork. **Threatened** Nationally (CR).

Black Stork *Ciconia nigra* 90–100cm

Rare winter visitor. **ID** Adult mainly glossy black, with white lower breast and belly, and red bill and legs; in flight, white underparts and axillaries contrast strongly with black neck and underwing. Juvenile has brown head, neck and upperparts flecked with white; bill and legs greyish-green. **Voice** Largely silent away from nest. **HH** Usually single birds or small parties. Often very shy and wary. Forages by walking with measured strides in shallow water. Soars for long periods on thermals with legs and neck extended. Mainly found along large rivers on sandy islands (chars). **Threatened** Nationally (VU).

Lesser Adjutant

ad

ad

Painted Stork

br

non-br

imm

ad

imm

br

non-br

br

Asian Woollyneck

ad

ad

Asian Openbill

ad

ad

imm

ad

Black Stork

Black-necked Stork *Ephippiorhynchus asiaticus*
129–150cm

Rare visitor. **ID** Large black-and-white stork with long red legs and huge black bill. In flight, wings white except for broad black band across coverts, and tail black. Male has brown iris, yellow in female. Juvenile has fawn-brown head, neck and mantle, mainly brown wing-coverts, and mainly blackish-brown flight feathers; legs dark. **Voice** Largely silent away from nest. **HH** Forages singly, in well-separated pairs within sight of each other, or in family parties after the breeding season. Usually very wary. When foraging, walks sedately in marshes or wades slowly while probing among aquatic vegetation and in shallow water with its bill open at the tip. Frequently soars for long periods, often at great height; legs are extended and neck is outstretched like other storks. Builds a solitary nest in large tree. Inhabits large wetlands and large rivers. **Threatened** Nationally (EN).

Eurasian Spoonbill *Platalea leucorodia*
80–90cm

Scarce and declining winter visitor. **ID** White, with spatulate-tipped bill. In flight, neck is outstretched, and flapping is rather stiff and interspersed with gliding. Adult has black bill with yellow tip; has crest and yellow breast patch when breeding. Juvenile has pink bill; in flight, shows black tips to primaries. **Voice** Usually silent away from breeding colonies. **HH** Often in small parties or flocks. Spends much of the day resting on one leg or sleeping with the bill tucked under a wing. Forages mainly in the mornings and evenings and at night. Wades actively in shallow water, making rhythmic side-to-side sweeps of its bill and sifting floating and swimming prey. Mainly on coastal mudflats and chars, also rarely along large rivers. **Threatened** Nationally (CR).

Black-headed Ibis *Threskiornis melanocephalus*
75cm

Scarce resident and winter visitor. **ID** Stocky, mainly white ibis with stout downcurved black bill. Adult breeding has naked black head, white lower-neck plumes, variable yellow wash to mantle and breast, and grey on scapulars and elongated tertials. In flight, shows stripe of bare red skin on underside of white forewing and on flanks. Adult non-breeding has all-white body and lacks neck plumes. Immature has grey feathering on head and neck, and black-tipped wings. **Voice** Usually silent away from breeding colonies. **HH** Seen mainly in parties. Forages in shallow water and marshes, often in well-scattered flocks. Walks about or often wades belly-deep while probing rapidly in water and mud. Readily perches and roosts in trees. Inhabits large wetlands including haors, major rivers and coastal chars. **Threatened** Nationally (VU).

Red-naped Ibis *Pseudibis papillosa*
68cm

Rare and localised resident restricted to extreme NW. **ID** Stocky, dark ibis with relatively stout downcurved bill. Has white shoulder patch and reddish legs. Appears bulky and broad-winged in flight, with only the feet extending beyond tail. Adult is mostly dark brown with green-and-purple gloss, but has naked black head with red nape. Immature dark brown, including feathered head. **Voice** Mainly silent away from breeding colonies. **HH** Habits are similar to those of Black-headed Ibis but prefers to feed in drier habitats and is less gregarious. Usually found singly or in small parties. Often nests singly, but sometimes in small, single-species colonies. Builds large platform nest in large tree; old nests of kites, storks and vultures are often used. Frequents riverbanks and open fields, including dry areas.

Glossy Ibis *Plegadis falcinellus*
55–65cm

Rare but increasing winter visitor mainly to NE. **ID** Small, dark ibis with rather fine downcurved bill. Graceful in flight, with extended slender neck, somewhat bulbous head, and legs and feet projecting well beyond tail. Breeding adult deep chestnut, glossed with purple and green; has narrow white surround to bare lores. Non-breeding adult duller, with white streaking on dark brown head and neck. Juvenile similar to adult non-breeding, but is dark brown with white mottling on head, and only faint greenish gloss to upperparts. **HH** Walks in shallow water or wades belly-deep while probing rapidly in water and mud. Usually found in flocks. Freshwater wetlands, especially shallow waters in haors and nearby wet paddy-fields.

Black-necked Stork
(not to scale)

♀

♂

ad

imm

imm

ad

imm

juv

ad

Black-headed Ibis

br

Eurasian Spoonbill

ad

br

non-br

juv

Red-naped Ibis

Glossy Ibis

Yellow Bittern *Ixobrychus sinensis* 38cm

Uncommon resident. **ID** Yellowish-buff wing-coverts contrast with dark brown flight feathers. Male has pinkish-brown mantle/scapulars, and face and neck-sides are vinaceous. Female is similar to male but has rufous streaking on black crown, variable rufous-orange streaking on foreneck and breast, and buff streaking to rufous-brown mantle and scapulars. Juvenile appears buff with bold dark streaking to upperparts including wing-coverts; foreneck and breast are heavily streaked. **Voice** Territorial call is low-pitched *ou-ou*. **HH** Habits are those of a typical bittern. Solitary. Most active at dusk; usually spends the day concealed in thick waterside vegetation but may be seen during the daytime in cloudy weather. Forages by creeping through dense vegetation or by standing and waiting at the edges of cover. If disturbed, often freezes with head and bill pointing vertically skywards. Nest is usually built in a dense reedbed. Inhabits freshwater wetlands with dense reed-swamp and emergent vegetation.

Cinnamon Bittern *Ixobrychus cinnamomeus* 38cm

Uncommon resident. **ID** Uniform-looking cinnamon-rufous flight feathers and tail in all plumages. Male has cinnamon-rufous crown, hindneck and mantle/scapulars. Female has browner crown and mantle, and brown streaking on foreneck and breast. Juvenile has buff mottling on dark brown upperparts, and is heavily streaked with dark brown on underparts. **Voice** Territorial call is a loud *kok-kok*. **HH** Habits very similar to those of Yellow; often found in same locality and same habitat as that species. Nest is built on bent-over reeds about 1m above water or mud. Inhabits freshwater wetlands and nearby dense vegetation, including smaller wetlands and drier scrubby thickets compared with Yellow.

Black Bittern *Ixobrychus flavicollis* 58cm

Rare and local resident, mainly in NE. **ID** Male has blackish upperparts, with yellowish malar and sides of neck and dark streaking on underparts. Female similar but browner upperparts and chestnut-streaked underparts. Juvenile similar but with distinct fringes to upperparts. **Voice** Territorial call is a loud booming. **HH** Habits are those of a typical bittern. Chiefly nocturnal and crepuscular; skulks in dense swamps during day. Most often seen flying at dawn and dusk and in cloudy weather. Nests in reeds or in dense thicket in a marsh. Inhabits forest pools, marshes and reed-edged wetlands. **TN** Sometimes included in genus *Dupetor*.

Malay Night-heron *Gorsachius melanolophus* 51cm

Rare resident in E, one record in Sundarbans. **ID** Stocky, with stout bill and short neck. Adult has black crown and crest, rufous sides to head and neck, and rufous-brown upperparts. Juvenile is greyish in coloration, finely vermiculated with white, black and rufous-buff, and has bold white spotting on crown and crest. **Voice** Usually silent. A sequence of 10–11 deep *oo* notes each about 1.5 seconds apart. **HH** Shy and mainly nocturnal. Skulks in damp places in forest undergrowth in dense forest during day. If flushed, flies off silently into nearby thickly foliaged tree. Builds nest in tree overhanging or close to stream in thick forest. Inhabits wet areas in dense forest. **AN** Malayan Night Heron.

Black-crowned Night-heron *Nycticorax nycticorax* 58–65cm

Locally uncommon resident. **ID** Stocky, with thick neck. Adult has black crown and mantle contrasting with grey wings and whitish underparts. Juvenile brownish and boldly streaked and spotted with white. Immature is similar to adult but has browner mantle/scapulars, and variable streaking on underparts. **Voice** A distinctive, deep and rather abrupt *wouck* in flight. **HH** Nocturnal and crepuscular except when feeding young. Usually spends the day perching hunched in a densely foliaged tree or in reedbed. Most often seen at dusk, flying singly or in small groups from its daytime roost. Like other herons, flies with leisurely flaps, with legs outstretched and projecting beyond tail, nearly always with head and neck drawn back. Mainly forages singly, sometimes in loose groups. Wetlands including ponds and beels (lakes).

Yellow Bittern

♂

juv

Cinnamon Bittern

♂

♀

juv

Black Bittern

♂

imm

ad

Malay Night-heron

juv

ad

juv

Black-crowned Night-heron

Green-backed Heron *Butorides striata* 40–48cm

Locally common resident along coast, scarce elsewhere. **ID** Small, stocky, short-legged heron. Adult has black crown and crest, dark greenish upperparts and greyish underparts. Juvenile has buff streaking and spotting on upperparts and dark-streaked underparts. Immature is similar to juvenile with uniform brown crown and mantle. **Voice** Usually silent but sometimes a *k-yow k-yow* or *k-yek k-yek.* **HH** Normally frequents the same area day after day. Hunts alone and in typical heron fashion. Often crepuscular although sometimes birds are active during day, especially in overcast weather. In daytime, mainly keeps to thick vegetation on banks of rivers and pools and is often seen perching on branches overhanging water. Mangrove forest and coastal wetlands, also large rivers and freshwater wetlands with dense shrubby vegetation on banks. **AN** Striated Heron.

Indian Pond-heron *Ardeola grayii* 42–45cm

Common and widespread resident. **ID** Whitish wings contrast with dark mantle/scapulars. Adult breeding has yellowish-buff head and neck, and maroon-brown mantle/scapulars. Head, neck and breast streaked/spotted in non-breeding and immature plumages. **Voice** A high, harsh squawk when flushed. **HH** Usually solitary when foraging but will gather in scattered flocks at drying-out pools to feed on stranded fish, and before roosting. Like other herons, typically hunts by standing motionless at water's edge, waiting for prey to swim within reach, or by slow stalking in shallow water or on land. Roosts communally. Tame and inconspicuous when perched but flies up with a startling flash of white wings. Marshes, flooded paddy-fields, lakes, village ponds, streams and ditches.

Chinese Pond-heron *Ardeola bacchus* 52cm

Rare visitor, mainly in NE and SE. **ID** In breeding plumage has maroon-chestnut head and neck and slaty-black mantle/scapulars. Non-breeding and immature plumages probably not separable from those of Indian Pond. **Voice** Presumably like that of Indian Pond. **HH** Similar to Indian Pond; found in spring (when in breeding plumage) in haors and smaller wetlands in NE and SE, status in other seasons unknown.

Grey Heron *Ardea cinerea* 90–98cm

Widespread, locally common winter visitor and rare resident. **ID** A large, mainly grey heron, lacking any brown or rufous in its plumage. In flight, black flight feathers contrast with grey upperwing- and underwing-coverts and prominent white leading edge to wing when head-on. Adult has yellow bill, whitish head and neck with black head plumes, and black patches on belly. In breeding season, has whitish scapular plumes, and bill and legs become orange or reddish. Immature is duller than adult, with grey crown, reduced black 'crest', greyer neck, less pronounced black patches on sides of belly, and duller bill and legs. Juvenile has dark grey cap with slight crest, dirty grey neck and breast, lacks black patches on belly-sides, lacks plumes, and has dark legs. See Appendix 1 for comparison with White-bellied Heron. **Voice** Often calls in flight, a loud *frarnk.* **HH** A typical diurnal heron. Usually feeds singly; occasionally gathers in loose parties at good feeding areas. Roosts communally in winter. Prefers to hunt in the open, unlike Purple Heron. Larger wetlands, including coastal mudflats, mangroves, large rivers, beels and haors.

Purple Heron *Ardea purpurea* 78–90cm

Local and uncommon winter visitor and resident. **ID** Rakish, with long, thin neck. In flight, compared with Grey Heron, bulge of recoiled neck is very pronounced, protruding feet large, underwing-coverts purplish (adult) or buff (juvenile), and lacks white leading edge to wing. Adult has chestnut head and neck with black stripes, grey mantle and upperwing-coverts, and dark chestnut belly and underwing-coverts. Juvenile has black crown, buffish neck, and brownish mantle and upperwing-coverts with rufous-buff fringes. See Appendix 1 for comparison with Goliath Heron. **Voice** Flight call similar to Grey, but higher-pitched and not so loud *frarnk.* **HH** Active mainly in early morning and evening; sometimes also feeds by day. Shyer than Grey, normally feeding out of sight among dense aquatic vegetation. Most often seen in flight. Hunts alone, usually by standing motionless and waiting; less often by slow stalking in shallow water. Dense aquatic vegetation particularly in NE haors; scarcer in coastal wetlands.

juv

Green-backed Heron

non-br

br

Indian Pond-heron

ad

br

Chinese Pond-heron

ad

ad

non-br

imm

ad

Grey Heron

Purple Heron

juv

Cattle Egret *Bubulcus ibis*

48–53cm

Common and widespread resident. **ID** Small and stocky with short yellow bill and short dark legs. Has orange-buff on head, neck and mantle in breeding plumage; base of bill and legs become reddish in breeding condition. All white in non-breeding plumage. **Voice** Usually silent away from breeding colony. **HH** Gregarious when feeding and roosting. Typically seen in flocks around domestic cattle and buffalo, feeding on insects disturbed by animals; often rides on animals' backs, picking parasitic insects and flies from their hides. Also forages in flooded fields. Unlike other egrets feeds mainly on insects, also takes tadpoles and lizards. Breeds colonially in large trees, not necessarily close to water; sometimes with other herons and egrets. Inhabits damp grassland, paddy-fields, lakes, pools and marshes.

Great White Egret *Ardea alba*

90–102cm

Common and widespread resident. **ID** Compared with Intermediate, larger and longer-billed, and looks thinner-necked with more angular and pronounced kink to neck. Black line of gape extends behind eye. In breeding plumage, bill is black, lores blue and tibia reddish, and has prominent plumes on mantle. In non-breeding plumage, bill yellow and lores pale green. **Voice** Normally silent, but occasionally utters low *kraak*; gives various deep guttural calls and softer notes during display. **HH** A typical diurnal heron. Like other herons, feeds by standing motionless at water's edge waiting for prey to swim within reach, or by slow stalking in shallow water or on land. Prey is normally grasped and killed by battering and is less often speared. Generally less sociable than other egrets, and is often solitary when foraging, although will feed communally at concentrated food sources. Roosts communally. Breeds colonially with other herons and cormorants. Builds nest in lone tree or grove, either standing in water or on dry land. Inhabits coastal and freshwater wetlands including rivers, lakes, marshes, pools and wet fields. **AN** Great Egret. **TN** Placed in *Egretta* or in monotypic genus *Casmerodius* by some authorities, but genetic studies link it more closely with *Ardea*.

Intermediate Egret *Ardea intermedia*

65–72cm

Common and widespread resident. **ID** Smaller than Great White, with shorter bill and neck. Black gape-line does not extend beyond eye. Shape can appear similar to Cattle but larger and usually found in water. Bill is black and lores yellow-green during courtship, and has pronounced plumes on breast and mantle. Has black-tipped yellow bill and yellow lores outside breeding season. **Voice** Normally silent; gives distinctive buzzing calls during display. **HH** A typical diurnal heron. Flies with leisurely flaps, with the legs outstretched and projecting beyond the tail, and nearly always with head and neck drawn back like other herons. Usually in small flocks which separate when foraging. Hunts chiefly by slow stalking. Roosts communally. Breeds colonially with other herons and cormorants. Builds nest in lone tree or grove, either standing in water or on dry land. Inhabits coastal and freshwater wetlands including mangroves, haors, marshes, flooded grassland, well-vegetated pools and rivers. **TN** Placed in *Egretta* or *Mesophoyx* by some authorities but genetic studies link it more closely with *Ardea*.

Little Egret *Egretta garzetta*

55–65cm

Common and widespread resident. **ID** Slim and graceful. Has black bill, black legs with yellow feet, and greyish or yellowish lores. In breeding plumage, has two elongated nape plumes and mantle plumes; lores and feet become reddish during courtship. Bill in non-breeding and immature can be paler and pinkish or greyish at base, or dull yellowish on some. Very rarely has some dark feathers in plumage. See Appendix 1 for comparison with Eastern Reef Egret. **Voice** Normally silent, except for throaty squawk when disturbed and various guttural calls at colonies. **HH** A typical diurnal heron. Often in flocks when foraging and more sociable than the two larger egrets; also found singly. Roosts communally. Breeds colonially with other herons and cormorants. Builds nest in lone tree or grove, either standing in water or on dry land. Inhabits coastal and freshwater wetlands including haors, lakes, rivers, pools, marshes and flooded paddy-fields.

br

non-br

Cattle Egret

non-br

br

Great White Egret

non-br

non-br

br

non-br

Intermediate Egret

non-br

br

non-br

Little Egret

Spot-billed Pelican *Pelecanus philippensis* 140cm

Former rare resident; presumed extirpated but with one recent vagrant record in NW. **ID** Much smaller than Great White Pelican (see Appendix 1), with dingier appearance, rather uniform pinkish bill and pouch (except in breeding condition), and black spotting on upper mandible (except juveniles). Pale circumorbital skin looks separated from bill (appears to be wearing goggles). Tufted crest/hindneck usually apparent even on young birds. Underwing pattern shows little contrast between wing-coverts and flight feathers, and paler greater coverts produce distinct central panel. Adult breeding has cinnamon-pink rump, underwing-coverts and undertail-coverts; head and neck appear greyish; has purplish skin in front of eye, and pouch is pink to dull purple and blotched with black. Adult non-breeding dirtier greyish-white, with paler pouch and facial skin. Immature has variable grey-brown markings on upperparts. Juvenile has brownish head and neck, brown mantle and upperwing-coverts (fringed with pale buff), and brown flight feathers; spotting on bill initially lacking (and still indistinct at 12 months). **Voice** Usually silent away from breeding colonies. **HH** Fishes alone or cooperatively by swimming forward in a semicircular formation, driving fish into shallow waters. A powerful flier, proceeding by steady flaps with the head drawn back between the shoulders. Ganges River.

Little Cormorant *Microcarbo niger* 51cm

Common resident. **ID** Much smaller than Great Cormorant. Has shortish bill, rectangular head (with steep forehead), short neck and tail appears long. Lacks yellow gular pouch. Breeding adult all black, with white plumes on head-sides. Bill, eyes, facial skin and pouch are black. Non-breeding browner (and lacks white head plumes), with whitish chin, and paler bill and pouch. Immature has whitish chin and throat, and foreneck and breast a shade paler than upperparts, with some pale fringes. **Voice** Usually silent except in vicinity of nest. **HH** Like other cormorants swims with body low in water, the neck straight and the head and bill pointing a little upwards. In flight, neck is extended and head is held slightly above horizontal. On smaller waters occurs singly or in small groups; on large waters often gathers in flocks. Breeds in trees in mixed colonies with other waterbirds. Inhabits all coastal and freshwater wetlands including mangroves, rivers, haors, lakes, reservoirs and ponds. **TN** Formerly in genus *Phalacrocorax*.

Great Cormorant *Phalacrocorax carbo* 80–100cm

Fairly common winter visitor. **ID** Much larger and bulkier than Little. Has thick neck, large and angular head, and stout bill. Adult breeding glossy black, with dark gular skin, red spot at base of bill, white cheeks and throat, extensive white plumes covering much of head, and white thigh patch. Non-breeding lacks white head plumes and thigh patch. Base of bill and gular skin are yellow. Immature similar but browner with underparts dark or extensively whitish or pale buff. Flies and swims like a typical cormorant; see Little. **Voice** Usually silent except in vicinity of nest. **HH** Like other cormorants it often perches for long periods in upright posture with spread wings and tail. Usually found singly or in small groups. Often roosts communally in winter. Large freshwater wetlands including large rivers, haors and large fish ponds.

Indian Cormorant *Phalacrocorax fuscicollis* 63cm

Rare presumed resident and wanderer. **ID** Smaller and slimmer than Great Cormorant, with thinner neck, slimmer oval-shaped head, finer-looking bill, and proportionately longer tail. In flight, looks lighter, with thinner neck and quicker wing action. Larger than Little Cormorant, with longer neck, oval-shaped head and longer bill. Adult breeding glossy black, with blue eyes, dark facial and gular skin, tuft of white behind eye, scattering of white filoplumes on neck. Non-breeding lacks white plumes; has whitish throat, yellowish gular pouch, and browner-looking head, neck and underparts. Immature has brown upperparts and whitish underparts. **HH** Flies and swims like a typical cormorant; see Little. Large freshwater wetlands: rivers and beels.

Oriental Darter *Anhinga melanogaster* 85–97cm

Scarce resident and visitor. **ID** Long, slim head and neck, dagger-like bill, and long tail. Adult breeding has dark brown crown and hindneck, white stripe down neck-side, blackish breast and underparts, lanceolate white scapular streaks, and white streaking on wing-coverts. Duller in non-breeding plumage. Immature browner with indistinct neck-stripe and buff fringes to coverts forming pale panel on upperwing. **Voice** Usually silent except in vicinity of nest. **HH** Spends much time drying its spread wings and tail. Often swims with body below water surface. Frequently nests colonially with other large waterbirds. Inhabits freshwater wetlands including haors, lakes, slow-moving rivers and marshes. **AN** Darter if African and Australasian forms are not split.

Spot-billed Pelican

non-br

br

br

imm

juv

Little Cormorant

non-br

imm

br

Great Cormorant

non-br

br

imm

Indian Cormorant

non-br

imm

br

br

Oriental Darter

br

imm

br

Indian Thick-knee *Burhinus indicus* 36–39cm

Scarce local resident in NW and SW. **ID** Sandy-brown and streaked. Has short yellow-and-black bill, striking yellow eye, and long yellow legs. In flight, has black flight feathers with patches of white in primaries. Is smaller, darker and more heavily streaked than extralimital Eurasian Thick-knee *B. oedicnemus*, with larger bill, shorter tail and longer tarsi. Bill is mainly black and shows more pronounced barring on wing-coverts (with broader pale panel). **Voice** Piercing calls recall Great Thick-knee. **HH** Typically spends day sitting in shade. Very wary; if suspicious runs off furtively with its head low, then squats and flattens itself on ground. When foraging (often at dusk), makes short runs, stopping to capture prey with swift snatch. River islands and other sandy ground and open dry fields near main rivers. **TN** Treated as conspecific with Eurasian Thick-knee *B. oedicnemus* by some authorities.

Great Thick-knee *Esacus recurvirostris* 49–54cm

Rare presumed resident in NW and SW. **ID** Has large, slightly upturned black-and-yellow bill, and yellow eye. At rest, most striking features are white forehead and 'spectacles' contrasting with black ear-coverts, and blackish and whitish bands across wing-coverts. In flight, has grey panel on wings and white patches on primaries. **Voice** A rising, wailing whistle of two or more syllables; a loud, harsh *see-eek* alarm call. **HH** Habits similar to Eurasian but usually rests in full sun during day. Sandbanks of large rivers and islands along coast.

Eurasian Oystercatcher *Haematopus ostralegus* 40–46cm

Rare winter visitor to coast; one exceptional nesting record from edge of Sundarbans. **ID** Black and white, with broad white wingbar. Bill and eyes reddish, and legs pinkish. White collar in non-breeding plumage. Shows broad white wingbar in flight. **Voice** A piping *pi... peep... peep... peep* and *pi-peep.* **HH** Coastal islands and intertidal mudflats. **Threatened** Nationally (VU).

Pied Avocet *Recurvirostra avosetta* 42–45cm

Uncommon winter visitor. **ID** Upward kink to black bill. Distinctive black-and-white pattern. Juvenile has brown and buff mottling on mantle and scapulars. **Voice** A throaty *quib... quib.* **HH** Characteristically feeds by sweeping bill and head from side to side in shallow water; often swims and upends like a dabbling duck, and picks prey from surface of water or mud. Coastal mudflats, rare in large rivers and haors.

Black-winged Stilt *Himantopus himantopus* 35–40cm

Locally common winter visitor to NE, scarce elsewhere. **ID** Slender appearance, with long pinkish legs, and fine straight bill. Black upperwing strongly contrasts with white back V in flight. Adult at rest shows mainly white head, neck and underparts, contrasting with upperparts, and reddish-pink legs. Both sexes can show variable amounts of black and/or dusky grey on crown and hindneck. Juvenile has browner upperparts with buff fringes. **Voice** Noisy and readily agitated; calls include *kek... kek* and a rather anxious *kikikikiki.* **HH** Gregarious throughout year. Graceful; walks slowly and deliberately. Forages on dry mud and by wading into shallows, sometimes belly-deep in water. Picks prey from surface, probes in soft mud and sweeps bill from side to side; sometimes immerses head and neck in water. Shallow waters in haors and other wetlands, rare on coast.

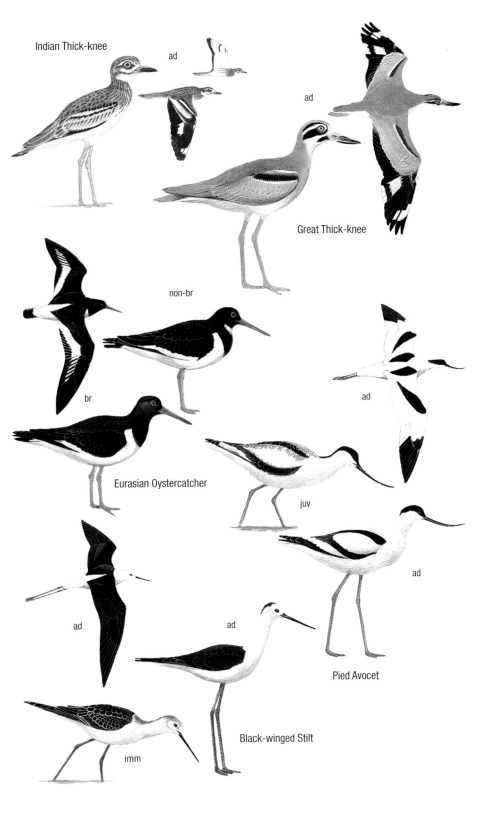

Indian Thick-knee

ad

ad

Great Thick-knee

non-br

br

ad

Eurasian Oystercatcher

juv

ad

ad

ad

Pied Avocet

Black-winged Stilt

imm

Grey Plover *Pluvialis squatarola* 27–30cm

Common winter visitor to coast. **ID** White underwing and black axillaries. Stockier, with stouter bill and shorter legs than Pacific Golden. Whitish rump and prominent white wingbar are important features from Pacific Golden. Has extensive white spangling to upperparts in breeding plumage; upperparts mainly grey in non-breeding (in all plumages lacks golden spangling of Pacific Golden). **Voice** A mournful *chee-woo-ee*. **HH** Habits similar to those of Pacific Golden but usually less gregarious; often in pairs or small groups and with other wader species. Sandy shores, mudflats and tidal creeks.

Pacific Golden Plover *Pluvialis fulva* 23–26cm

Common winter visitor and passage migrant. **ID** A slim-bodied, long-necked and long-legged plover with short dark bill. In all plumages, has golden-yellow markings on upperparts, and dusky-grey underwing-coverts and axillaries. In flight, shows narrower white wingbar and dark rump. Adult breeding has black on face, foreneck, breast and belly, which is strikingly bordered by white. Adult non-breeding and juvenile have yellowish-buff wash to supercilium, cheeks and neck; usually shows pronounced supercilium, which often curves down as diffuse crescent behind ear-coverts, and also a dark patch on rear ear-coverts. **Voice** An abrupt disyllabic *chi-vit* recalling Spotted Redshank and a more plaintive *tu-weep* or *chew you*. **HH** Very wary and, if disturbed, groups rise almost simultaneously in a compact flock, twisting and turning rapidly in unison. Has a typical plover feeding behaviour: running in short spurts, pausing, then stooping to pick up prey. Mudbanks of coastal and freshwater wetlands, short wet grass and ploughed fields.

Little Ringed Plover *Charadrius dubius* 14–17cm

Common and widespread resident and winter visitor, scarcer on coast. **ID** Small, elongated and small-headed appearance, and uniform upperwing with only a very narrow wingbar. Bill small and mainly dark. Legs yellowish or pinkish. Adult breeding has black mask and frontal band, striking yellow eye-ring, and black breast band. Adult non-breeding and juvenile have rather uniform head pattern with pale forehead and supercilium. See Appendix 1 for differences from Long-billed Plover. **Voice** Clear, descending *pee-oo* or shorter *peeu* in flight. **HH** In pairs or small flocks which scatter over a wide area when feeding. Carries the head low and drawn into the shoulders. Has typical plover feeding action: makes short runs, then pauses and stoops stiffly to pick up prey. Sand and mudbanks of rivers, pools and beels.

Kentish Plover *Charadrius alexandrinus* 15–17cm

Common winter visitor. **ID** Small and stocky. White hind-collar and usually small, well-defined patches on sides of breast. Upperparts paler and sandier than Little Ringed. Legs usually appear blackish but may be tinged brown or olive-yellow. Male has rufous cap and black eyestripe and forecrown. **Voice** Flight call is a soft *pi... pi... pi*, recalling Little Stint, or a rattling trill *prrr* or *prrtut* (harsher than similar call of Lesser Sandplover); and a plaintive *whoheet*. **HH** Typical plover gait and feeding behaviour (see Little Ringed) but runs more rapidly. Sand and mud along main rivers and coast.

Lesser Sandplover *Charadrius mongolus* 19–21cm

Common winter visitor to coast, were it is the most abundant shorebird. **ID** Larger and longer-legged than Kentish, lacking white hind-collar. Very difficult to identify from Greater but is smaller with stouter bill (equal to or shorter than distance between bill-base and rear of eye, with blunt tip), and shorter dark grey or dark greenish legs (with tibia shorter than tarsus). In flight, feet do not usually extend beyond tail and white wingbar is narrower across primaries. Breeding male typically shows full black mask and forehead, or white patch on forehead, and more extensive rufous on breast compared with Greater (although these characters vary). **Voice** A hard *chitik*, *chi-chi-chi*, and *kruit-kruit* in flight, all rather short and sharp. **HH** Gait and feeding behaviour typical of plovers. Coastal mudflats and sandbanks, also rarely along large rivers.

Greater Sandplover *Charadrius leschenaultii* 22–25cm

Common winter visitor to coast. **ID** Larger and lankier than Lesser Sandplover, with longer and larger bill, usually with pronounced gonys and more pointed tip (longer than distance between bill-base and rear of eye). Longer legs are paler, with distinct yellowish or greenish tinge. In flight, feet project noticeably beyond tail, has more pronounced dark subterminal band to tail, and has broader white wingbar across primaries. **Voice** In flight, a trilling *prrrirt* or *kyrrrr... trrr*, softer and longer than that of Lesser. **HH** Gait and feeding behaviour are typical of other plovers. Coastal mudflats and wetlands.

Grey Plover

br

non-br

non-br

Pacific Golden
Plover

br

non-br

non-br

Little Ringed Plover

br

non-br

♂ br

non-br

Kentish Plover

non-br

♀ br

non-br

non-br

non-br

♂ br

non-br

♂ br

♀ br

Lesser Sandplover

Greater Sandplover

Northern Lapwing *Vanellus vanellus* 28–31cm

Rare winter visitor. **ID** Black crest, white (or buff) and black face pattern, black breast-band, and dark green upperparts. Juvenile has prominent buff fringes to mantle, scapulars and wing-coverts. Has very broad, rounded wingtips. Shows whitish rump and blackish tail-band in flight. **Voice** Mournful *eu-whit*. **HH** Found singly or in small flocks. Feeding behaviour is similar to that of Red-wattled. Wet grassland, marshes, beel margins, riverbanks; sometimes fallow fields and dry stubbles.

River Lapwing *Vanellus duvaucelii* 29–32cm

Rare resident. **ID** Black crest, face and throat, grey sides to neck, and black bill and legs. Black patch on belly. In flight, shows broad white greater-covert wingbar contrasting with black flight feathers, and black tail. Juvenile is similar to adult, but black of head is partly obscured by white tips, and has buff fringes and dark subterminal marks to feathers of upperparts. **Voice** Sharp insistent, high-pitched *did, did, did*, sometimes ending with *did-did-do-weet*. **HH** Usually occurs singly, in pairs or in small groups. Feeding behaviour is similar to that of Red-wattled. Often has a hunched posture with head drawn in. Declining; frequents sandbanks on coast and along main rivers.

Yellow-wattled Lapwing *Vanellus malabaricus* 26–28cm

Local and uncommon resident, mainly in C and NW. **ID** Yellow wattles and legs. White eyestripe joining at nape, dark cap, and brown breast-band. In flight, shows white greater-covert wingbar contrasting with black flight feathers, and white tail with black subterminal band. Juvenile has small and dull yellow wattles, white chin, brown cap, and prominent buff fringes and dark subterminal bars to feathers of upperparts. **Voice** Strident *chee-eet* and hard *tit-tit-tit*. **HH** Singly, in pairs or sometimes in small flocks outside breeding season. Foraging behaviour is similar to that of Red-wattled. Flight is buoyant with rather slow wingbeats. Dry fields, open dry country and dry riverbeds.

Grey-headed Lapwing *Vanellus cinereus* 34–37cm

Common winter visitor in NE, uncommon elsewhere. **ID** Yellow bill with black tip, and yellow legs. Grey head, neck and breast, latter with diffuse black border, and black tail-band. Secondaries are white. Juvenile has brownish head and neck, lacks the dark breast-band, and has prominent buff fringes to feathers of upperparts. **Voice** Plaintive *chee-it, chee-it*. **HH** Behaviour is similar to that of Red-wattled. Usually found in small parties or flocks of up to 300 birds. Wet fields and short wet grassland, also muddy fringes of beels and rivers.

Red-wattled Lapwing *Vanellus indicus* 32–35cm

Common and widespread resident. **ID** Black head and breast with white cheek patch, red bill with black tip, and yellow legs. In flight, shows white greater-covert wingbar and black tail-band. Juvenile is duller than adult, with whitish throat. Race in Bangladesh is *atronuchalis*, which differs from widespread nominate subspecies of rest of Indian Subcontinent by lack of white neck-sides, and in having a white collar dividing black of neck from brown mantle. **Voice** Agitated and penetrating *did he do it, did he do it* and a less intrusive *did did did*. **HH** In pairs or small flocks of up to about 12 birds. A vigilant and noisy bird; when alarmed calls loudly and frantically while circling overhead. Forages by walking or running in short spurts, then stops and probes with body tilted forward; also vibrates its foot rapidly on surface to flush invertebrates. Feeds mainly at night and in early morning and evening. Usually flies slowly with deep flaps. Open flat ground near water.

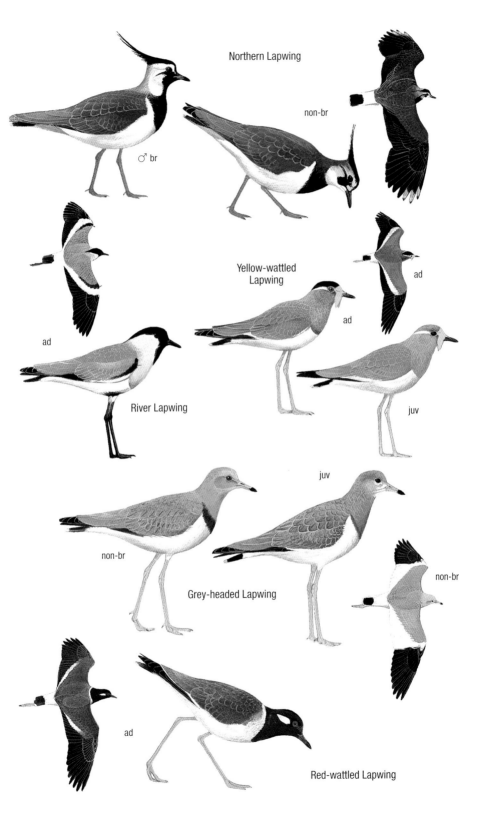

Northern Lapwing

♂ br

non-br

Yellow-wattled
Lapwing

ad

ad

ad

juv

River Lapwing

juv

non-br

Grey-headed Lapwing

non-br

ad

Red-wattled Lapwing

Greater Painted-snipe *Rostratula benghalensis* 25cm

Uncommon resident. **ID** Rail-like wader with broad, rounded wings and longish, downcurved bill. White or buff 'spectacles' and 'braces'. Adult female has maroon head and neck, and dark greenish wing-coverts. Adult male and juvenile duller and have buff spotting on wing-coverts; juvenile lacks dark breast-band, and throat and breast are finely streaked. **Voice** Occasionally an explosive *kek* when flushed; female has a soft *koh koh* in display. **HH** Found singly and in small flocks. Chiefly crepuscular or nocturnal. Skulking, and reluctant to fly if approached; rises heavily with legs trailing, and lands in cover again a short distance away. Feeds by probing in mud, or by sweeping bill from side to side in shallow water while bobbing rear body. Movements are slow and deliberate. Makes a nocturnal display flight along a regular route (roding). Marshes, wet overgrown paddy-fields, pools and ditches, thickly vegetated and with mud patches.

Pheasant-tailed Jacana *Hydrophasianus chirurgus* 31cm

Uncommon resident, more numerous in NE. **ID** Extensive white on upperwing, and white underwing. Yellowish patch on sides of neck. Breeding adult has brown underparts and long, curved tail. Non-breeding adult and juvenile lack elongated tail and have white underparts, with dark line down side of neck and dark breast-band (less distinct than in plate). Juvenile has buff fringes to upperparts and barring across breast. **Voice** A distinctive *me-e-ou* or *me-oup* in breeding season, and a nasal *tewu*. **HH** Gregarious outside breeding season. Walks or rests on floating vegetation; usually found in the open. Forages actively throughout day. Swims well and floats buoyantly. Flight is slow and flapping, with large feet dangling behind. Beels and pools with floating aquatic plants.

Bronze-winged Jacana *Metopidius indicus* 28–31cm

Uncommon, widely distributed resident. **ID** Dark upperwing and underwing. Adult has white supercilium, bronze-green upperparts and blackish underparts. Juvenile has orange-buff wash on breast, short white supercilium and yellowish bill. **Voice** A short, harsh grunt and wheezy, piping *seek-seek-seek.* **HH** Habits and habitat similar to those of Pheasant-tailed, but more frequent in smaller waterbodies and less gregarious. If alarmed, will partially submerge its body among aquatic vegetation.

Oriental Pratincole *Glareola maldivarum* 23–24cm

Scarce passage migrant, winter and summer visitor to coast and large wetlands. **ID** Breeding adult has black-bordered creamy-yellow throat and peachy-orange wash to underparts (pattern much reduced in non-breeding plumage). Shows red underwing-coverts in flight. **Voice** Sharp *kyik, chik-chik* or *chet* calls. **HH** Usually crepuscular, also active in overcast conditions; rests during hottest part of day, squatting on ground. Hawks insects with mouth wide open in powerful swallow-like flights; also feeds on ground like a plover, making short dashes to capture prey. Large migrating flocks can occur; at other times, single birds or small flocks. Dried-out bare silt and sand along larger rivers, marshes and coast, also flies high over wetlands and adjacent fields.

Little Pratincole *Glareola lactea* 16–19cm

Locally common resident. **ID** Small size, with sandy-grey coloration and square-ended tail (or with shallow fork). White panel across secondaries, blackish underwing-coverts and black tail-band in flight. Breeding adult has black lores and buff wash to throat; non-breeding lacks these features and has streaked throat. Juvenile has indistinct buff fringes and brown subterminal marks on feathers of upperparts. **Voice** A high-pitched, rattling *tiririt.* **HH** Habits similar to Oriental but often hawks insects later in the evening. Breeds colonially. Large rivers with sand and silt banks. **AN** Small Pratincole.

juv

Greater Painted-snipe ♂ ♀

br

Pheasant-tailed Jacana br non-br

ad imm

Bronze-winged Jacana ad

br non-br br

br juv

Oriental Pratincole br non-br

Little Pratincole

Whimbrel *Numenius phaeopus* 40–46cm

Common winter visitor to coast. **ID** Smaller than Eurasian Curlew, with shorter bill, which often has more marked downward kink. Has prominent whitish supercilium and crown-stripe, contrasting with blackish eyestripe and sides to crown, resulting in more striking head pattern. Juvenile as adult. Eastern *variegatus* (possible but unconfirmed in Bangladesh) has back and rump, as well as underwing, marked heavily with brown (nominate subspecies has white back V and whiter underwing). **Voice** Flight call distinctive, *he-he-he-he-he-he-he*, flat-toned and laughter-like. **HH** Feeds chiefly by picking from surface of open mud, also by probing. Mainly estuarine mudflats, tidal creeks and mangroves, including Sundarbans.

Eurasian Curlew *Numenius arquata* 50–60cm

Common winter visitor to coast. **ID** From Whimbrel by larger size, much longer bill, more uniform head pattern (lacking prominent supercilium and crown-stripe). Juvenile has shorter bill. **Voice** Has distinctive mournful rising *cur-lew* call and an anxious *were-up* when alarmed. Its song, a sequence of bubbling phrases, accelerating and rising in pitch, is often heard in winter and on passage. Frequently calls in flight. See Appendix 1 for differences from Far Eastern Curlew. **HH** Usually singly or in small groups. Generally wary and difficult to approach. Feeds by walking on mud or grassland and probing deeply, also by picking from surface. Coastal mudflats and wet grass fields, rarely along main rivers.

Bar-tailed Godwit *Limosa lapponica* 37–41cm

Scarce winter visitor to coast. **ID** Lacks wingbar, has barred tail and white V on back. At rest, stockier than Black-tailed, with shorter legs and shorter, more upturned bill. Breeding male has chestnut-red head, neck and underparts; mantle and scapulars are more uniformly streaked than Black-tailed. Breeding female has pale chestnut underparts, although mainly same as non-breeding. Non-breeding has dark streaking on breast and streaked appearance to upperparts. Juvenile similar to non-breeding, but with buff wash to underparts and buff edges to mantle/scapulars. **Voice** Gives a barking *kak-kak* and deep *kirruc*. **HH** Habits similar to Black-tailed but often feeds in shallower water. Coastal mudflats.

Black-tailed Godwit *Limosa limosa* 36–44cm

Locally common winter visitor and passage migrant, mainly to coast and NE. **ID** White wingbar and white rump with black tail-band. Long straight bill is mainly pinkish with darker tip; long dark legs. In breeding plumage, male has rufous-orange neck and breast, with blackish barring on underparts and white belly; breeding female larger and duller than male. In non-breeding, plumage is uniform grey on neck, upperparts and breast. Juvenile has cinnamon underparts and cinnamon fringes to dark-centred upperparts. *L. l. melanuroides* ('Eastern Black-tailed Godwit') occurs and is more similar in shape to Bar-tailed Godwit, showing narrower wingbar in flight; breeding male has deeper and more extensive red on underparts, and in non-breeding plumage is darker grey on upperparts and breast, compared with nominate. **Voice** Yapping *kek-kek* occasionally uttered in flight. **HH** Feeds mainly by walking slowly and probing in open mud or shallows; also, by picking prey from surface. Feeds in deeper water than most waders; sometimes wades up to belly and probes with head and neck almost completely submerged. Mainly shallows, wet grass and mudbanks along coast and in haors. **AN** Western Black-tailed Godwit if eastern form *melanuroides* is split.

Asian Dowitcher *Limnodromus semipalmatus* 34–36cm

Rare winter visitor to coast. **ID** From Bar-tailed by straight, broad-based all-black bill, with swollen tip; also by smaller size and stouter appearance, and square-shaped head with steeply rising forehead. Has distinctive 'sewing-machine' feeding action. In flight, shows diffuse pale wingbar, greyish tail, and dark markings on lower back and rump (lacking white V of Bar-tailed). Underparts brick-red in breeding plumage, with chestnut fringes to dark feathers of mantle and scapulars. Upperparts and underparts heavily streaked in non-breeding plumage. Juvenile has buff fringes to upperparts and buffish wash to breast. See Appendix 1 for differences from Long-billed Dowitcher. **Voice** Gives a yelping *chep-chep* or *chowp* and a soft moaning *kiaow*. **HH** Intertidal mudflats and mudbanks. **Threatened** Nationally (EN).

ad
phaeopus

Whimbrel

ad
phaeopus

ad
variegatus

Eurasian Curlew

ad

ad

non-br

♂ br

♂ br

♀ br

non-br

juv

Bar-tailed Godwit

non-br

♂ br

♂ br

melanuroides

non-br

Black-tailed
Godwit

juv

juv

br

non-br

Asian Dowitcher

Ruddy Turnstone *Arenaria interpres*　　　　23cm

Widespread but uncommon winter visitor to coast. **ID** Short bill and orange legs. In flight, shows white stripes on wings and back, and black tail-band. In breeding plumage, has complex black-and-white neck and breast pattern and much chestnut-red on upperparts; duller and less strikingly patterned in non-breeding plumage. Juvenile has buff fringes to upperparts, and blackish breast. **Voice** Utters a rolled *trik-tuk-tuk-tuk* or *tuk-er-tuk*; a sharp *chick-ik* or *kuu* when flushed. **HH** Runs actively on shore, turning over shells to catch small invertebrate prey sheltering beneath; also probes into soft sand and pokes among detritus at high-water mark in search of food. Flight is strong and direct, usually low over shore. Often feeds in small parties. Tidal mudflats and sandy shores.

Great Knot *Calidris tenuirostris*　　　　26–28cm

Rare winter visitor to coast. **ID** Larger than Red Knot, and often has slightly downcurved bill. Also, is less neatly proportioned, with head looking proportionately smaller and neck and body longer, at times recalling Ruff. In flight, compared with Red Knot shows more clearly defined white rump contrasting with grey tail. Breeding adult heavily marked with black on breast and flanks, and chestnut pattern on scapulars. Non-breeding adult typically more heavily streaked on upperparts and breast than Red Knot. Juvenile has darker centres and contrasting white fringes to feathers of mantle, scapulars and wing-coverts, and more heavily marked breast, compared with juvenile Red Knot. **Voice** Low, disyllabic *nyut nyut*. **HH** Usually found in small parties, often with godwits and other waders. Feeds slowly, chiefly by probing deeply into mud or sand. Flight is strong and direct. Intertidal flats and tidal creeks. **Threatened** Globally (EN) and nationally (EN).

Red Knot *Calidris canutus*　　　　23–25cm

Rare winter visitor to coast. **ID** Stocky, with short, straight bill. Breeding adult has brick-red underparts. Non-breeding adult has whitish underparts and uniform grey on upperparts. Juvenile has buff fringes and dark subterminal crescents on upperparts, and buff wash on breast and flanks. **Voice** Utters low short *knutt... knutt*; often silent. **HH** Highly gregarious outside breeding season. Feeds chiefly by probing in soft mud and picking from surface. Mainly intertidal mudflats.

Ruff *Calidris pugnax*　　　　♂26–32cm, ♀20–25cm

Common winter visitor especially to NE. **ID** Confusingly variable but distinctive shape, with long neck, small head, short and slightly downcurved bill, and long yellowish or orangey legs. In all plumages, lacks prominent supercilium and, in flight, shows narrow white wingbar and prominent white sides to uppertail-coverts. Male is considerably larger than female. Non-breeding and juvenile have neatly fringed scaly upperparts, juvenile also buff underparts. Breeding birds typically have black-and-chestnut markings on upperparts, male with striking ruff. **Voice** Generally silent. **HH** Forages by walking purposefully, picking prey from surface and vegetation and probing into mud, sometimes wading into shallow water. When stationary, has upright posture with head raised. Marshes, wet fields and mudbanks of wetlands, particularly haors where flocks of more than 1,000 are regular. **TN** Formerly in monotypic genus *Philomachus*.

Sanderling *Calidris alba*　　　　20cm

Uncommon winter visitor to coast. **ID** Stocky, with short bill. Very broad white wingbar. Adult breeding variable in appearance; initially mottled grey and black, with head and breast becoming more rufous with wear. Rufous birds possibly confusable with Little and Red-necked Stints, but Sanderling is considerably larger, with broader wingbar, has rufous centres to scapulars and coverts, patterned tertials, and lacks hind toe. Sides of head are distinctly streaked compared with Red-necked. Non-breeding is pale grey above and very white below. Has blackish lesser wing-coverts, which are especially noticeable in flight but also show at rest as black patch at bend of wing (unless concealed by breast feathers). Juvenile chequered black and white above with buff wash to streaked sides of breast. **Voice** Call is a liquid *plit*. **HH** Extremely active; runs swiftly after retreating waves, stopping suddenly to catch minute prey or probe in sand. Forages in small parties on shoreline, and prefers sandy beaches.

Ruddy Turnstone

non-br

juv

br

non-br

non-br

Great Knot

br

juv

non-br

non-br

juv

Red Knot

juv

non-br

br

♀ br

♀ juv

♀ juv

♂ br

♂ non-br

♂ br

♂ br

♂ juv

Ruff

Sanderling

br

juv

non-br

br

non-br

Broad-billed Sandpiper *Calidris falcinellus*

16–18cm

Uncommon winter visitor to coast. **ID** Distinctive shape: stockier than Dunlin with legs set well back and downward-kinked bill. In all plumages, has more prominent supercilium than Dunlin, with 'split' before eye, and contrasting with dark eyestripe. Adult breeding has bold streaking on neck and breast contrasting with white belly, and rufous-fringed mantle and scapular feathers with whitish mantle and scapular lines. Birds in worn breeding plumage can appear uniformly very dark on upperparts. Non-breeding has dark patch at bend of wing (sometimes obscured by breast feathers) and strong streaking on breast; dark inner wing-coverts show as dark leading edge to wing in flight. Juvenile has buff mantle/scapular lines and streaked breast. **Voice** Flight call is a buzzing *chrrreet* and a shorter *tzit* or *trr*. **HH** Feeds in similar way to Curlew Sandpiper. Mudbanks of creeks and intertidal mudflats. **TN** Formerly placed in monotypic genus *Limicola*.

Curlew Sandpiper *Calidris ferruginea*

18–23cm

Common winter visitor, mainly to coast, also inland. **ID** White rump. More elegant than Dunlin, with longer, more downcurved bill and longer legs. Breeding adult has chestnut-red head and underparts. Non-breeding adult paler grey than Dunlin, with more distinct supercilium. Juvenile has strong supercilium, buff wash to breast and buff fringes to upperparts. **Voice** Flight call is a low, purring *prrriit*. **HH** Feeds in wet mud and silt in similar fashion to Dunlin, also by wading in deeper water. Mainly coastal, on intertidal mudflats and saltpans; also, occasionally on silty islands in large rivers.

Dunlin *Calidris alpina*

16–22cm

Rare winter visitor. **ID** Shorter legs and bill compared with Curlew Sandpiper, with dark centre to rump. Breeding adult has black belly. Non-breeding adult darker grey-brown than Curlew Sandpiper, with less distinct supercilium. Juvenile has streaked belly, rufous fringes to mantle and scapulars, and buff mantle V. **Voice** Flight call is a distinctive slurred *screet*. **HH** Makes short runs over wet mud and wades near water's edge. Intertidal mudflats along coast and sandbanks in large rivers.

Red-necked Stint *Calidris ruficollis*

13–16cm

Common winter visitor to coast. **ID** Very similar to Little Stint; very subtle structural differences include stouter, deeper-tipped bill, shorter legs, and longer wings, together giving rise to more elongated appearance. Adult non-breeding is almost identical to Little, but is cleaner and paler grey above, with clearer and less extensive dark centres to mantle and scapulars, and markings on sides of breast are more clearly defined. Adult breeding typically has unstreaked rufous-orange throat, foreneck and upper breast, white sides of lower breast with dark streaking, and greyish-centred tertials and wing-coverts (with greyish-white fringes). Juvenile lacks or has indistinct mantle V; has different coloration and pattern to lower scapulars (grey with dark subterminal marks and whitish or buffish fringes; typically blackish with rufous fringes in Little), and grey-centred tertials (usually blackish with rufous edges in Little); supercilium does not usually split in front of eye. **Voice** Call is high-pitched rasping *chriit*. **HH** Habits very similar to Little. Mainly coastal.

Little Stint *Calidris minuta*

13–15cm

Common winter visitor. **ID** Very similar to Red-necked Stint (see that species for differences). More rotund and upright than Temminck's, with dark legs. In flight, shows grey sides to tail. Adult breeding has pale mantle V, rufous wash to face, neck-sides and breast, and rufous fringes to upperpart feathers. Non-breeding has untidy, mottled/streaked appearance (Temminck's is more uniform), with grey breast-sides. Juvenile has whitish mantle V, greyish nape, prominent white supercilium, which typically splits above eye (not visible in plate), and rufous fringes to upperparts. **Voice** Flight call is weak *pi, pi, pi*. **HH** An active wader, rapidly picks at surface and frequently darts about to catch very tiny prey. Muddy edges of large rivers, beels and wetlands, and coastal mudflats.

br

non-br

juv

juv

br

non-br

Broad-billed Sandpiper

non-br

non-br

juv

br

br

Curlew Sandpiper

br

non-br

Dunlin

non-br

br

br

juv

br fresh

non-br

juv

br worn

Red-necked
Stint

non-br

br fresh

non-br

non-br

br worn

Little Stint

juv

juv

Temminck's Stint *Calidris temminckii*   13–15cm

Common winter visitor. **ID** More elongated than Little, with more horizontal stance, and tail extends noticeably beyond closed wings at rest. In flight, shows white sides to tail. Legs yellowish. In all plumages, lacks mantle V and is usually rather uniform, with complete breast-band and indistinct supercilium. Breeding adult has irregular dark markings on upperparts and juvenile has regular buff fringes (pattern very different from Little). **Voice** A trilling, cicada-like *trrrrrit*. **HH** Unobtrusive. Forages more among vegetation at wetland edges than other stints. Favours vegetated freshwater habitats: marshes, pools, lakes, wet fields and riverbanks; scarce on coastal wetlands and mudflats.

Long-toed Stint *Calidris subminuta*   13–15cm

Rare winter visitor, mainly to SE coast. **ID** Long yellowish legs, longish neck and upright stance recall miniature Wood Sandpiper. In all plumages, has prominent supercilium and heavily streaked foreneck and breast. Adult breeding and juvenile have prominent rufous fringes to upperparts, and rufous crown; juvenile also has very striking mantle V. In winter, upperparts more heavily marked than Little. In all plumages is more heavily marked above and on breast than Temminck's and has more upright stance (although note both species have yellow legs). **Voice** Call is soft *prit* or *chirrup*, similar to but less purring than that of Curlew Sandpiper. **HH** Often feeds with other stints; runs about energetically to pick up tiny invertebrates. Freshwater and brackish marshes, and intertidal mudflats.

Spoon-billed Sandpiper *Calidris pygmaea*   14–16cm

Rare winter visitor to coast, mainly in SE. **ID** Stint-sized. Spatulate tip to bill (although bill shape can be difficult to see side-on). Adult non-breeding has paler grey upperparts than Little Stint, with more pronounced white supercilium, forehead and cheeks; underparts appear cleaner and whiter. Adult breeding more uniform rufous-orange on face and breast compared with Little (recalling Red-necked). Juvenile very similar to Little Stint but shows more white on face and darker eyestripe and ear-coverts (masked appearance). **Voice** Flight call is quiet, rolled *preep* or shrill *wheet*. **HH** Feeds mostly by sweeping bill from side to side in shallow water while walking. Intertidal mudflats. **TN** Formerly placed in monotypic genus *Eurynorhynchus*. **Threatened** Globally (CR) and nationally (CR).

Pintail Snipe *Gallinago stenura*   25–27cm

Fairly common winter visitor. **ID** Virtually indistinguishable in field from Swinhoe's Snipe (see Appendix 1). Compared with Common, has more rounded wings, and slower and more direct flight. Lacks well-defined white trailing edge to secondaries, and has densely barred underwing-coverts and pale upperwing-covert panel. Feet project beyond tail in flight. At rest, shows little or no tail projection beyond wings. Usually shows bulging supercilium in front of eye, with little contrast between buff supercilium and cheeks, and eyestripe often narrow in front of eye and poorly defined behind. Width and colour of edges to lower large scapulars similar on inner and outer webs, creating scalloped appearance. **Voice** If flushed gives short, rasping *tetch*, deeper than Common. **HH** Flushes with little or no zigzagging; usually drops into cover more quickly than Common. Marshy pool edges, wet paddy-fields; sometimes dry ground, unlike Common. **AN** Pin-tailed Snipe.

Common Snipe *Gallinago gallinago*   25–27cm

Common winter visitor. **ID** Compared with Pintail Snipe, wings more pointed, and flight faster and more erratic. In flight, shows prominent white trailing edge to wing, white banding on underwing-coverts and more extensive white belly patch. At rest, shows noticeable projection of tail beyond wings, poorly defined median-covert panel, buff supercilium contrasts with white cheek-stripe, and broad buff edges to outer webs of lower scapulars contrast with narrower, browner inner webs. **Voice** If flushed, gives an anxious, rising, grating *scaaap*, which usually sounds higher-pitched and more urgent than Pintail. **HH** When flushed, rises steeply with rapid zigzagging, circles high and lands some distance away. Marshes, wet fields and muddy edges to beels and rivers.

non-br

br

juv

Temminck's Stint

non-br

Long-toed Stint

br

non-br

non-br

juv

non-br

br

juv

Spoon-billed Sandpiper

flight

juv

ad

outer tail

Pintail Snipe

ad

Common Snipe

juv

flight

outer tail

juv

Terek Sandpiper *Xenus cinereus* 22–25cm

Fairly common winter visitor to coast. **ID** Longish, upturned bill and short yellowish legs. In flight, shows prominent white trailing edge to secondaries and grey rump and tail. Adult breeding has blackish scapular lines. Juvenile is similar to breeding adult but has buff fringes and dark subterminal marks on upperparts feathers. **Voice** Flight call is a soft pleasant whistle *hu-hu-hu* and sharper *twit-wit-wit-wit*, recalling Common Sandpiper. **HH** A very active feeder, running erratically here and there to chase prey; also probes deeply, and sometimes feeds in shallow water. Solitary or in scattered groups when feeding; often gathers in small flocks to roost. Coastal mudflats.

Common Sandpiper *Actitis hypoleucos* 19–21cm

Common and widespread winter visitor. **ID** Horizontal stance, long tail projecting well beyond closed wings. White wingbars, and brown rump and centre of tail show in flight. In breeding plumage, has irregular dark streaking and barring on upperparts, absent in non-breeding. Juvenile has buff fringes and dark subterminal crescents to upperparts. **Voice** Anxious *wee-wee-wee* when flushed or alarmed. **HH** Characteristically rocks rear end of body and bobs head constantly when feeding. Runs along the water's edge and picks prey from the ground or vegetation. Flies low over water, with rapid, shallow wingbeats alternating with brief glides on stiff downcurved wings. Scattered individuals typically found along banks of small and large rivers, including estuarine rivers; less frequent in marshes.

Green Sandpiper *Tringa ochropus* 21–24cm

Uncommon winter visitor. **ID** From Wood Sandpiper by shorter greenish legs and stockier appearance, darker and less heavily spotted upperparts, and supercilium indistinct or absent behind eye. In flight, shows very dark underwing, strongly contrasting with white belly and vent, and striking white rump is distinctive. Breeding adult has white streaking on crown and neck, heavily streaked breast, and prominent whitish spotting on upperparts. Non-breeding adult is more uniform on head and breast, and is less distinctly spotted on upperparts. Juvenile has browner upperparts with buff spotting. **Voice** Ringing *tluee-tueet* and *tuee-weet-weet* calls. **HH** Favours small waters: ditches, small pools and streams; also marshes, wet fields and banks of rivers and lakes.

Wood Sandpiper *Tringa glareola* 18–21cm

Common winter visitor. **ID** From Green by longer yellowish legs and slimmer appearance, heavily speckled upperparts, and prominent supercilium behind eye; in flight by call, slimmer body and narrower wings, toes projecting clearly beyond tail, paler underwing contrasting less with white underparts, and paler brown upperparts contrasting less with smaller white rump. Adult breeding has heavily streaked breast and barred flanks; upperparts barred and spotted pale grey-brown and white. Adult non-breeding has more uniform grey-brown upperparts, spotted whitish, and breast brownish and lightly streaked. Juvenile has warm brown upperparts speckled warm buff, and lightly streaked buff breast. **Voice** Soft *chiff-if* or *chiff-if-if* flight call. **HH** Favours shallow water with emergent vegetation: marshes, pools, beels and often wet paddy-fields.

Grey-tailed Tattler *Tringa brevipes* 24–27cm

Rare spring passage migrant to SE coast. **ID** A stocky grey-and-white wader with stout straight bill and short yellow legs. In all plumages, shows prominent white supercilium contrasting with dark eyestripe, uniform grey wings lacking prominent wingbar, grey rump and tail, and grey underwing contrasting with white belly. Adult breeding has barring on breast and flanks. Adult non-breeding uniform grey on upperparts and breast. Juvenile has indistinct white spotting on upperparts. **Voice** A plaintive *too-weep*. **HH** Coastal wetlands; small numbers roosted with other waders on shrimp ponds at Patenga (Chittagong) in Apr–May during the 1990s.

Terek Sandpiper

br

non-br

br

non-br

juv

br

Common Sandpiper

juv

br

Green Sandpiper

br

non-br

non-br

br

non-br

juv

non-br

Wood Sandpiper

br

non-br

non-br

Grey-tailed Tattler

non-br

juv

Spotted Redshank *Tringa erythropus* 29–32cm

Uncommon winter visitor, mainly to NE, rarer on coast. **ID** Red at base of bill, and red legs. Longer bill and legs than Common Redshank, lacking broad white trailing edge to wing. Non-breeding plumage is paler grey above and whiter below than Common with more prominent white supercilium. Underparts black in breeding plumage. In first-summer plumage has dark barring on underparts and dark-mottled upperparts; legs can be black. Juvenile similar to non-breeding adult, but has darker grey upperparts more heavily spotted with white, and underparts are finely barred with grey. **Voice** A distinctive *tu-ick* in flight and shorter *chip* alarm call. **HH** Usually solitary or in small groups. Feeds by picking from the surface, often after a short dash; frequently forages in water, either sweeping the bill from side to side or probing rapidly, often submerging the head and neck completely. Swims readily and upends like a surface-feeding duck to reach bottom mud. Muddy banks and shallow water of haors (where it can form dense flocks), beels, rarely along rivers and coast.

Common Redshank *Tringa totanus* 27–29cm

Common winter visitor, mainly to coast. **ID** Orange-red at base of bill, orange-red legs, and broad white trailing edge to wing. Non-breeding plumage is grey-brown above, with grey breast. Neck and underparts heavily streaked in breeding plumage; upperparts with variable dark brown and cinnamon markings. Juvenile quite different from juvenile Spotted, with brown upperparts entirely fringed and spotted with buff, underparts heavily streaked with dark brown, and dull orange legs and base to bill. **Voice** Very noisy, often giving its alarm call, an anxious *teu-hu-hu*, in flight; also a mournful *tyuuu* on ground. **HH** Mainly found singly or in small groups. Generally wary. Feeds by walking briskly and picking from surface; also probes and wades in shallow water. Coastal mudflats, rarely inland on muddy edges of rivers and beels.

Common Greenshank *Tringa nebularia* 30–34cm

Common winter visitor. **ID** Stocky, with long, stout (and slightly upturned) bill and long, stout greenish legs. Upperparts grey and foreneck and underparts white in non-breeding plumage. In breeding plumage, foreneck and breast streaked, upperparts untidily streaked. Juvenile has dark-streaked upperparts with fine buff or whitish fringes. **Voice** Loud, ringing *tu-tu-tu* flight call; sometimes a more throaty *kyoup-kyoup-kyoup*. **HH** Usually forages singly. Generally wary and when alarmed bobs head and body nervously up and down. Feeds actively, chiefly in shallow water or at the water's edge; detects prey mainly by sight and makes frequent rapid runs to seize fast-moving prey. Flies strongly and sometimes erratically. Wide range of wetlands including coast.

Spotted Greenshank *Tringa guttifer* 29–32cm

Rare winter visitor to coast, mainly Nijhum Dwip and Sonadia areas. **ID** Stockier than Common Greenshank, with shorter, yellowish legs, and deeper bill with blunt tip. Breeding adult has black spotting on breast and prominent white notching on scapulars and tertials. Non-breeding is paler and more uniform upperparts than Common. Juvenile has rather uniform upperparts, with paler fringes to wing-coverts, and strongly bicoloured bill. **Voice** Flight call is *kwork* or *gwaak*, very different from Common Greenshank. **HH** Behaviour similar to Common. Coastal mudflats. **AN** Nordmann's Greenshank. **Threatened** Globally (EN) and nationally (CR).

Marsh Sandpiper *Tringa stagnatilis* 22–25cm

Uncommon winter visitor, mainly to haors of NE; less numerous on coast. **ID** Smaller and daintier than Common Greenshank, with proportionately longer legs and finer bill. Legs greenish or yellowish. Upperparts grey and foreneck and underparts white in non-breeding plumage, when the pale lores, forehead and chin create a pale-faced appearance. In breeding plumage, foreneck and breast are streaked and upperparts blotched and barred. Juvenile upperparts appear streaked blackish, the feathers notched and fringed with buff, and head-sides, hindneck and upper mantle are streaked dark grey and white. **Voice** An abrupt, dull *yup* flight call; also a rapid, excitable series of *kiu-kiu-kiu* notes. **HH** A particularly graceful wader. Mainly singly or in small groups. Forages actively, often in water or at water's edge; picks delicately from surface, making frequent rapid darts to seize prey; probes occasionally. Shallow water and mud in haors and beels, also coastal mudflats.

Spotted Redshank

br

juv

non-br

non-br

Common Redshank

non-br

br

non-br

juv

Common Greenshank

br

non-br

non-br

br

non-br

Marsh Sandpiper

non-br

non-br

juv

Spotted Greenshank

br

Indian Skimmer *Rynchops albicollis* 40cm

Locally common winter visitor and rare breeder. **ID** Adult has large, drooping orange-red bill (with lower mandible projecting noticeably beyond upper), black cap, and black mantle and wings contrasting with white underparts. In flight, shows broad white trailing edge to upperwing, white underwing with blackish primaries, and white rump and tail with black central tail feathers. Juvenile has whitish fringes to browner mantle and upperwing-coverts, diffuse cap and dull orange bill with black tip. **Voice** Nasal *kap kap.* **HH** Feeds chiefly in early morning or near dusk. Flight is fast, powerful and graceful, resembling a *Sterna* tern. Systematically quarters water surface and has characteristic skimming feeding method. Much of the global population winters around islands and mudflats on the central coast, occasionally on passage along main rivers and recently found nesting on sandbanks in the Ganges River. **Threatened** Globally (EN) and nationally (CR).

Slender-billed Gull *Larus genei* 43cm

Rare winter visitor. **ID** Gently sloping forehead, longish neck and longer bill compared with Black-headed. In flight, both neck and tail appear longer than Black-headed. Adult has white head throughout year (may show grey ear-covert spot in winter), deep red bill (often looks blackish, paler in winter), pale iris (dark in Black-headed) and variable pink flush on underparts. First-winter/first-summer from Black-headed by paler and less distinct dark eye-crescent and ear-spot (sometimes completely lacking), pale iris, paler orange bill (with dark tip smaller or absent), paler legs, less prominent dark trailing edge to inner primaries, and more extensive white on outer primaries resulting in more prominent white 'flash' on wing. Juvenile has grey-brown mantle and scapulars, with pale fringes (generally paler and lacking ginger-brown coloration of juvenile Black-headed). **Voice** Slightly deeper than that of Black-headed. **HH** Small numbers recently found to be regular at Sonadia Island in SE, on coastal waters and mudflats. **TN** Some authorities place in genus *Chroicocephalus.*

Brown-headed Gull *Larus brunnicephalus* 42cm

Common winter visitor. **ID** Slightly larger than Black-headed, with more rounded wingtips and broader bill, which is dark-tipped in all ages. Adult has broad black wingtips (broken by white 'mirrors') and white patch on outer primaries and primary coverts; underside to primaries largely black; iris pale yellow (brown in adult Black-headed). In breeding plumage, hood paler brown than Black-headed. Juvenile and first-winter have broad black wingtips contrasting with white patch on primary coverts and base of primaries. **Voice** As Black-headed, but deeper and gruffer. **HH** Mainly coastal waters and mudflats, also around large wetlands and principal rivers. **TN** Some authorities place in genus *Chroicocephalus.*

Black-headed Gull *Larus ridibundus* 38cm

Common winter visitor. **ID** Smaller than Brown-headed, with finer bill and narrower and more pointed wings. In all plumages has distinctive white 'flash' and less black on wingtips than Brown-headed. Bill blackish-red and hood dark brown in breeding plumage. In non-breeding and first-winter plumages, bill tipped black and head white with dark ear-covert patch. **Voice** Nasal *kyaaar,* short *keck* and deeper *kuk.* **HH** Coastal islands and waters, large rivers and haors. **TN** Some authorities place in genus *Chroicocephalus.*

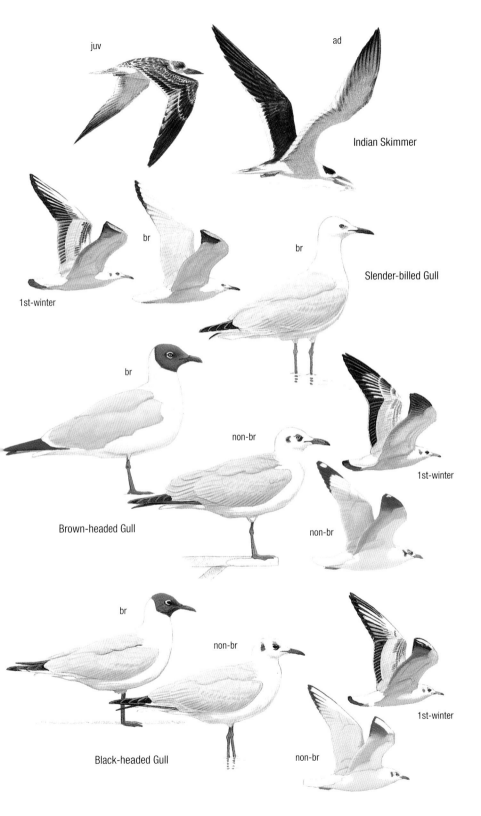

juv

ad

Indian Skimmer

br

Slender-billed Gull

1st-winter

br

br

non-br

1st-winter

Brown-headed Gull

non-br

br

non-br

1st-winter

Black-headed Gull

non-br

PLATE 37: LARGE GULLS AND POMARINE JAEGER

Pallas's Gull *Larus ichthyaetus*   69cm

Common winter visitor to coast, scarce elsewhere. **ID** Largest gull in Bangladesh. Head is angular, with sloping forehead, and crown peaks behind eye; bill is long and strikingly dark-tipped (except in juvenile), with bulging gonys. Eyes always dark. Breeding adult has black hood, white eye-crescents, and yellow bill with black and red at tip. White primary tips contrast with black subterminal marks; white primary coverts and base of outer primaries form white patch on outer wing. Non-breeding adult has white head with black mask and white eye-crescents. Juvenile has brown mantle and scapulars with pale fringes. First-winter/first-summer has grey mantle, pronounced dark mask and streaking across hindcrown, and clear-cut dark tail-band. May acquire partial hood in first summer. Second-winter has largely grey upperwing, with dark lesser-covert bar and extensive black on primaries. Third-winter as adult non-breeding, but with more black on primaries. **Voice** Corvid-like *kra-ah*. **HH** Coastal mudflats and waters, also haors and large rivers. **AN** Great Black-headed Gull. **TN** Some authorities place in genus *Ichthyaetus*.

Lesser Black-backed Gull *Larus fuscus*   58–65cm

Rare winter visitor to coast; two records in NE. **ID** Form occurring is *L. f. heuglini* ('Heuglin's Gull'), and due to uncertainty about the status of large gulls in Bangladesh comparison is made here with Caspian Gull *L. cachinnans* and 'Steppe Gull' *L. f. barabensis*, neither of which has been confirmed. Generally stockier and squarer-headed than Caspian and 'Steppe'. Adult has darker grey upperparts than Caspian and 'Steppe' (although form *'taimyrensis'*, is more similar to 'Steppe'), and head more heavily streaked in non-breeding plumage. Eyes yellow (appear small and dark in Caspian). Adult shows more black on wingtips than Caspian, typically with smaller white 'mirror' at tip of longest primary, and no white 'mirror' on second-longest primary (latter present in *'taimyrensis'*). Juvenile and first-winter from Caspian by darker inner primaries, greater coverts and underwing-coverts. Retains neat, pale-fringed juvenile mantle feathers for longer than Caspian (to Dec). Reaches adult plumage one year ahead of Caspian. First-summer has grey feathers in mantle, coverts and tertials. By second-winter, mantle and coverts are largely grey (coverts of Caspian retain many immature feathers). Form *'taimyrensis'*, which may occur, is bulkier and broader-winged than *L. f. heuglini*, and upperparts of adult are a shade paler. **Voice** A braying *ka-yaow-owowow-ow-ow ow*. **HH** Coastal mudflats and estuarine waters, also large haors. **TN** Heuglin's Gull *L. heuglini* (including subspecies *L. h. taimyrensis*) and Steppe Gull *L. barabensis* are regarded as separate species by some authorities.

Pomarine Jaeger *Stercorarius pomarinus*   56cm

Rare winter visitor. **ID** Larger and stockier than Arctic Jaeger (see Appendix 1), with heavier bill (and more pronounced dark tip) and broader-based wings. In flight, appears slower and heavier, with deeper chest and broader-based wings. Adult breeding has long, broad central tail feathers twisted at end to form swollen tip (although tips may be broken off); when present this is the best feature from Arctic. Occurs in both pale and dark morphs. Prominent pink base to bill, black chin, brighter yellow neck, more prominent breast-band, and dark flank barring are additional features to help separate it from pale-morph Arctic. Adult non-breeding (pale morph) has indistinct cap, and barring to breast and upper- and undertail-coverts; uniform dark underwing-coverts distinguish it from birds in first- and second-winter plumage. Broader, round-tipped central tail feathers best distinction from Arctic. Juvenile variable, typically dark brown with broad pale barring on uppertail- and undertail-coverts and underwing-coverts. Combination of strongly barred uppertail-coverts and dark uniform head is diagnostic of Pomarine, and head, neck and underparts never appear rufous-coloured as on some juvenile Arctic. Other juvenile plumages appear virtually identical to Arctic. Additional finer features from juvenile Arctic are the second pale crescent at base of primary coverts on underwing on most birds (in addition to pale base to underside of primaries), diffuse vermiculations (never streaking) on nape and neck, lack of (or very indistinct) pale tips to primaries, and blunt-tipped or almost non-existent projection of central tail feathers (more prominent and pointed in juvenile Arctic). Dark-morph juveniles occur, some being uniform sooty-black with white only at base of primaries; they differ little in subsequent immature plumages and are not distinguishable by plumage from juvenile dark-morph Arctic. **HH** Often associates with feeding flocks of terns. Flight is direct and powerful, with steadily flapping wingbeats. Coastal waters, with most records well offshore in the Bay of Bengal, over the Swatch of No Ground marine canyon. **AN** Pomarine Skua.

Pallas's Gull

non-br

non-br

br

1st-year

Lesser
Black-backed Gull

non-br

1st-winter

Steppe
Gull

Caspian
Gull

non-br
pale morph

br
pale morph

juv
dark

juv
intermediate

Pomarine Jaeger

br
dark morph

Little Tern *Sternula albifrons* 22–24cm

Local and declining uncommon resident. **ID** Fast flight with rapid wingbeats on narrow-based wings. Feeds by hovering and plunge-diving. Breeding adult has white forehead and black lores, black-tipped yellow bill, orange legs and feet, and black outer primaries. Non-breeding adult and immature have blackish bill, black mask and nape band, dark lesser-covert bar, and dark legs. Juvenile has dark subterminal marks to feathers of upperparts, and whitish secondaries form broad pale trailing edge to wing. **Voice** A *ket* or *ket-ket*. **HH** Hovers more frequently and for longer periods than other terns and with faster-fluttering beats before plunge-diving steeply after prey; also dips steeply to surface, skimming prey from water. Lakes and rivers. **TN** Formerly placed in genus *Sterna*.

Common Gull-billed Tern *Gelochelidon nilotica* 35–38cm

Common winter visitor mainly to coast. **ID** Has stout gull-like black bill, and broader-based, less pointed wings compared with other terns (except Caspian, which has huge red bill). Flight is steady and more gull-like, less graceful, with shallower wingbeats. Rump and tail are grey and concolorous with back in all plumages. Breeding adult has black cap and darker grey upperparts than in other plumages. Black of head reduced to black mask in non-breeding and immature plumages. Juvenile has sandy tinge to crown and mantle and indistinct dark fringes to tertials and some wing-coverts, but is less heavily marked on upperparts than other juvenile terns. **Voice** Guttural *gek-gek-gek* or *gir-vit*. **HH** Frequently feeds by hawking, swooping or dipping to pick up food, often skimming close to the surface, or seizing insects in mid-air. Coastal waters, also large rivers and rarely haors. **AN** Gull-billed Tern.

Caspian Tern *Hydroprogne caspia* 47–54cm

Uncommon winter visitor to coast. **ID** Large, with broad-winged, short-tailed appearance. Huge red bill and black underside to primaries. Breeding adult has complete black cap. Non-breeding adult has black-streaked crown and black mask; bill is duller, with more black at tip. First-winter and first-summer plumages are similar to adult non-breeding, but show faint dark lesser-covert and second bars, and dark-tipped tail. Juvenile has narrow dark subterminal bars to scapulars and wing-coverts; forehead and crown are more heavily marked, almost forming dark cap. **Voice** Loud and far-carrying, hoarse *kretch*. **HH** Fishes by patrolling high above water, bill pointing downwards, hovering occasionally before plunging into water to seize prey. Coastal waters and islands. **TN** Formerly placed in genus *Sterna*.

Whiskered Tern *Chlidonias hybrida* 23–25cm

Common winter visitor and rare resident. **ID** In breeding plumage, white cheeks contrast with black cap and grey underparts. In non-breeding and juvenile plumages, from White-winged by larger bill, grey rump concolorous with back and tail, and different head pattern (see White-winged). Head markings can be limited to dark mask, recalling small Common Gull-billed. Compared with White-winged, juvenile generally lacks pronounced dark lesser-covert and second bars, and has black and buff markings on mantle/scapulars that appear more chequered (more uniformly dark in White-winged). **Voice** Hoarse *eirchk* or *kreep*. **HH** Feeds mainly on insects by hawking or picking from water surface; unlike White-winged, occasionally plunge-dives for fish. Coastal waters and islands, large rivers and other large wetlands such as haors.

White-winged Tern *Chlidonias leucopterus* 20–23cm

Rare spring passage migrant and winter visitor. **ID** In breeding plumage, black head and body contrast with pale upperwing-coverts. Black underwing-coverts are last part of plumage to be lost during moult into non-breeding plumage (always white in Whiskered). In non-breeding and juvenile plumages, smaller bill, whitish rump contrasting with grey tail, and different head pattern are distinctions from Whiskered. Black ear-covert patch is bold and reaches below eye, and usually has well-defined black line down nape. First-year birds show dark lesser-covert and second bars, and by late winter these contrast strongly with pale (worn) median and greater coverts, which form pale panel in wing, while mantle also can appear noticeably darker than pale coverts, giving rise to 'saddled' appearance as in juvenile; birds in this plumage are distinct from non-breeding and first-year Whiskered, which have more uniform mantle and wings. **Voice** Hoarse, dry *kirsch*. **HH** Hawks insects or swoops down to pick up small prey from water surface; flies with great agility. Most records from coastal waters and islands; rarer in haors of NE.

juv

Little Tern

non-br

br

br

br

juv

1st-winter

non-br

br

Common
Gull-billed Tern

br

non-br

juv

br

Caspian Tern

1st-winter

br

non-br

br

non-br

juv

br

Whiskered Tern

1st-summer

juv

non-br

non-br

br

White-winged Tern

River Tern *Sterna aurantia*
38–46cm

Uncommon localised resident along main rivers and on coastal islands. **ID** Adult breeding has orange-yellow bill, black cap, greyish-white underparts and long greyish-white outer tail feathers; whitish primaries contrast with otherwise grey wing to form striking 'flash' on outer wing in flight. In non-breeding plumage lacks elongated outer tail feathers, and has blackish mask and mainly grey crown. Large size, stocky appearance and stout yellow bill (with dark tip) help separate adult non-breeding and immature plumages from Black-bellied. Juvenile has dark fringes to upperparts, black streaking on crown and nape, whitish supercilium, and dark mask extending as dark streaking onto ear-coverts and sides of throat. **Voice** Fairly short, shrill, staccato *kiuk-kiuk* in flight, quite high-pitched and melodious. **HH** Mainly feeds by plunge-diving from the air; also by dipping to surface and picking up prey. Large rivers and estuarine islands. **Threatened** Globally (VU).

Common Tern *Sterna hirundo*
31–35cm

Uncommon winter visitor and passage migrant to coast, rare elsewhere. **ID** In breeding plumage has orange-red bill with black tip, orange-red legs, pale grey wash to underparts and elongated outer tail feathers, which reach wingtip at rest. In non-breeding and first-winter plumages has mainly dark bill, darker grey upperparts and shorter tail. Juvenile has orange legs and bill-base (bill becomes black with age). *S. h. longipennis* may occur; in breeding plumage, it has mostly black bill, with greyer upperparts and underparts than nominate subspecies, and more distinct white cheek-stripe; legs are dark reddish-brown. **Voice** Drawn-out and harsh *krri-aaah* and short *kik*. **HH** Fishes mainly by plunge-diving from the air. Coastal and estuarine waters, also large rivers.

Black-bellied Tern *Sterna acuticauda*
33cm

Rare and declining resident. **ID** Smaller than River, with orange bill (and variable black tip) in all plumages. Breeding adult has grey breast, black belly and vent, and long outer tail feathers. Like River, whitish primaries contrast with grey rest of wing to form striking 'flash' on outer wing in flight. Long orange bill and deeply forked tail are best features from Whiskered. Non-breeding adult and immature have white underparts, shorter tail, and black mask and streaking on crown. Confusingly, can occur with black cap and white underparts, when most similar to River, but structural differences and orange bill are diagnostic. Juvenile has dark mask and streaking on crown and nape, sandy coloration to head and mantle, and brown fringes to upperparts. **Voice** Clear, piping *peuo*. **HH** Feeds by plunge-diving from the air, also by dipping to surface to pick up prey. Large rivers; formerly regular along Jamuna River, feared extirpated after 2000, but recently rediscovered. **Threatened** Globally (EN) and nationally (CR).

Lesser Crested Tern *Thalasseus bengalensis*
35–37cm

Rare winter visitor to SE coast. **ID** From Greater Crested by smaller and slimmer yellowish-orange to orange bill, smaller size and lighter build, and paler grey upperparts. From Sandwich Tern (see Appendix 1) by bill colour, and grey rump and tail. Breeding adult has black crown and crest, and forehead, although black of latter is quickly lost (forehead is never black on Greater Crested). Non-breeding adult has black nape-band. Juvenile has dark centres to lesser and greater coverts and secondaries, which show as diffuse dark bars across wing, and has dark centres to feathers of mantle, scapulars and tertials. Upperwing pattern is similar to Greater Crested, but dark bars are typically paler and less contrasting. First-winter and first-summer plumages are similar to adult non-breeding but show darker grey primaries and dark lesser-covert and second bars. **Voice** Gives an upward-inflected *kree-it*, much as Sandwich Tern. **HH** A marine tern, usually found in offshore waters and often far out to sea; most sightings in coastal waters from Sonadia to St. Martin's Islands. **TN** Formerly placed in genus *Sterna*.

Greater Crested Tern *Thalasseus bergii*
46–49cm

Scarce winter visitor to coast. **ID** From Lesser Crested by broader, slightly drooping, cold yellow to lime-green bill, and by larger size and stockier build. In adult plumage shows well-defined whitish fringes to tertials, and has darker grey coloration to upperparts than Lesser Crested. Immature plumages are similar but generally more strongly patterned than in Lesser Crested. **Voice** A harsh *kerrer* or *kerrak*. **HH** Mainly offshore waters and often at considerable distances out at sea; also, coastal waters and islands from outer edge of the Sundarbans east to St. Martin's Island. **AN** Swift Tern. **TN** Formerly placed in genus *Sterna*.

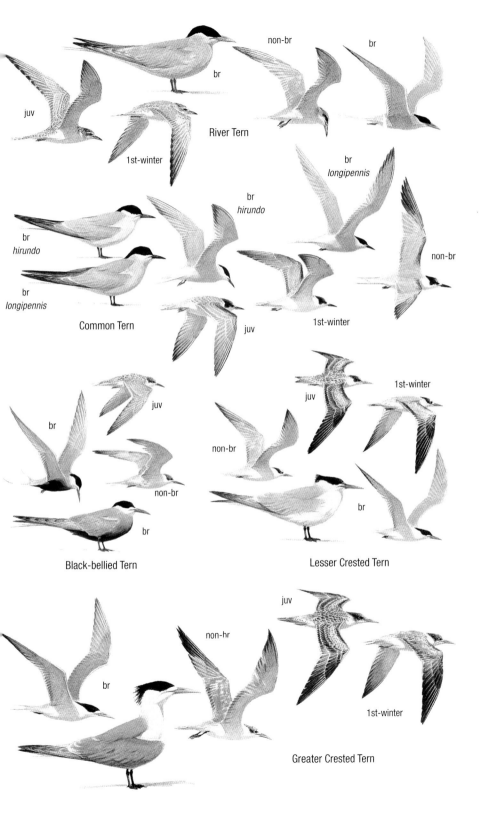

juv

non-br

br

br

River Tern

1st-winter

br
longipennis

br
hirundo

br
hirundo

non-br

br
longipennis

Common Tern

juv

1st-winter

juv

br

non-br

br

Black-bellied Tern

juv

1st-winter

non-br

br

Lesser Crested Tern

juv

br

non-hr

1st-winter

Greater Crested Tern

Common Barn Owl *Tyto alba* 36cm

Uncommon resident. **ID** Readily identified by combination of unmarked white face and contrasting black eyes, white to golden-buff underparts finely spotted with black, and golden-buff and grey upperparts, which are finely spotted with black and white. Wings and tail appear very uniform in flight, lacking any prominent tail barring or wing patches. **Voice** Variety of eerie screeching and hissing notes. **HH** Mainly nocturnal. Hunts by quartering open country a few metres above ground. Roosts and nests in large, old buildings in cities, towns and villages, also in ruins.

Brown Boobook *Ninox scutulata* 32cm

Common and widespread resident. **ID** Hawk-like profile (with slim body, long tail and narrow head). Has uniform brown upperparts showing variable amounts of white spotting on scapulars, all-dark face except for variable white patch above bill (lacking pale facial discs shown by many owl species), and bold rufous-brown streaking and spotting on underparts. **Voice** A soft *oo... ok, oo... ok, oo... ok* in runs of 6–20 calls. **HH** Crepuscular and nocturnal, typically spending day concealed in top of forest tree. Flight is hawk-like: a series of rapid wingbeats followed by a glide. Well-wooded areas, including village groves and all forest types. **AN** Brown Hawk-Owl.

Collared Owlet *Glaucidium brodiei* 17cm

Local resident restricted to Chittagong Hill Tracts in SE. **ID** Most similar to Asian Barred Owlet, but is smaller and has distinct buff (or rufous) 'spectacles' on upper mantle, which frame blackish patches (creating pattern resembling an owl's face), and crown appears spotted rather than neatly and finely barred. Occurs in rufous, grey and brown morphs. **Voice** Pleasant four-note bell-like whistle *toot... tootoot... toot*, uttered in runs of three or four and repeated at intervals. **HH** Diurnal and crepuscular, often hunting in daylight. Usually seeks prey by watching and listening from prominent perch. Calls persistently by day and night in breeding season. Evergreen forest.

Asian Barred Owlet *Glaucidium cuculoides* 23cm

Fairly common in E, rare in C. **ID** From Spotted Owlet by barred rather than spotted appearance, with finely barred crown, upperparts and breast. White eyebrows are narrow and facial discs are barred. From Collared Owlet by larger size, barred rather than spotted appearance to crown and absence of owl-like face pattern on nape. Juvenile has buff spotting on crown, nape and mantle; breast barring and flank streaking more diffuse. **Voice** Crescendo of harsh squawks. In breeding season, a continuous bubbling whistle lasting up to *c.* 7 seconds. **HH** Mainly diurnal; scans and listens for prey from prominent perch. Puffs itself into ball before starting its bubbling call, then gradually subsides to its normal size. Evergreeen and semi-evergreen forests and wooded areas such as tea estates.

Spotted Owlet *Athene brama* 21cm

Common and widespread resident. **ID** From Asian Barred by spotted rather than barred appearance (with prominent white spotting on crown, mantle and wing-coverts, and brown spotting rather than close dark barring on underparts). In addition has pale facial discs and pale hind-collar. **Voice** Harsh screechy *chirurr-chirurr-chirurr*, followed by or alternated with *cheevak, cheevak, cheevak* and a variety of other discordant screeches and chuckles. **HH** Mainly crepuscular and nocturnal. Hides by day in tree hollow, chimney or roof. Frequently hunts insects around street lights. Usually in pairs or family groups. Open wooded groves, also around habitation and cultivation.

Common Barn Owl

ad

ad

Brown Boobook

ad
turning away

juv

ad

ad

Collared Owlet

Asian Barred Owlet

ad

Spotted Owlet

Collared Scops-owl *Otus lettia* 23–25cm

Common, widespread resident, except NW where status uncertain (see Indian Scops-owl). **ID** From Oriental Scops-owl by larger size, prominent buff nuchal collar edged with dark brown, more sparsely streaked underparts, and buffish (less distinct) scapular spots. Eyes are dark orange or brown (yellow in Oriental). See also Indian Scops-owl. **Voice** Call is subdued, frog-like *broou* with falling inflection, usually repeated at irregular intervals. **HH** Very similar to Oriental Scops-owl. Forest, village groves and well-wooded areas. **TN** Considered conspecific with Indian Scops-owl as Collared Scops-owl *O. bakkamoena* by some authorities.

Indian Scops-owl *Otus bakkamoena* 23–25cm

Rare presumed resident discovered in NW in 21st century. **ID** From Collared Scops-owl mainly by call, although this is not considered diagnostic; upperparts and underparts are less heavily and more regularly marked with finer, longer dark streaks; dark tip to yellowish bill. Variable in coloration with buff, through to brown morphs. **Voice** Call is frog-like *whuk* (stonger than Collared Scops-owl), with rising inflection. **HH** Same as Collared.

Mountain Scops-owl *Otus spilocephalus* 20cm

Scarce local resident in SE hills, one old record from NE. **ID** From similar species by unstreaked underparts, which are indistinctly spotted with buff and barred with brown, and by unstreaked upperparts, which are mottled with buff and brown (crown and nape usually the most heavily marked). Has poorly defined facial discs and stubby ear-tufts. Bill and claws are pale. Often shows a paler band on upper mantle (forming diffuse 'collar'), and head can appear slightly paler than rest of upperparts. **Voice** Repeated double whistle, *toot-too*. **HH** Nocturnal like other scops-owls. Roosts during day in tree hollow. Calls intermittently during night. Evergreen and second forest.

Oriental Scops-owl *Otus sunia* 19cm

Fairly common, widespread resident. **ID** Highly variable, with grey, brown and rufous morphs. Smaller than Collared Scops-owl with prominent white scapular spots, profusely streaked underparts, and absence of prominent nuchal collar. Rufous morph distinct from Collared Scops-owl. Eyes are yellow (dark in Collared Scops-owl). **Voice** Repeated, resonant, rhythmic, frog-like *wut-chu-chraaii*, first note musical, last two more rasping. **HH** Hides by day in dense foliage. Forest.

Short-eared Owl *Asio flammeus* 37–39cm

Rare but widespread winter visitor. **ID** Has heavily streaked head, upperparts and breast, whitish facial discs and yellow eyes. In flight, has buffish patches on primaries, pronounced dark carpal patches and dark barring to buff tail. **Voice** Usually silent. **HH** Diurnal and crepuscular. Hunts by quartering low over ground. Open grasslands along main rivers, coastal islands and large wetlands.

Brown Wood-owl *Strix leptogrammica* 47–53cm

Uncommon and local resident. **ID** Has uniform dark brown crown and upperparts (with patch of white barring on scapulars), and greyish-white underparts finely barred with brown; flight feathers are dark brown and narrowly barred paler brown. Also has dark brown face, with prominent white eyebrows and striking white band across foreneck. **Voice** Calls include a double hoot *tu-whoo*, a sonorous squawk *hoo-hoohoohoo (hoo)*, and a variety of eerie shrieks and chuckles. **HH** Nocturnal. Roosts by day in large trees. Very shy; if disturbed will compress itself to resemble a stump or fly a long distance through forest. Dense evergreen, deciduous and mangrove forest.

Collared Scops-owl

ad

ad
buff morph

ad
intermediate morph

ad
brown morph

Indian Scops-owl

Mountain Scops-owl

ad

Oriental Scops-owl

ad
grey morph

ad
rufous morph

ad

Short-eared Owl

ad

Brown Wood-owl

Spot-bellied Eagle-owl *Bubo nipalensis* 63cm

Rare and local resident in E and formerly in C. **ID** Very large, with bold chevron-shaped spots on whitish underparts, whitish facial discs, buff-barred dark brown upperparts, large pale bill and brown eyes. Juvenile is very distinctive: crown, mantle, coverts, rump and underparts are white and buff, with brown spotting and barring, and face is off-white. **Voice** Low, deep, moaning hoot, and far-carrying mournful scream. **HH** Largely nocturnal, usually hiding by day among dense foliage in deep forest. Very bold and powerful owl, able to overcome large prey such as Kalij Pheasant. Hunts in forest. Nests in tree hollow, deserted raptor nest, or cliff cave or fissure. Inhabits dense evergreen and sal forest.

Dusky Eagle-owl *Bubo coromandus* 58cm

Rare and local resident, more regular near Sundarbans. **ID** Upperparts are greyish-brown, finely vermiculated whitish, and with diffuse darker brown streaking; underparts are greyish-white, finely vermiculated and more strongly streaked with brown. From Brown Fish-owl by more upright ear-tufts, more pronounced facial discs, more uniform grey-brown upperparts, and absence of any rufous tones. Legs are feathered to toes (largely unfeathered in Brown Fish-owl). **Voice** Deep, resonant *wo, wo, wo, wo-o-o-o-o.* **HH** Crepuscular. Generally roosts by day in a shady tree, sometimes a thicket, emerging about an hour before sunset. Frequently begins calling in early afternoon and then continues intermittently, but can call at any time. Usually nests in deserted raptor nest high in tree near water. Inhabits well-watered areas with extensive cover of thickly foliaged trees, in or near mangrove or sal forest.

Brown Fish-owl *Ketupa zeylonensis* 56cm

Uncommon, widespread resident. **ID** From Tawny Fish-owl by combination of duller brown upperparts, finer dark brown streaking on crown, mantle and scapulars, finer streaking on dull buff underparts (with close cross-barring, lacking on other species) and absence of white above bill. **Voice** Calls include soft, rapid, deep *hup-hup-hu,* maniacal laugh *hu-hu-hu-hu... hu ha* and mournful scream similar to Spot-bellied Eagle-owl. **HH** Usually lives in pairs, generally roosting in thickly foliaged tree. Partly diurnal, emerging from roost long before sunset, when pair members begin to call to each other. Sometimes hunts during day, especially in dull weather. Nests in tree or deserted raptor nest. Inhabits forest and well-wooded areas near water.

Tawny Fish-owl *Ketupa flavipes* 61cm

Very rare presumed resident; two records in 1990s. **ID** From smaller Brown Fish-owl by pale orange upperparts (much more richly coloured than on Brown) with bolder and more distinct black streaking, bold orange-buff barring on wing-coverts and flight feathers, and broader and more prominent black streaking on pale rufous-orange underparts, which lack black cross-barring; also, often shows prominent whitish patch on forehead. **Voice** Deep *whoo-hoo* and cat-like mewing. **HH** Habits similar to those of Brown Fish-owl. Banks of streams and rivers in dense broadleaved forest.

Buffy Fish-owl *Ketupa ketupu* 50cm

Rare resident in Sundarbans. **ID** From much larger Tawny Fish-owl by finer streaking on underparts, with only very fine shaft streaking on belly and flanks, and duller buffish-white to greyish-white barring on flight feathers and tail (more orange-buff on Tawny). From Brown by white forehead (although not always apparent), more rufous-orange upperparts with thicker and more prominent black streaking, orange-buff underparts with more clearly defined black streaking on breast, and lack of black cross-barring on feathers of underparts. **Voice** Call is monotonous *bup-bup-bup-bup-bup-bup-bup.* **HH** Forested creeks in mangroves.

Spot-bellied Eagle-owl

juv

ad

ad

Dusky Eagle-owl

ad

Brown Fish-owl

ad

Tawny Fish-owl

ad

Buffy Fish-owl

Osprey *Pandion haliaetus*
55–58cm

Scarce winter visitor. **ID** Long wings, typically angled at carpals, and short tail. Head is often held downwards when hunting. Has whitish head with black stripe through eye, dark and uniform upperparts with pale barring on tail, white underparts and underwing-coverts, and black carpal patches. Juvenile similar to adult but with buff tips to feathers of upperparts. **Voice** Gives a shrill cheeping whistle, but mostly silent away from nest. **HH** Usually solitary. Often perches on bamboo poles or stakes in or near water. Feeds entirely on fish, captured in powerful shallow dive feet first to grasp prey. Sundarbans, large wetlands including haors and large rivers.

Black-winged Kite *Elanus caeruleus*
31–35cm

Uncommon resident. **ID** Small size. Grey and white with black 'shoulders' and black eye patch. Wings pointed and tail rather short. Flight buoyant, with much hovering. Juvenile has brownish-grey upperparts with pale fringes, and less distinct shoulder patch. **Voice** Weak whistling notes. **HH** Usually found singly or in widely separated pairs, and in same area day after day. Spends much time perched on prominent vantage points. Hunts by quartering open ground, hovering at intervals with wings held high over back and beaten rather slowly, with feet often trailing. Open country: cultivation, grassland and open scrubland.

Jerdon's Baza *Aviceda jerdoni*
46cm

Scarce local resident in NE and SE, recent records in C. **ID** Long and erect white-tipped crest. Broad wings (pinched in at base) and fairly long tail. Greyish (male) or pale rufous (female) head, indistinct gular stripe, rufous-barred underparts and underwing-coverts, and bold barring across primary tips. At rest, closed wings extend well down tail. Juvenile has dark streaking on head and breast, and narrower dark barring on tail. **Voice** A *kip-kip-kip* or *kikiya, kikiya* uttered in display flight; also, plaintive mewing *pee-ow.* **HH** Crepuscular and elusive, keeping mostly within cover. Rather sluggish. Spends long periods perched in tree on lookout for prey. Evergreen and semi-evergreen forest.

Black Baza *Aviceda leuphotes*
33cm

Locally uncommon winter visitor to NE and SE, rare in C. **ID** Largely black, with long crest, white breast-band, rufous barring on underparts, and greyish underside to primaries contrasting with black underwing-coverts. Wings are broad and rounded and tail medium length. Flight like corvid, interspersed with short glides on flat wings. Male has more extensive patch of white on upperwing (extending onto secondaries) compared with female. **Voice** Loud, shrill, high-pitched *tcheeoua*, often repeated. **HH** Perches upright high in canopy of tall forest tree, joins mixed-species foraging flocks of larger birds such as laughingthrushes. Makes short flights to capture prey. Evergreen and semi-evergreen forest.

Crested Serpent-eagle *Spilornis cheela*
56–74cm

Fairly common and widespread resident. **ID** Broad, rounded wings. Soars with wings held forward and in pronounced V. At rest, has floppy black-and-white crest, yellow cere and lores, and unfeathered yellow legs. Adult has broad white bands across wings and tail and white spotting and barring on brown underparts. Juvenile has blackish ear-coverts, whitish head and underparts, narrower barring on tail than adult, and largely white underwing with fine dark barring and dark trailing edge. **Voice** Variety of loud, ringing, musical whistling or screaming calls in flight. **HH** Often quite tame. Has a characteristic habit of soaring over forest in pairs, the birds screaming to each other and sometimes rising to great heights. Spends hours perched very upright on forest tree. Raises crest if alarmed. Usually hunts by dropping almost vertically onto prey from perch. Stick and twig nest is built high in tree, often near stream. Forests and well-wooded country.

Osprey

ad

juv

ad

ad

Black-winged Kite

♀

juv

ad

Black Baza

Jerdon's Baza

juv

juv

ad

soaring

Crested
Serpent-eagle

Red-headed Vulture *Sarcogyps calvus*   85cm

Former rare resident recorded in 19th and early 20th centuries in NW, SE and C; now extirpated. **ID** Comparatively slim and pointed wings. Adult mainly black with bare reddish head and cere, white patches at base of neck and upper thighs, and reddish legs and feet; in flight, greyish-white bases to secondaries show as broad panel (particularly across underwing). Juvenile is browner with white down on head; pinkish coloration to head and feet, white patch on upper thighs and whitish undertail-coverts are best features. **Voice** Mostly silent, but squeaks, hisses and grunts like other vultures. **HH** Usually singly or in pairs. Frequently feeds on carcasses of small animals, which are overlooked by other vultures; also feeds timidly with other vultures at carcasses of larger animals. Open country near habitation, and well-wooded hills. **Threatened** Globally (CR).

Himalayan Griffon *Gyps himalayensis*   115–125cm

Rare but apparently increasing winter visitor to northern half of country, particularly NE; so far only juveniles and immatures recorded in Bangladesh. **ID** Larger than White-rumped and Griffon Vultures (see Appendix 1), with broader body and slightly longer tail. In unrecorded adult, wing-coverts and body plumage pale buffish, contrasting strongly with dark flight feathers and tail, and ruff is buffish. Underparts lack pronounced streaking. Legs and feet pinkish with dark claws, yellowish bill, and pale blue cere and facial skin (blackish in Griffon). Juvenile has brown-feathered ruff, with bill and cere initially black, dark brown body and upperwing-coverts boldly and prominently streaked with buff (wing-coverts almost concolorous with flight feathers), and back and rump also dark brown. Streaked upperparts and underparts and pronounced white banding across underwing-coverts are best distinctions of juvenile from Cinereous Vulture; very similar in plumage to juvenile White-rumped, but much larger and more heavily built, with broader wings and longer tail, underparts more heavily streaked, and streaking on mantle and scapulars. **Voice** A variety of grunts and hisses. **HH** Occurs in small groups or as individuals, roosts gregariously in large trees. Aggressive when feeding and dominates all other vultures except Cinereous. Nests in Himalayas. Birds wandering to Bangladesh sometimes perch on village houses or near habitation and are vulnerable to capture, mainly due to starvation or sometimes (if they have found a carcass) due to difficulty taking off after gorging. **AN** Himalayan Vulture.

White-rumped Vulture *Gyps bengalensis*   75–85cm

Rare resident, now largely restricted to NE and SW; declined sharply in late 20th and early 21st centuries; remaining colonies are protected as part of two 'vulture safe zones'. **ID** Smallest of the *Gyps* vultures. Adult mainly blackish, with white neck-ruff, white rump and back, and white underwing-coverts. Juvenile dark brown with streaking on underparts and upperwing-coverts, dark rump and back, whitish head and neck, and all-dark bill. In flight, underparts and underwing-coverts of juvenile darker than Slender-billed. Juvenile similar in colour to juvenile Himalayan, but smaller and less heavily built, with narrower wings and shorter tail; underparts less heavily streaked, and lacks prominent streaking on mantle and scapulars. **Voice** Croaks, grunts, hisses and squeals at nest colonies, roosts and carcasses. **HH** Gregarious. Often soars at great height in search of carrion. Roosts communally at traditional sites in groves. Nests colonially at traditionally used sites. Formerly found around human habitation, cultivation and open country, but remaining colonies are in forest. **Threatened** Globally (CR) and nationally (CR).

Slender-billed Vulture *Gyps tenuirostris*   93–100cm

Rare visitor, former resident, to NE and C, always uncommon but affected by sharp decline in late 20th and early 21st centuries. **ID** Bill, head and neck, and body are more slender than in Griffon (see Appendix 1). Adult has dark bill and cere with pale culmen, lacks any down on black head and neck, and has dirty white ruff that is rather small and ragged (by comparison, Griffon has extensive white down on head and neck, mainly yellowish bill, and extensive white ruff). Upperparts are colder grey-brown; white thighs. In flight, from below, shows prominent white thigh patches, trailing edges of the wings appear rounded and pinched in at the body, outer primaries noticeably longer than inner primaries, undertail-coverts dark (pale in Griffon) and in flight feet reach tip of tail (falling short of tail-tip in Griffon). Juvenile is similar to adult but with mainly dark bill, some white down on head and neck, and pale streaking on underparts. **Voice** Hissing and cackling sounds. **HH** Habits like White-rumped and often associates with that species. Nest is platform of twigs, sometimes with green leaves attached, in large leafy tree. Recorded in open cultivated areas, wetlands, forest edge, and around villages. **Threatened** Globally (CR).

Red-headed
Vulture

Himalayan
Griffon

White-rumped
Vulture

Slender-billed
Vulture

Cinereous Vulture *Aegypius monachus*   100–110cm

Rare winter visitor to NW, C and NE. **ID** Very large vulture with broad, parallel-edged wings. Soars with wings flat (*Gyps* vultures soar with wings held in shallow V). At distance, typically appears uniformly dark, except for pale areas on head and bill. Adult blackish-brown with paler brown ruff; may show paler band across greater underwing-coverts, but underwing darker and more uniform than on *Gyps* species. Juvenile is blacker and more uniform in colour than adult. **Voice** Usually silent. **HH** Normally solitary by day. Usually roosts communally on ground. Most birds recorded are wandering juveniles and prone to capture by local people, possibly due to shortage of carcasses. Dominates all other vultures at carcasses. Open country.

Changeable Hawk-eagle *Nisaetus cirrhatus*   63–77cm

Uncommon but widely distributed resident. **ID** Wings are slightly narrower and more parallel-edged compared with Mountain Hawk-eagle and has proportionately longer tail. Soars with wings held flat (except in display, when both wings and tail raised). Best told from Mountain Hawk-eagle by lack of prominent crest, paler sides to head, boldly streaked underparts (any barring confined to flanks, thighs and vent), and narrower dark tail barring. Occurs as dark morph when confusable with Black Eagle; best told by structural differences, greyish underside to tail and greyish bases to underside of flight feathers. Juvenile has pale fringes to upperparts (some with white forewing), buff underparts and underwing-coverts, and narrower and more numerous tail-bands than adult. Difficult to tell from juvenile Mountain but has paler head and lacks crest. Has pale crescent across outer primaries, absent in Mountain. **Voice** Silent except in breeding season; gives an ascending series of shrill whistles, *kri-kri-kri-kri-kree-ah* and *kreeee-krit* and a loud high-pitched *ki-ki-ki-ki-ki-ki-ki-ki-kee*, rising in crescendo and ending in a long, drawn-out scream. **HH** All forest types including open sal, mangroves and evergreen forest. **TN** Considered here to be conspecific with Crested Hawk-eagle of peninsular India and not a separate species *N. limnaeetus* as recognised by some authorities.

Mountain Hawk-eagle *Nisaetus nipalensis*   72cm

Five or more records from Chittagong Hill Tracts in SE, where possibly resident. **ID** Wings broader than on Changeable Hawk-eagle, with squarer wingtips and more pronounced curve to trailing edge, and has proportionately shorter tail. Soars with wings in shallow V. Distinguished from Changeable by prominent crest, dark crown and ear-coverts (with variable pale supercilium), heavily barred underparts and underwing-coverts, whitish-barred uppertail-coverts and stronger dark barring on tail. Juvenile from juvenile Changeable by more extensive dark streaking on crown and sides of head, white-tipped black crest, and fewer, more prominent tail-bars. Has patch of light and dark barring on inner primaries, lacking in Changeable. **Voice** Three shrill notes *tlueet-weet-weet*; also, a rapid, bubbling call in display. **HH** Forested hills. **TN** Formerly placed in genus *Spizaetus*. Threatened Nationally (VU).

Rufous-bellied Eagle *Lophotriorchis kienerii*   53–61cm

Six records from SE, where possibly resident. **ID** Smallish, with buzzard-like wings and tail. Glides and soars with flat wings. Adult has blackish hood and upperparts, white throat and breast, and (black-streaked) rufous rest of underparts. Upperwing is uniformly dark except for pale patches at base of primaries. At rest, shows short crest. Juvenile has white underparts and underwing-coverts, dark mask and white supercilium, dark patch on sides of upper breast, and dark patch on flanks. When head-on, shows striking white leading edge to wing; upperwing dark with paler base to primaries. **Voice** Silent except in breeding season, when utters piercing or plaintive scream. **HH** Evergreen forest in hills. **TN** Formerly placed in genus *Hieraaetus*. Threatened Nationally (VU).

Black Eagle *Ictinaetus malaiensis*   69–81cm

Five records from Chittagong Hill Tracts in SE, where it may be a rare resident. **ID** Distinctive wing shape and long tail. Flies with wings raised in V, with primaries upturned. At rest, long wings extend to tip of tail. Adult dark brownish-black, with striking yellow cere and feet; in flight, shows whitish barring on uppertail-coverts, and faint greyish barring on tail and underside of remiges (compared with dark morph of Changeable Hawk Eagle). Juvenile as adult; may show indistinct pale streaking to head and underparts. **Voice** Normally silent, shrill yelping cries in aerial courtship. **HH** Invariably seen on wing. Frequently soars over forest, often reaching considerable heights. Hunts by sailing buoyantly and slowly very low over forest. Hill forests. **TN** Former specific name *malayensis* recently shown to be pre-dated by *malaiensis*.

Cinereous Vulture

ad

juv

ad

soaring

ad
le morph

ad
dark
morph

juv

ad
pale
morph

ad
dark
morph

Changeable Hawk-eagle

ad

juv

ad

juv

ad

soaring

ad

juv

Mountain
Hawk-eagle

juv

ad

juv

soaring

soaring

ad

Black Eagle

Rufous-bellied
Eagle

Booted Eagle *Hieraaetus pennatus*   45–53cm

Rare winter visitor. **ID** Smallish rather kite-like eagle. Wings comparatively long and narrow, and tail long and square-ended. When gliding and soaring, wings are held slightly forward and are flat or slightly angled down at carpal. Twists tail in kite-like fashion. In all plumages, shows small white shoulder patches, pale panel across median coverts, pale wedge on inner primaries, pale scapulars, white crescent across uppertail-coverts and greyish underside to tail. Head, underparts and underwing-coverts are whitish, brown or rufous respectively in pale, dark and rufous morphs. Juvenile as adult but shows broad white trailing edge to wings and tail. **Voice** High-pitched double whistle, *ki-keee*. In display, very noisy, *pi-peee*, *pi-pi-pi-pi-peee*; also, at times a longer scream, *kleeek-kleeek*. **HH** Open and well-wooded country.

Indian Spotted Eagle *Clanga hastata*   59–67cm

Scarce but widespread resident, probably rarest in SE. **ID** As Greater Spotted, Indian Spotted is a stocky, medium-sized eagle with rather short and broad wings, buzzard-like head with comparatively fine bill (particularly so in Indian Spotted), long and closely feathered tarsi, and rather short tail. Wings are angled down at carpals when gliding and soaring. Adult is similar in overall appearance to Greater Spotted although warmer brown in coloration. Has wider gape than Greater Spotted, with thick 'lips' (gape flanges) visible at distance; gape-line extends to back of or behind eye (in Greater Spotted, reaches level to centre of eye). Lacks spiky nape feathers of Greater Spotted and has shorter thigh feathering. Underwing-coverts are paler or same colour as flight feathers (darker in Greater Spotted). Juvenile is more distinct from juvenile Greater Spotted. Spotting on upperwing-coverts is less prominent, tertials are pale brown with diffuse white tips (dark with bold white tips in Greater Spotted), uppertail-coverts are pale brown with white barring (white in Greater Spotted), and underparts are paler light yellowish-brown with dark streaking. In some plumages, can resemble Steppe Eagle – differences mentioned below for Greater Spotted are likely to be helpful for separation (although gape-line is also long in Steppe). **Voice** Very high-pitched cackling laugh. **HH** Like other *Clanga* and *Aquila* eagles, an aggressive and powerful predator that can soar well, often at considerable heights. Hunts by quartering with slow glides over areas within and near woodland, usually flying above treetop level; seizes most prey on ground. Hunts wide range of prey, mainly small mammals, but also amphibians, medium-sized or small birds, reptiles and insects. Nest is large platform of sticks and twigs, some with leaves attached, built in large tree. Favours mixed habitats: hunts in open wetlands and floodplains with woodland nearby; nests in lightly wooded areas. **TN** Formerly considered conspecific with Lesser Spotted Eagle as *Aquila pomarina*. **Threatened** Globally (VU) and nationally (EN).

Greater Spotted Eagle *Clanga clanga*   65–72cm

Scarce winter visitor, mainly to haors in NE. **ID** Medium-sized eagle with rather short and broad wings, stocky head and short tail. Wings distinctly angled down at carpals when gliding, almost flat when soaring. See account for Indian Spotted for differences from that species. Compared with Steppe Eagle, has less protruding head in flight, with shorter wings and less deep-fingered wingtips; at rest, trousers less baggy, and bill smaller with rounded (rather than elongated) nostril and shorter gape; lacks adult Steppe's barring on underside of flight and tail feathers, and dark trailing edge to wing, and has dark chin. Pale variant *'fulvescens'* distinguished from juvenile Imperial Eagle by structural differences, lack of prominent pale wedge on inner primaries on underwing, and unstreaked underparts. Juvenile has bold whitish tips to dark brown coverts. **Voice** A barking *kluck-kluck* or *tyuck-tyuck*. **HH** Often seen perched on treetop, bush or bank near water. Hunts on wing or on ground, generally taking slow-moving prey, especially frogs. Also takes faster-moving waterbirds by swooping low and scattering flock, isolating an individual; also eats stranded or dead fish and lizards. Large wetlands (haors), main rivers and (rarely) coastal islands. **TN** Formerly placed in genus *Aquila*. **Threatened** Globally (VU) and nationally (VU).

soaring

ad
dark morph

ad
pale morph

Booted Eagle

ad
dark morph

ad

ad

juv

Indian Spotted Eagle

juv

gliding

gliding

ad

ad

juv

juv
'*fulvescens*'

juv

Greater Spotted Eagle

juv

Tawny Eagle *Aquila rapax*
63–71cm

Very rare winter visitor, possibly former resident, with no records since 1980s. **ID** Compared with Steppe Eagle, hand of wing does not appear so long and broad, tail slightly shorter, and looks smaller and weaker at rest; gape-line ends level with centre of eye (extends to rear of eye in Steppe), and adult has yellowish iris (usually brown in Steppe). Differs from Greater and Indian Spotted in more protruding head and neck in flight, baggy trousers, yellow iris and oval nostril. Adult extremely variable, from dark brown through rufous to pale cream, and unstreaked or streaked with rufous or dark brown. Dark morph very similar to adult Steppe (which shows much less variation); distinctions include less pronounced barring and dark trailing edge on underwing, dark nape and dark throat. Rufous to pale cream Tawny Eagle is uniformly pale from uppertail-coverts to back, with undertail-coverts same colour as belly (contrast often apparent on similar species). Pale adults also lack prominent whitish trailing edge to wing, tip to tail and greater-covert bar (present on immatures of similar species). Characteristic, if present, is distinct pale inner-primary wedge on underwing. Juvenile also variable, with narrow white tips to unbarred secondaries; otherwise as similar-plumaged adult. Immature/subadult birds can show dark throat and breast contrasting with pale belly, and dark banding across underwing-coverts; whole head and breast may be dark. **Voice** A variety of loud, raucous cackles; distinctive guttural *kra* while in pursuit and harsh grating *kekeke* in display flight. **HH** Perches on treetops in cultivation or near village rubbish dumps. Feeds on carrion and refuse; also, small mammals, birds and reptiles, mainly stolen from smaller birds of prey, but also captures small mammals by pouncing on them from small bush. Inhabits open dry country. **Threatened** Globally (VU).

Steppe Eagle *Aquila nipalensis*
76–80cm

Scarce winter visitor. **ID** Broader and longer wings than Greater and Indian Spotted, with more pronounced and spread fingers, and more protruding head and neck; wings flatter when soaring, and less distinctly angled down at carpals when gliding. When perched, clearly bigger and heavier than Greater and Indian Spotted, with heavier bill and baggy trousers. Adult separated from adult spotted eagles by underwing pattern (dark trailing edge, distinct barring on remiges, indistinct or non-existent pale crescents in carpal region), pale rufous nape patch and pale chin. Juvenile has broad white bar across underwing, double white bar on upperwing and white crescent across uppertail-coverts; prominence of bars on upperwing and underwing much reduced on older immatures (when more similar in appearance to Indian Spotted; see that species). **Voice** Undescribed in region. **HH** Pirates food and eats carrion, also takes easily available prey, such as injured birds, stranded fish, rodents, lizards and snakes. Mainly grassy and sandy islands in main rivers and along coast, also large wetlands in NE and open country. **Threatened** Globally (EN).

Eastern Imperial Eagle *Aquila heliaca*
72–83cm

Rare winter visitor. **ID** Large, stout-bodied eagle with long and broad wings, longish tail, and distinctly protruding head and neck. Wings flat when soaring and gliding. Adult has almost uniform upperwing, small white scapular patches, golden-buff crown and nape, and two-toned tail. Juvenile has pronounced curve to trailing edge of wing, pale wedge on inner primaries, streaked buffish body and wing-coverts, uniform pale rump and back (lacking distinct pale crescent shown by other *Aquila* species, except Tawny), and white tips to median and greater upperwing-coverts. **Voice** A sonorous barking *owk-owk-owk*; rather silent outside breeding season. **HH** A solitary, majestic eagle. Rather inactive, typically spends much of day perched on good vantage point, such as tree, or often on ground. In the region, mainly feeds by robbing other birds of prey in flight; also eats carrion and small mammals, birds and reptiles taken on ground. Soars high and can fly at great speeds. Mainly on sandbanks in main rivers, also in vicinity of large wetlands. **Threatened** Globally (VU) and nationally (VU).

subad

ad

ad

ad

juv

Tawny Eagle

gliding

juv

ad

ad

Steppe Eagle

imm

juv

subad

ad

subad

ad

juv

gliding

ad

juv

gliding

juv

Eastern Imperial Eagle

Western Marsh-harrier *Circus aeruginosus*

42–54cm

Fairly common and widespread winter visitor. **ID** A broad-winged, stout-bodied harrier; in common with other harriers, glides and soars with wings held in noticeable V. Adult male is distinguished from other male harriers by combination of chestnut-brown mantle and upperwing-coverts contrasting with grey secondaries/inner primaries and black outer primaries, pale head (variably streaked with brown), pale leading edge to wing, and brown streaking on breast and belly, becoming uniformly brown on lower belly and vent. Shows variable amount of brown on underwing-coverts, with rest of underwing being white except black tips to primaries. All-dark melanistic morphs occur, which are sooty-grey above and blackish below, with grey patch at base of underside of primaries. Adult female is mainly dark brown except for creamy crown, nape and throat, creamy leading edge to wing, and paler patch at base of underside of primaries. Melanistic female (and juvenile) much as melanistic male, but with grey or whitish colour extending across underside of all flight feathers (not just on primaries). Juvenile is similar to female, but head and wing-coverts may be entirely dark. **Voice** Usually silent outside breeding season. **HH** Like other harriers has characteristic method of hunting: systematically quarters the ground a few metres above it, gliding slowly on raised wings and occasionally flapping rather heavily several times; on locating prey, drops quickly with claws held out to catch it. Freshwater and coastal wetlands and marshes, including haors and riverine grasslands. **AN** Eurasian Marsh-harrier if Eastern Marsh-harrier is not split.

Eastern Marsh-harrier *Circus spilonotus*

43–54cm

Uncommon winter visitor, mainly to NE, also C and NW. **ID** Differs markedly from Western Marsh, and in some plumages could be confused with Pied and Hen Harriers. Compared with Pied, it is larger, has broader wings and tail, and has stouter body, with more prominent head. Adult male has black streaking on whitish head and breast, and has black mantle and median coverts with feathers boldly edged with white. Head and mantle can appear mainly black on some, and is then not unlike adult male Pied, but is never as clean-cut in appearance as that species (i.e. breast is streaked, mantle and back show some white fringes, and lacks clear-cut black bar across median coverts and broad white leading edge to wing). Young male can appear similar to female Pied; small size, finer build and prominent barring across tail are useful features for Pied. Adult female has white uppertail-coverts, greyish flight feathers and tail with dark barring, heavily streaked head and breast, and broad, diffuse rufous streaking on underparts. Secondaries on underwing appear grey and diffusely barred, with paler patch at base of primaries. Structural differences and heavily marked underparts and underwing are best features for separation from female Pied. White 'rump'-band recalls female Hen, but pattern and coloration of upperwing, underwing and underparts is otherwise rather different. Juvenile is rather dark, with cream breast-band and pale patch at base of underside of primaries; head usually mainly cream, with variable dark streaking, and some have pronounced cream streaking on mantle. Can show white band across uppertail-coverts. **HH** Habits similar to those of Western Marsh. Large wetlands, marshes, reed-swamps, flooded fields and cultivation in vicinity of water. **TN** Split from *C. aeruginosus* is not recognised by some authorities.

Pied Harrier *Circus melanoleucos*

41–46.5cm

Widespread but uncommon winter visitor, most numerous in NE. **ID** Most likely to be confused with Eastern Marsh-harrier. Adult male is distinctive, with black head and breast contrasting with white underparts; black upperparts, median coverts and primaries contrast with grey of rest of wing, and white leading edge to wing. Yellow cere and iris are especially striking against black of head. Adult female superficially resembles female Hen Harrier but has paler underwing with narrow dark barring on flight feathers and sparse streaking on underwing-coverts, greyer primaries and secondaries with more pronounced dark banding, and greyer tail with narrower dark barring. Juvenile has dark brown head and upperparts, with white supercilium and patch below eye, white uppertail-coverts, rufous-brown underparts and underwing-coverts, and pale underside to primaries (with dark fingers) contrasting with dark underside of secondaries. **Voice** Usually silent in winter quarters. **HH** Flight and hunting behaviour like Western Marsh, but more graceful on wing. Wetlands, also open grassland and cultivation.

Western Marsh-harrier

♂ ♂ ♀ ♀ juv ♀

Eastern Marsh-harrier

♂ imm ♀ ♀ juv

♂ ♀ juv

Pied Harrier

♂ ♂ imm juv

juv ♀ ♀

Hen Harrier *Circus cyaneus* 44–52cm

Rare winter visitor. **ID** Compared with Pallid and Montagu's Harriers (see Appendix 1) has slower and more laboured flight and is stockier, with broader wings and more rounded hand, normally with five (rather than four) visible primaries at tip. Adult male is mainly dark grey with whiter belly, with extensive black at tips of wings and prominent white uppertail-covert patch. Adult female has boldly streaked underparts. White band across uppertail-coverts is broader than on Pallid and Montagu's. Has narrow pale neck-collar, but otherwise head pattern is plain compared with those species, usually lacking dark ear-covert patch. Juvenile recalls female but has rufous-brown wash to underparts and underwing-coverts, noticeably streaked with dark brown (unlike on juvenile Pallid and Montagu's). **Voice** Usually silent. **HH** Flight and hunting behaviour like Western Marsh-harrier, but more graceful on wing. Open grassland, wetlands and cultivation.

Crested Goshawk *Accipiter trivirgatus* 30–46cm

Rare to uncommon resident in E. **ID** Larger size and crest are best distinctions from Besra. Short, broad wings, pinched in at base. Wingtips barely extend beyond tail-base at rest. Yellow (rather than grey) cere and shorter crest help separate from Jerdon's Baza (Plate 43). Compared with other *Accipiter* species has short, stoutish legs. Soars low above canopy in fluttering display flight, with white undertail-coverts prominent. Narrow white line along uppertail-coverts diagnostic if present. Male has dark grey crown and paler grey ear-coverts, well-defined black submoustachial and gular stripes, and rufous-brown streaking on breast and barring on belly and flanks. Female larger with browner crown and ear-coverts, and browner streaking and barring on underparts. Juvenile has paler brown upperparts with pale fringes, buffish fringes to crown and crest, streaked ear-coverts and buffish wash to underparts, which are mainly streaked brown (with barring to flanks and thighs). **Voice** Calls include passerine-like whistles, a shrill scream, *he he, hehehehe* and shrill whistle *chewee...chewee...* **HH** Evergreen and semi-evergreen forest.

Shikra *Accipiter badius* 30–36cm

Fairly common resident. **ID** Adult paler than Besra and Eurasian Sparrowhawk. Underwing pale, with fine barring on remiges, and slightly darker wingtips. Head cuckoo-like compared with other *Accipiter* species. Male has pale blue-grey upperparts with contrasting dark primaries, indistinct grey gular stripe, fine brownish-orange barring on underparts, unbarred white thighs, and unbarred or lightly barred central tail feathers. Female upperparts brownish-grey. Female has pale brown upperparts, more prominent gular stripe and streaked underparts; from juvenile Besra by paler upperparts and narrower tail-barring, and from Eurasian Sparrowhawk by streaked underparts. **Voice** Loud, harsh *titu-titu*, similar to Black Drongo call; also, long drawn-out screams, *iheeya, iheeya* and constantly repeated *ti-tui* in display. **HH** Open forest and wooded country, including village groves.

Besra *Accipiter virgatus* 29–36cm

Rare resident. **ID** Small, with short primary projection (less than one-third down tail). Upperparts are darker and underwing is strongly barred compared with Shikra. Prominent gular stripe and streaked breast separate it from Eurasian Sparrowhawk. In all plumages, resembles Crested Goshawk but is considerably smaller, lacks crest, and has longer and finer legs. Adult male has dark slate-grey upperparts lacking strong contrast with primaries, broad blackish gular stripe, bold rufous streaking on breast and barring on belly, flanks and thighs, and broad dark tail-barring. Adult female has browner upperparts, with blackish crown and nape, and yellow iris (red in male). Juvenile has brown upperparts with rufous fringes, prominent gular stripe, and streaked/spotted underparts with barred flanks. From immature Shikra by darker, richer brown upperparts, broad gular stripe and broader tail-barring. From Eurasian Sparrowhawk by streaked underparts. **Voice** Loud squealing *ki-weeer* and rapidly repeated *tchew-tchew-tchew* during displays. **HH** Mainly in denser forests of E, rarer in sal and mangrove forests.

Eurasian Sparrowhawk *Accipiter nisus* 31–36cm

Rare winter visitor. **ID** Upperparts darker than Shikra, with prominently barred underparts and underwing. Uniform barring on underparts and absence of prominent gular stripe should separate from Besra. Male has dark slate-grey upperparts and reddish-orange barring on underparts. Female dark brown on upperparts, with brown-barred underparts and yellow iris (red in some males). Juvenile has dark brown upperparts and barred underparts. See Appendix 1 for comparison with Japanese Sparrowhawk and larger Northern Goshawk. **Voice** Usually silent in winter. **HH** Well-wooded country and open forest, including mangroves.

Hen Harrier

♂

♂ imm

♀

♂

Crested
Goshawk

♂

♂

juv

juv

Shikra

♂

♀

Eurasian
Sparrowhawk

♂

♀

juv

♂

♀

juv

Besra

juv

♂

White-bellied Sea-eagle *Haliaeetus leucogaster* 66–71cm

Localised coastal resident. **ID** Soars and glides with wings pressed forward in pronounced V. Distinctive shape, with slim head, bulging secondaries and short, wedge-shaped tail. Adult has white head and underparts, grey upperparts, white underwing-coverts contrasting with black remiges, and mainly white tail. Juvenile has pale head, dark breast-band, whitish tail with dark subterminal band and pale wedge on inner primaries. Immature shows mixture of juvenile and adult features. **Voice** Loud, far-carrying and goose-like honking, *ank... ank... ank*, and faster and more duck-like *ka, ka-kaaa*. **HH** Frequently soars, circling at great heights. Often seen on prominent perch, usually near water. Chiefly in Sundarbans and other coastal islands with mangrove forest.

Pallas's Fish-eagle *Haliaeetus leucoryphus* 76–84cm

Rare winter breeding visitor mainly to NE. **ID** Soars and glides on flat wings. Long, broad wings and protruding (rather small) head and neck. Adult has pale head and neck, dark brown upperwing and underwing, and mainly white tail with broad black terminal band. Juvenile less bulky, looks slimmer-winged, longer-tailed and smaller-billed than juvenile White-tailed Eagle (see Appendix 1); has dark mask, pale band across underwing-coverts, pale patch on underside of inner primaries, all-dark tail (lacking pale inner webs of White-tailed) and pale crescent on uppertail-coverts. Older immature has more uniform underwing and whitish tail with mottled dark band. Compare also with Steppe and Eastern Imperial Eagles. **Voice** Commonest call is hoarse, barking gull-like, *kvo kvok kvok*. **HH** Rather sluggish, perching for long periods on tree, post or sandbank close to water, but also patrols slowly over water and soars. Feeds mainly on fish snatched near surface. Nest is huge construction of sticks built over several years in large tree bordering wetland, sometimes close to villages. Immatures are regular in large wetlands away from nest sites. Haors and rarely large rivers. **Threatened** Globally (EN) and nationally (EN).

Grey-headed Fish-eagle *Icthyophaga ichthyaetus* 69–74cm

Rare but widespread breeding resident. **ID** Only really confusable with Lesser Fish-eagle *I. humilis* although this species has not been recorded in Bangladesh. Larger than Lesser, with longer tail. Wingtips fall short of tail-tip at rest. Adult from Lesser by largely white tail with broad black subterminal band. Also, darker and browner upperparts and deeper rufous-brown breast. Juvenile has boldly streaked head and underparts, diffuse brown tail-barring, and pale underwing with dark barring to flight feathers and pronounced dark trailing edge. **Voice** A squawk *kwok, kuwok* or *kuwonk*, or harsh shrieks. In display flight, gives eerie series of far-carrying, dreamy *tiu-weeeu* notes. **HH** Singly or in pairs. Usually perched in tree overlooking water; flies only short distances and rarely soars. Feeds almost entirely on fish. Well-wooded areas with abundant rivers and waterways.

Brahminy Kite *Haliastur indus* 48cm

Common and widespread resident. **ID** Small size and kite-like flight but with rounded tail. Wings usually angled at carpals. Tail rounded. Adult mainly chestnut, with white head, neck and breast. Juvenile mainly brown, with pale streaking on head, mantle and breast, large pale patch at base of primaries on underwing, and pale brown and unmarked underside to tail. **Voice** A nasal, drawn-out, slightly undulating *kyerrh* or a squeal. **HH** Frequently perches on tall tree overlooking water or flies slowly above ground looking for prey. All open habitats but mostly found in vicinity of water: coastal and freshwater wetlands including mangroves, ponds and flooded paddy-fields.

Black Kite *Milvus migrans* 58–66cm

Common and widespread resident (*M. m.govinda*); common and widespread winter visitor (*M. m. lineatus*). **ID** Shallow tail-fork. Much manoeuvring of arched wings and twisting of tail in flight. Dark rufous-brown, with variable whitish crescent at primary bases on underwing, and pale band across median coverts on upperwing. Juvenile has broad whitish or buffish streaking on head and underparts. *M. m. lineatus* is larger than *govinda*, with broader wings and generally more prominent whitish patch at base of primaries on underwing. Shows more pronounced dark mask, with paler crown and throat. Belly and vent are also paler. Juvenile is more heavily streaked. **Voice** Shrill, almost musical whistle *ewe-wir-r-r-r-r*. **HH** Gregarious throughout year, birds often soaring together and roosting communally, sometimes in large numbers. Closely associated with human habitation. Feeds mainly on refuse and offal but is omnivorous. Mainly occurs around cities, towns and villages, and wetlands; also around fishing camps. **TN** *M. m. lineatus* is treated as separate species 'Black-eared Kite' by some authorities.

ad

ad

juv

Pallas's Fish-eagle

ad

ad

juv

White-bellied
Sea-eagle

imm

imm

Grey-headed
Fish-eagle

juv

ad

Brahminy Kite

ad

juv

qovinda

juv

lineatus

ad

juv

ad

ad

juv

Black Kite

Oriental Honey-buzzard *Pernis ptilorhynchus* 57–60cm

Scarce resident. **ID** Tail long and broad, and has narrow neck, small head and bill, and short bare tarsi. Soars with wings flat. Has pronounced crest. Underparts and underwing-coverts range from dark brown through rufous to white, and unmarked, streaked or barred; often shows dark moustachial stripe and gular stripe, and gorget of streaking across lower throat. Lacks dark carpal patch. Male has grey face, two black tail-bands, usually three black underwing bands and brown iris. Female has browner face and upperparts, three black tail-bands, four narrower black underwing bands and yellow iris. Juvenile has narrower underwing banding, three or more tail-bands and extensive dark tips to primaries; cere yellow (grey on adult) and iris dark. **Voice** High-pitched screaming whistle *wheeew.* **HH** Spends long periods perched within tree foliage. Forests, open woodland and groves near cultivation and villages.

White-eyed Buzzard *Butastur teesa* 43cm

Scarce localised resident mainly in C. **ID** Longish, rather slim wings, long tail, and buzzard-like head. Pale median-covert panel. Adult has black gular stripe, white nape patch, barred underparts, dark wingtips and rufous tail; iris white and cere yellow. Juvenile has buffish head and breast streaked with dark brown, and throat-stripe indistinct or absent; rufous uppertail more strongly barred; iris brown. **Voice** Plaintive mewing *pit-weer, pit-weer* in breeding season. **HH** Sluggish, spends long periods perched very upright on vantage point such as isolated tree, post or mound. Hunts chiefly by dropping to ground to seize prey. Open sal forest, occasionally other open country, including woodland.

Japanese Buzzard *Buteo japonicus* 51–57cm

Thought to be scarce but widespread winter visitor, although specific identity of birds occurring in Bangladesh is uncertain, and confusion is compounded by varying taxonomic treatment of birds occurring in Indian Subcontinent. **ID** Stocky, with broad rounded wings and moderate-length tail. Plumage very variable. Typically, has variable streaking on breast and brown patches on sides of belly; tail is grey-brown to greyish-white with diffuse dark barring (appearing pale and unbarred from below); shows prominent dark carpal patches on underwing. Elsewhere, also occurs as an all-dark melanistic form. Birds similar in appearance to Long-legged Buzzard, but different structurally, occur and could be Eurasian Buzzard *B. buteo vulpinus*, with rufous on underparts and underwing-coverts, and narrowly barred rufous tail. **Voice** A loud, repeated mew *peee-oo.* **HH** Sandy islands in large rivers and on coast, tea estates and other open country. **TN** Treatment here recognises Japanese as separate from Eurasian Buzzard and Himalayan Buzzard *B. refectus* (= *burmanicus*), based on these three forms being well differentiated genetically (but not morphologically), and that Japanese is a winter visitor to the region from NE Asia. Genetic studies of birds occurring in Bangladesh are needed to clarify the situation.

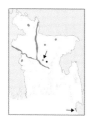

Long-legged Buzzard *Buteo rufinus* 61cm

Scarce but widespread winter visitor. **ID** Larger and longer-necked than Japanese and Eurasian Buzzards, with longer wings and tail; soars with wings in deeper V. Most distinctive morph has combination of paler head and upper breast, rufous-brown lower breast and belly, more uniform rufous underwing-coverts, more extensive black carpal patches, larger pale primary patch on upperwing, and unbarred pale orange uppertail. Intermediate and black morphs much as some plumages of Japanese and Eurasian Buzzards. Juvenile generally less rufous, with narrower and more diffuse trailing edge to wing, and lightly barred tail which on many is pale greyish-brown. **Voice** Rather silent; when anxious a short, sharp, high-pitched *mew.* **HH** Mostly on sandy islands in large rivers.

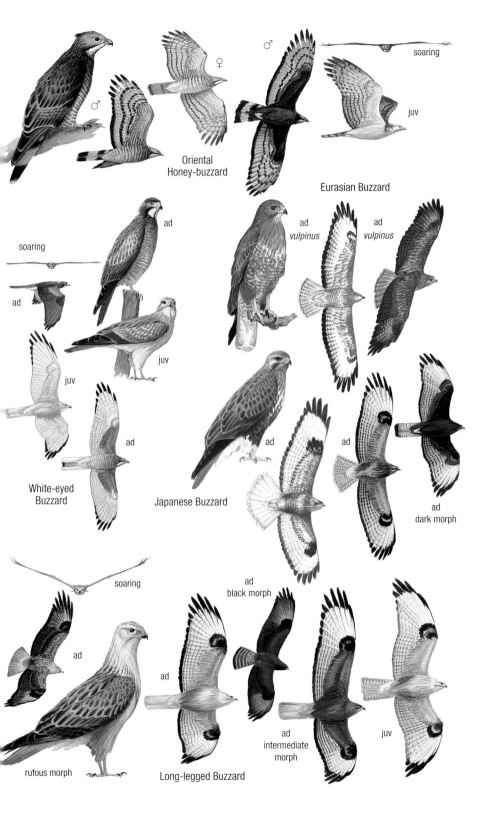

Oriental
Honey-buzzard

♂

♀

♂

soaring

juv

Eurasian Buzzard

soaring

ad

ad
vulpinus

ad
vulpinus

ad

juv

juv

ad

White-eyed
Buzzard

Japanese Buzzard

ad

ad

ad
dark morph

soaring

ad
black morph

ad

ad

ad

juv

rufous morph

Long-legged Buzzard

ad
intermediate
morph

Red-headed Trogon *Harpactes erythrocephalus* 35cm

Uncommon resident in E , rare in C. **ID** Eye-ring and bill can be bright blue. Male has crimson head and breast, white breast-band, pinkish-red underparts, and black-and-grey vermiculated wing-coverts. Female is similar but with dark cinnamon head and breast, and brown-and-buff vermiculated coverts. Immature resembles female. **Voice** Descending sequence, *tyaup, tyaup, tyaup, tyaup, tyaup.* **HH** Usually singly or in widely separated pairs. Perches almost motionless in upright posture for long periods. Captures flying insects on the wing, twisting and turning like a flycatcher, or snatches prey while hovering in front of foliage, occasionally by swooping to ground. Middle storey of evergreen and semi-evergreen forest.

Common Hoopoe *Upupa epops* 31cm

Uncommon, widespread resident and winter visitor. **ID** Mainly rufous-orange to orange-buff, with striking black-and-white wings and tail, black-tipped fan-like crest (usually held flat) and downcurved bill. Broad, rounded wings reminiscent of giant butterfly in flight. **Voice** Repetitive *poop, poop, poop.* **HH** Usually seen singly or in pairs. Searches for insects on ground, running, walking, probing and pecking. Flight is undulating, slow and butterfly-like, with irregular wingbeats. Open country, cultivation and around villages.

Great Hornbill *Buceros bicornis* 95–105cm

Rare resident in Chittagong Hill Tracts, one vagrant record in NE. **ID** Huge size, massive yellow casque and bill, and white tail with black subterminal band. Has broad black face-band contrasting with yellowish-white nape, neck and upper breast. In flight, shows broad whitish wingbar and trailing edge to wing. Neck, breast, white wingbars and base of tail typically stained yellow with preen-gland oils. Male has red iris, black circumorbital skin and black at each end of casque. Female smaller, with smaller bill, white iris and red circumorbital skin, and lacks black at ends of casque. Immature lacks casque until at least six months old. **Voice** Loud, deep, retching calls, often in short series and frequently given as duet. Loud *ger-onk* flight call. **HH** Quite wary. Keeps to regular schedule of feeding circuits and roosting flights. Often flies high over forest for long distances. Chiefly arboreal, but sometimes descends to ground to feed. Mature evergreen forest with fruiting trees. **Threatened** Globally (VU) and nationally (VU).

Oriental Pied Hornbill *Anthracoceros albirostris* 55–60cm

Uncommon local resident in E. **ID** Has yellow-and-black bill, black head and neck, and mainly black tail with white tips to outer feathers. Upperwing is black except for narrow white trailing edge. Sexes are similar, but female is smaller with less convex casque lacking projecting tip and has black at tip of bill and casque. Both sexes have pale blue circumorbital skin. Immature has smaller bill and casque with less black; orbital skin and throat patch are whitish. **Voice** Variety of loud, shrill, nasal squeals and raucous chucks, including loud cackling. **HH** Feeds in treetops, particularly on figs and other fruits; often detected by loud wingbeats when flying over canopy. Mature evergreen and semi-evergreen forest with fruiting trees.

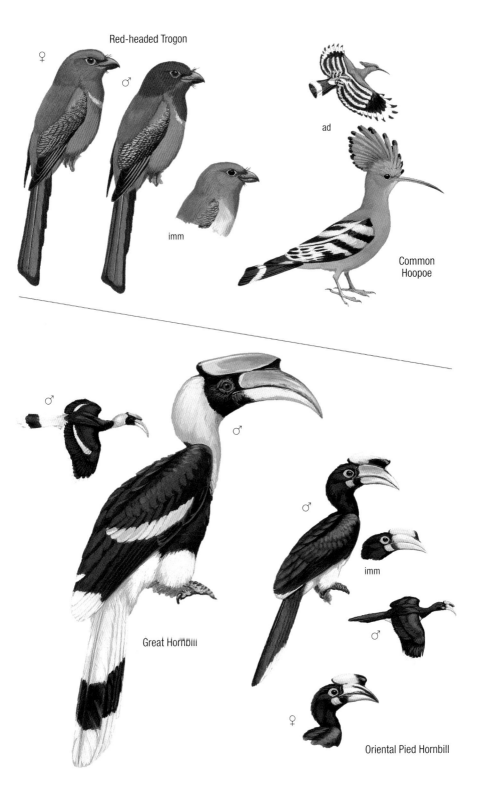

Red-headed Trogon

♀

♂

imm

ad

Common
Hoopoe

♂

♂

♂

imm

♂

Great Hornbill

♀

Oriental Pied Hornbill

Blue-bearded Bee-eater *Nyctyornis athertoni*   31–34cm

Uncommon local resident in E. **ID** Large green bee-eater with broad, square-ended tail. Adult has blue forehead and 'beard', green upperparts, broad greenish streaking on yellowish-buff belly and flanks, and yellowish-buff undertail-coverts and underside to tail. Shows yellowish-buff underwing-coverts in flight. Juvenile is similar to adult and has blue 'beard' even when very young. **Voice** A gruff *gga gga ggr gr* or *kor-r-r kor-r-r.* **HH** Spends much time perched inactively and inconspicuously among foliage in middle or upper forest storey; sometimes in open on treetops. Typically, perches in hunched posture with its tail hanging vertically. Makes aerial sallies for insects from vantage point like other bee-eaters. Flight is laboured and deeply undulating. Edges and clearings of evergreen and semi-evergreen forest.

Asian Green Bee-eater *Merops orientalis*   16–18cm

Common and widespread resident. **ID** Small with elongated central tail feathers, blue throat with black gorget, golden-brown crown and nape, and green tail. Juvenile has square-ended tail; crown and nape are green, and lacks black gorget. **Voice** Pleasant throaty trill, *tree-tree-tree.* **HH** Often in small, loose parties. Perches on vantage point such as small tree, dead branch or post and sometimes on backs of cattle; then launches sallies for insects and circles gracefully back to base. Open country with scattered trees. **AN** Green Bee-eater.

Chestnut-headed Bee-eater *Merops leschenaulti*   18–20cm

Locally common resident. **ID** From Asian Green by combination of bright chestnut crown, nape and mantle, yellow throat, turquoise rump and broad tail with shallow fork. Juvenile is duller, with chestnut of upperparts absent or reduced to a wash on crown (crown and nape are uniform dark green on some). **Voice** A *pruik* or *churit*, briefer or less melodious than calls of larger bee-eaters. **HH** Similar to Blue-tailed, but a forest bird; often sallies from treetops or feeds high above canopy. Evergreen, deciduous and mangrove forest.

Blue-tailed Bee-eater *Merops philippinus*   23–26cm

Locally common summer visitor; rarely overwinters. **ID** Mainly green bee-eater with blue rump and tail. Has green crown and nape (with touch of blue on supercilium in front of eye). Throat and ear-coverts are chestnut (with touch of blue below eye), and upperparts and underparts are washed with rufous and turquoise. Juvenile is similar to adult but duller. Shows strong blue cast to rump, uppertail-coverts and tail. **Voice** A rolling *diririp.* **HH** In loose flocks when foraging. Darts out from exposed perches to seize prey; also hawks insects in continuous flight. Sometimes hunts from treetops at forest edge and clearings. Nest burrows in sandbanks and sandy ground. Mainly along rivers and other wetlands with sandy islands, also in open wooded country.

Oriental Dollarbird *Eurystomus orientalis*   28cm

Uncommon, local migrant and summer visitor in E. **ID** Dark greenish to bluish, appearing black at distance, with red bill and eye-ring. In flight, turquoise patch on primaries. Flight is buoyant, with broad wings. Juvenile is similar but has dull pinkish bill. **Voice** Raucous *check check.* **HH** Spends long periods in daytime perched inactively, often on tops of dead trees, only occasionally making short flights for insects. Feeds chiefly from late afternoon until dark. Hawks insects on the wing; small parties gather and hawk in circles where insects are swarming. Agile in flight when chasing prey. Evergreen and semi-evergreen forest and forest clearings. **AN** Dollarbird.

Blue-bearded Bee-eater

ad

Asian Green Bee-eater

ad

ad

Chestnut-headed Bee-eater

ad

juv

ad

Blue-tailed Bee-eater

ad

juv

juv

ad

Oriental Dollarbird

ad

Indian Roller *Coracias benghalensis* 33cm

Common and widespread resident in W but status needs further investigation. **ID** Has rufous-brown on nape and underparts, white streaking on ear-coverts and throat, and greenish mantle. Has turquoise band across primaries and dark blue terminal band to tail. Intergrades with Indochinese Roller occur. **Voice** Raucous *chack-chack-chack*, with discordant screeches. **HH** Spends most of day on prominent perch such as dead tree, post or telegraph wire in open country, scanning for prey. Swoops leisurely down to capture prey on ground. Often hunts until dusk is well advanced. Aggressively territorial throughout most of year. Cultivation, open woodland, groves and gardens.

Indochinese Roller *Coracias affinis* 33cm

Common and widespread resident in C and E, but status needs further investigation. **ID** Darker than Indian Roller. Has brownish-green upperparts, purplish-brown underparts and blue streaking on throat. In flight, turquoise band across primaries and turquoise on outer tail with dark corners. **Voice** As Indian. **HH** Habits as Indian. Cultivation, open woodland, tea estates. **TN** Often considered conspecific with Indian Roller.

Oriental Dwarf-kingfisher *Ceyx erithaca* 14cm

Rare and local summer visitor. **ID** Tiny forest kingfisher with coral-red bill. Has orange head, with variable violet iridescence on crown and nape, blue-black forehead and 'ear patch', black upperparts with coverts and scapulars boldly marked with blue, pale orange underparts, and orange rump and tail also with violet iridescence. Juvenile is duller; bill is initially dark, becoming pale yellow-orange. Underparts are whitish with orange breast-band, crown is more orange (less violet), and has less blue in upperparts. **Voice** Call a weak, thin *seet*, thinner and higher-pitched than Blue-eared; contact calls include weak, shrill *tit- sreet* and *tit-tit*. **HH** Shady streams in moist evergreen forest. **Threatened** Nationally (EN).

Blue-eared Kingfisher *Alcedo meninting* 17cm

Rare local resident, mainly in coastal regions. **ID** From Common by blue ear-coverts (except in juvenile plumage), darker blue upperparts (lacking greenish tones to crown, scapulars and wings), darker brilliant blue back and rump, and deeper orange underparts. Female has red on lower bill. Juvenile has rufous-orange ear-coverts as on Common Kingfisher but has darker blue upperparts (similar in coloration to adult) and lacks the broad blue moustachial stripe of that species (although can show short black moustachial, which does not extend beyond ear-coverts). **Voice** Call is shrill, single *seet* or *tsit*, higher-pitched and less strident than Common. Contact calls are thin, shrill *striiiiit, trrrrrt tit... trrrreu* etc. **HH** Habits very similar to Common. Streams and creeks in mangrove and evergreen forest.

Common Kingfisher *Alcedo atthis* 16cm

Common and widespread resident. **ID** From similar forest-dwelling Blue-eared by orange ear-coverts, paler greenish-blue upperparts and paler orange underparts. Note however that juvenile Blue-eared has rufous ear-coverts. Female has red on lower mandible. Juvenile is similar to adult but duller and greener above, with dusky scaling on breast. **Voice** Call is high-pitched, shrill *chee*, usually repeated, and *chit-it-it* alarm call. **HH** Uses post, reed or bank at water's edge as vantage point, perching 1–2m above surface. Plunges headlong into water to catch prey; may submerge completely, and sometimes hovers before diving. Wetlands and village ponds, usually away from forest.

Pied Kingfisher *Ceryle rudis* 25cm

Common and widespread resident. **ID** Crested black-and-white kingfisher. Has white-streaked black crown and crest, white supercilium contrasting with broad black eyestripe, white underparts with black breast-band, and black-and-white wings and tail. Male has double breast-band. Female has single, usually broken, breast-band. **Voice** Sharp *chirruk chirruk*. **HH** Characteristically hunts by hovering over water with fast-beating wings and bill pointing down, before plunging vertically downwards when it sees a fish. Large rivers, also wetlands and floodplains.

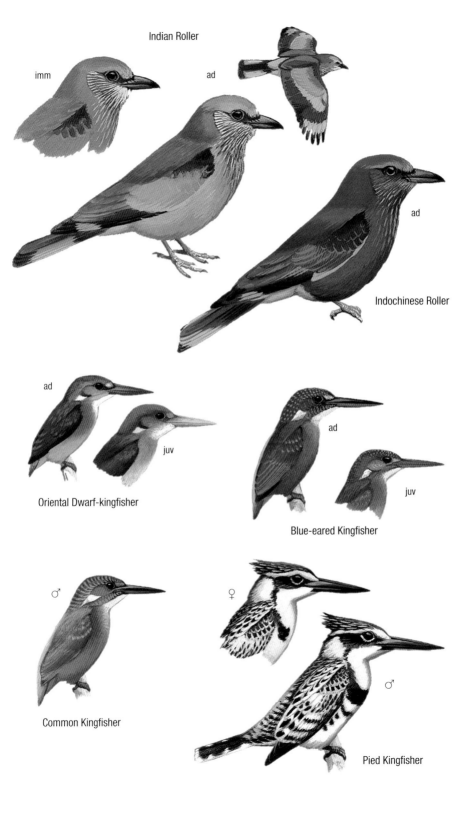

Indian Roller

imm

ad

ad

Indochinese Roller

ad

juv

Oriental Dwarf-kingfisher

ad

juv

Blue-eared Kingfisher

♂

Common Kingfisher

♀

♂

Pied Kingfisher

Stork-billed Kingfisher *Pelargopsis capensis* 35cm

Uncommon but widespread resident. **ID** Very large with huge coral-red bill, brownish cap, pale orange-buff collar and underparts, and blue-green upperparts. In flight, shows turquoise rump and lower back. Juvenile has dusky barring on underparts, especially on breast (forming a broad band). **Voice** Explosive, shrieking *ke-ke-ke-ke-ke* and pleasant *peer, peer, pur.* Song a long series of often paired melancholy whistles *iuu-iuu...iuu-iuu...iuu-iuu* etc. **HH** Rather sluggish and heard more often than seen. Sits, often half-hidden, on branch overhanging water, occasionally diving to catch prey. Flight is strong, direct and steady like other kingfishers. Beels, rivers and fish ponds bordered by large trees. **TN** Formerly placed in genus *Halcyon.*

Brown-winged Kingfisher *Pelargopsis amauroptera* 35cm

Locally common resident in SW. **ID** Very large with huge red bill. From Stork-billed by brown mantle, wings and tail, which contrast strongly with turquoise rump and lower back (especially in flight). Lacks the brown cap of Stork-billed, and the entire head and underparts are rich brownish-orange. Juvenile has pale fringes to mantle and wing-coverts, and fine dusky barring on nape and underparts. **Voice** A deep laughing *cha-cha-cha-cha,* descending in pitch, and mournful *chow-chow-chow* whistle. **HH** Sundarbans mangrove forest and nearby planted mangroves. **TN** Formerly placed in genus *Halcyon.* **Threatened** Nationally (VU).

Ruddy Kingfisher *Halcyon coromanda* 25cm

Rare local resident restricted to Sundarbans. **ID** Medium-sized forest-dwelling kingfisher, with large coral-red bill, rufous-orange upperparts with brilliant violet gloss and paler rufous underparts. In flight, shows striking bluish-white rump. Juvenile is darker and browner on upperparts and has bluer rump, faint blackish barring on rufous underparts and blackish bill. **Voice** Call is descending, high-pitched *titititititititi,* rather like White-breasted but more musical. Song is soft, trilling *tyuur-rrrr* incessantly repeated at *c.* 1-second intervals from perch. **HH** Secretive, shy and most easily detected by call. Mangrove forest.

White-breasted Kingfisher *Halcyon smyrnensis* 27–28cm

Common and widespread resident. **ID** Large kingfisher with large red bill, chocolate-brown head and underparts, white throat and centre of breast, and brilliant turquoise-blue upperparts including rump and tail. In flight, shows prominent white patches at base of black primaries. Juvenile is duller, with brown bill and dark scalloping on breast. **Voice** Call is loud, rattling laugh. Song is drawn-out musical whistle *kililili.* **HH** Bold and noisy. Like most kingfishers spends long periods perched singly or in well-separated pairs, watching intently for prey. Typically perches on fenceposts, telegraph wires or branches. Occasionally jerks tail, and on spotting quarry often bobs head and body to help judge distance before dropping swiftly downwards to seize prey in bill. Wide-ranging habitat, and can be found far from water in cultivation, forest edge, gardens and wetlands. **AN** White-throated Kingfisher.

Black-capped Kingfisher *Halcyon pileata* 28cm

Common winter visitor to coast, rare vagrant elsewhere. **ID** Large, mainly coastal kingfisher with coral-red bill. Has black cap, white collar, deep purplish-blue upperparts, black coverts contrasting with blue secondaries, white breast, and pale orange-buff belly and flanks. In flight, shows bright blue rump and prominent white patches at base of primaries. Juvenile has dusky scalloping on collar and breast. **Voice** Distinctive ringing cackle, *kikikikikiki,* similar to, but higher-pitched than, White-throated. **HH** Perches in open at forest edge or on telegraph wires. Dives down obliquely to catch prey; rarely plunges into water. Mangroves and coastal woodland.

Collared Kingfisher *Todiramphus chloris* 23–25cm

Locally common resident along coast. **ID** Medium-sized coastal kingfisher with stout, mainly blackish bill, blue-green crown and ear-coverts with short white supercilium, white collar, blue-green mantle, blue wings and tail, and white underparts. Female is duller with scaling on breast-sides. Juvenile has scaling on buffish-white underparts and buff supercilium. Lacks white in wing of Black-capped. **Voice** Raucous *krerk-krerk-krerk-krerk.* **HH** Mangrove forest and other coastal wetlands, including shrimp farms.

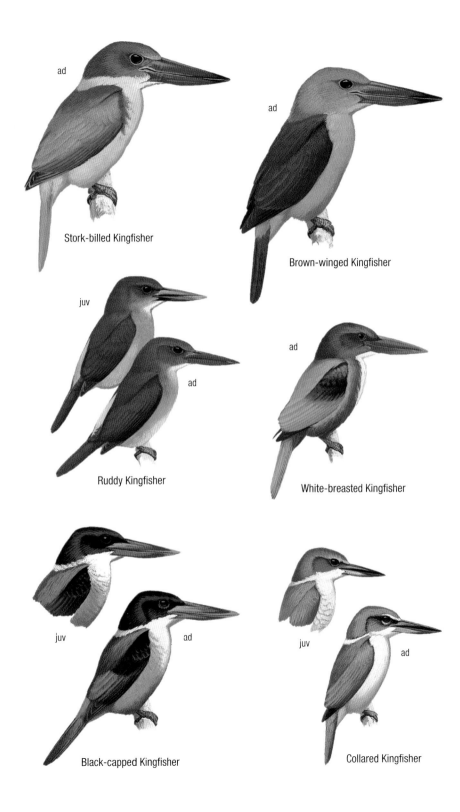

ad

Stork-billed Kingfisher

ad

Brown-winged Kingfisher

juv

ad

Ruddy Kingfisher

ad

White-breasted Kingfisher

juv

ad

Black-capped Kingfisher

juv

ad

Collared Kingfisher

Coppersmith Barbet *Psilopogon haemacephalus* 17cm

Common and widespread resident. **ID** A small, brightly coloured barbet, with crimson forehead and patch on breast, yellow patches above and below eye contrasting with blackish hindcrown and sides of head, yellow throat, dark streaking on belly and flanks, and bright red legs and feet. Juvenile lacks red on forehead and breast; has prominent pale yellow patches above and below eye (surrounded by dark olive sides of head and moustachial stripe), whitish throat, olive-green breast-band, and broad olive-green streaking on belly and flanks. **Voice** Call is loud, metallic, monotonous, repetitive *tuk, tuk, tuk* etc. **HH** Usually found singly or in small parties. Particularly vocal in heat of day. Village groves, open wooded areas and deciduous forest; scarcer in mangroves and evergreen forest edge. **TN** Previously in genus *Megalaima*.

Blue-eared Barbet *Psilopogon cyanotis* 17cm

Uncommon local resident in E and C. **ID** A small barbet with black forehead, blue throat and complex black, blue, red and yellow pattern on sides of head. Easily told from larger Blue-throated by blackish forehead and blue crown, multicoloured face and smaller, all-dark bill. Juvenile is mainly green, with blue wash to sides of head and throat, and lacks black and red head markings. **Voice** Disyllabic, repetitive *tk-trrt*, repeated about 120 times per minute, and throaty whistle. **HH** Habits similar to Coppersmith. Usually found singly, perched in treetop. Evergreen and semi-evergreen forest. **TN** Previously in genus *Megalaima*. Treated here as separate from extralimital Yellow-eared *P. australis* and Black-eared Barbets *P. duvaucelii*.

Great Barbet *Psilopogon virens* 33cm

Rare resident restricted to Chittagong Hill Tracts. **ID** Largest barbet and unmistakable, with large pale yellow bill, violet-blue head, brown breast and mantle, olive-streaked yellowish underparts, and red undertail-coverts. Juvenile is duller with greener head. **Voice** Monotonous, incessant and far-reaching *piho piho* uttered throughout day; also, repetitive *tuk tuk tuk* often given in duet (presumably with female). **HH** Rather silent in winter, but noisy in hot weather. Usually singly or in groups of up to five or six, but congregates in larger, feeding parties in fruit-laden trees. Evergreen forest and second growth. **TN** Previously in genus *Megalaima*.

Lineated Barbet *Psilopogon lineatus* 28cm

Common and widespread resident. **ID** Has bold white streaking on head, upper mantle and breast (extending to centre of belly), with usually whitish chin and throat (although can be dusky brown on some). Underparts can look white, streaked with brown. In addition, has naked yellowish patch around eye and yellowish bill. Juvenile has less prominent streaking. **Voice** Monotonous *kotur, kotur, kotur*. **HH** Sings persistently from high in canopy. Otherwise, found singly or in small groups, sometimes with other frugivorous birds. Forest and well-wooded areas. **TN** Previously in genus *Megalaima*.

Blue-throated Barbet *Psilopogon asiaticus* 23cm

Common and widespread resident. **ID** A medium-sized barbet, with red forehead, black band across centre of crown, red hindcrown, and blue 'face', throat and upper breast. Possibly confusable with Blue-eared Barbet but is larger, and has red on crown, uniform blue sides of head and larger pale bill (with variable dark culmen and tip). Juvenile is similar to adult, but head pattern is duller and poorly defined (with red of crown intermixed with green and black). **Voice** Loud, harsh *took-a-rook, took-a-rook* uttered very rapidly. **HH** Habits similar to Lineated. Mainly forest but also well-wooded villages, especially where there are figs. **TN** Previously in genus *Megalaima*.

Coppersmith Barbet

ad

juv

Blue-eared Barbet

ad

juv

Great Barbet

ad

Lineated Barbet

ad

paler variant

Blue-throated Barbet

juv

ad

Eurasian Wryneck *Jynx torquilla* 16–17cm

Uncommon but widespread winter visitor. **ID** Cryptically patterned with grey, buff and brown. Has dark stripe through eye, irregular dark stripe down nape and mantle, buff to pale rufous throat and breast, which are finely barred with black, and long, barred tail. **Voice** Piping *kek-kek-kek*. **HH** Rather sluggish and unobtrusive. Feeds mainly on ground, hopping along with tail slightly raised, picking up ants and other insects. Can cling to tree trunks. Scrub, second growth and edges of cultivation, tea estates and swamp-thickets.

White-browed Piculet *Sasia ochracea* 9–10cm

Scarce localised resident in E. **ID** Tiny and appears tail-less. Has greenish-olive upperparts (variably washed with rufous) and rufous underparts, very short black tail, fine white supercilium behind eye, and red iris and orbital skin. Male has golden-yellow on forehead, which is rufous on female. **Voice** A short, sharp *chi*, sometimes followed by rapid trill. Rapid, tinny drumming; also taps loudly on bamboo. **HH** Forages on bamboo stems and other twigs in undergrowth. Sometimes hops among leaf litter on forest floor, cocking its tail like wren. Bamboo and dense low vegetation in evergreen and semi-evergreen forest.

Speckled Piculet *Picumnus innominatus* 10cm

Scarce localised resident. **ID** Tiny. Whitish face is broken by blackish ear-covert patch and malar stripe, and white to yellowish-white underparts heavily spotted with black. Also, greyish crown, yellowish-green upperparts, and short, square-ended blackish tail has white on central and outer feathers. Male has dull orange forehead and forecrown, barred with black. Female has uniform forehead and crown. **Voice** Sharp *tsick*, high-pitched *sik-sik-sik* a *ti-ti-ti* and loud drumming. **HH** Usually with mixed feeding parties of insectivorous species. Energetic and agile when feeding; small enough to forage on slender twigs of trees and shrubs, often in canopy. Creeps along branches, clings upside-down and perches sideways across branches. Evergreen, semi-evergreen and mangrove forest.

Bay Woodpecker *Blythipicus pyrrhotis* 27cm

Rare resident in E. **ID** From Rufous by long yellowish-white bill. It is also larger, with more angular head shape, and has more broadly barred and brighter rufous upperparts, diffuse streaking on forehead and crown, darker brown underparts, and largely unbarred tail. Male has prominent scarlet patch on sides of neck, which extend onto nape. Female lacks scarlet patch. Juvenile has more prominent barring on mantle, diffuse rufous and dark brown barring on underparts, and more prominent pale streaking on head. **Voice** Loud, descending laughter, *keek, keek-keek-keek-keek-kerere-kerere*; also, loud chattering *kerere-kerere-kerere*. **HH** Forages mostly within a few metres of ground on moss-covered trunks, dead stumps and fallen logs; also, on ground among roots, and sometimes higher, keeping close to or on the trunk. Shy and elusive, heard much more often than seen. Eats ants and beetle larvae. Dense evergreen forests.

Pale-headed Woodpecker *Gecinulus grantia* 25cm

Rare resident in E. **ID** A smallish, mainly unbarred woodpecker with small, pale bill. Easily distinguished by golden-olive head and neck, dull crimson to crimson-brown upperparts, brown primaries barred with buffish-pink, and dark olive underparts. Male has crimson-pink on crown. **Voice** Territorial call is reminiscent of Bay, a loud, strident *yi-wee-wee-wee*; other calls are harsh, high-pitched and quickly repeated *grrritj-trrit-grrit* etc. Drumming is loud, evenly pitched and of fairly short duration. **HH** Forages noisily and often low on large bamboos in evergreen forest.

Rufous Woodpecker *Micropternus brachyurus* 25cm

Widespread, fairly common resident. **ID** A medium-sized, rufous-brown woodpecker with shaggy crest and short black bill. Heavily barred with black on mantle, wings, flanks and tail. Male has small scarlet flash on ear-coverts. Female has pale buff ear-coverts. **Voice** A high-pitched, nasal *keenk keenk kenk*. Diagnostic drumming like stalling engine *bdddd-d-d-d-dt*. **HH** Shy, forages in trees at all heights; often seen digging into tree-ant nests; also feeds on fallen logs, termite nests and cow dung. Forests, second growth and denser village groves.

Speckled Piculet

ad

Eurasian Wryneck

♀

♂

♀

♂

White-browed Piculet

♂

♀

♂ juv

Bay Woodpecker

♂

♂ juv

♀

♀

♂

Rufous Woodpecker

Pale-headed Woodpecker

Greater Flameback *Chrysocolaptes guttacristatus* 33cm

Locally common resident. **ID** From Himalayan by larger size and longer S-shaped neck, longer bill, white or black-and-white spotted hindneck, pale eyes and four (not three) toes. In addition, has clearly divided moustachial stripe (with obvious white oval centre), clean single black line down centre of throat and white spotting on black breast. Female has white spotting on crown (white streaking in Himalayan). **Voice** Sharp, metallic, monotone, *di-di-di-di-di-di-di*, like large cicada, and lower-pitched, more rapid *di-i-i-i-i-t* in flight. **HH** Frequently associates with other woodpeckers, drongos and larger insectivorous species. Chiefly visits large trees. Noisy. Forages at all levels, especially on dead wood, sometimes on ground. Mainly eats insects and grubs, also nectar. Evergreen, semi-evergreen, mangrove and mixed deciduous forest. **AN** Greater Goldenback. **TN** Previously *C. lucidus* but now recognised as separate species.

Himalayan Flameback *Dinopium shorii* 30–32cm

Rare resident, unrecorded since 2000 probably due to loss of large trees at its main site. **ID** Smaller size and bill than Greater Flameback, with black hindneck and brownish-buff centre of throat (and breast on some) with black spotting forming irregular border. Has indistinctly divided moustachial stripe (centre is brownish-buff, with touch of red on male). Has reddish or brown eyes and three toes. Breast is irregularly streaked and scaled with black, and on some is almost unmarked. Female has white streaking to black crest (white spotting in female Greater). **Voice** Rapid, repeated tinny *klak-klak-klak-klak-klak*, slower and softer than Greater. **HH** Often associates with other woodpeckers and small birds. Mixed deciduous forest. **AN** Himalayan Goldenback.

Common Flameback *Dinopium javanense* 28–30cm

Scarce resident, largely restricted to Sundarbans. **ID** Smaller size and bill than Greater Flameback, lacking cleanly divided moustachial, and has black hindneck. Has reddish or brown eyes and three toes. Smaller size and bill compared with Himalayan. Moustachial stripe lacks clear dividing line (usually solid black, although can appear divided on some and similar to Himalayan). Also, Common has irregular line of black spotting down centre of throat (brownish-buff line in Himalayan), and breast of Common is more heavily marked with black. **Voice** Rattle call faster and higher-pitched than Greater. **HH** Mangrove forest; status of isolated records in evergreen forests unclear. **AN** Common Goldenback.

Black-rumped Flameback *Dinopium benghalense* 26–29cm

Common and widespread resident. **ID** Smaller than other flamebacks. Best told by combination of black lower back and rump, different head pattern (has white-spotted black throat and black stripe through eye, but lacks moustachial stripe), barred primaries and (variable) white or buff spotting on blackish lesser wing-coverts. Female from females of other flamebacks by red hindcrown and crest. **Voice** Single strident *klerk* contact call and whinnying *kyi-kyi-kyi*. **HH** Bold and noisy. Often found in small parties with other species. Forages at all levels in trees and on ground. Chiefly eats ants. Village groves, open wooded areas, second growth, deciduous and mangrove forests. **AN** Lesser Goldenback.

Greater Yellownape *Chrysophlegma flavinucha* 33cm

Locally fairly common resident. **ID** From Lesser Yellownape by larger size and bill, brown on crown, and lack of red-and-white markings on head. Further differences include dark olive sides of neck adjoining dark-spotted white foreneck, uniform underparts, black barring on primaries and yellow (male) or rufous-brown (female) throat. **Voice** A plaintive, descending *pee-u... pee-u* and single metallic *chenk*. **HH** Often with mixed-species feeding flocks and other woodpeckers. Feeds at all forest levels, from floor to highest branches. Evergreen, deciduous and mangrove forest. **TN** Previously in genus *Picus*.

Lesser Yellownape *Picus chlorolophus* 27cm

Locally fairly common resident in E and SW. **ID** From Greater Yellownape by smaller size and bill, red-and-white head markings, rufous panel in wing, white barring on primaries and white barring on underparts. Male has red moustachial and line above eye. **Voice** A buzzard-like, drawn-out *pee-oow* and descending series of shrill notes, *kee kee kee kee kee*. **HH** Often accompanies roving mixed parties of insectivorous birds or other woodpecker species. Feeds chiefly in smaller trees, occasionally in understorey shrubs or fallen logs. Evergreen, semi-evergreen and mangrove forest.

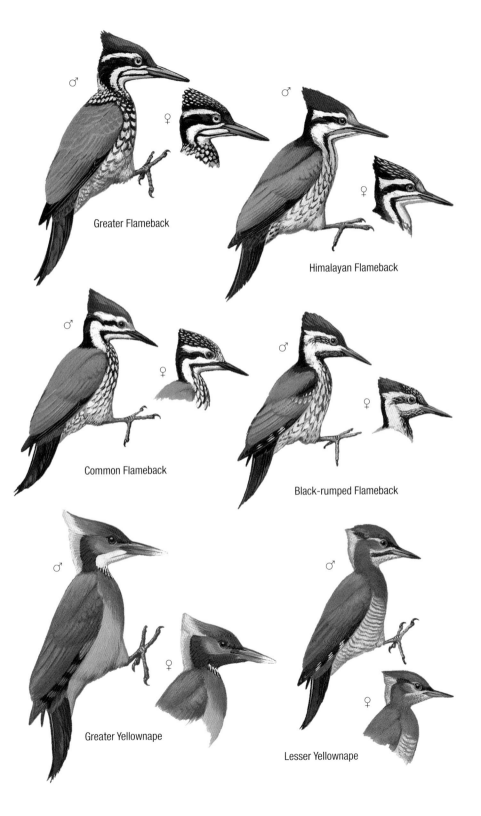

Greater Flameback

Himalayan Flameback

Common Flameback

Black-rumped Flameback

Greater Yellownape

Lesser Yellownape

Streak-throated Woodpecker *Picus xanthopygaeus*   30cm

Uncommon resident in W and C. **ID** Has bold scaling on underparts. From Streak-breasted by olive streaking on throat and upper breast. See account for Streak-breasted for other useful features. Male has red crown; female has black crown, streaked with grey. Juvenile has grey bases to feathers of mantle and scapulars, creating mottled appearance to upperparts. **Voice** Rather silent, although can give sharp, single *queemp.* **HH** Mainly feeds on ground on ants and termites; also pecks at cattle dung for beetle larvae and takes flower nectar. Open deciduous forest, mangroves and second growth.

Streak-breasted Woodpecker *Picus viridanus*   30–33cm

Rare resident restricted to Sundarbans. **ID** From Streak-throated by unmarked or indistinctly marked olive foreneck and upper breast. Other features mentioned in literature (plainer grey ear-coverts, red eye, more pronounced blackish moustachial, blacker tail) appear to be variable in both Streak-breasted and Streak-throated and may be unreliable in the field. **Voice** Calls include squirrel-like *kyup.* **HH** Mangroves.

Black-naped Woodpecker *Picus guerini*   32cm

Locally common resident. **ID** Has plain grey face, black nape and moustachial, dark bill and uniform greyish-green underparts. Male has red forehead and forecrown (black in female). Juvenile is duller, with greyer upperparts, less pronounced moustachial and whitish barring on underparts. **Voice** A high-pitched *peeek, peeek, peeek, peeek,* usually in runs of four or five notes and fading at end; also, chattering alarm cry. **HH** Usually solitary and somewhat shy species, edging up and down tree trunk out of sight from observer. Often feeds on ground. Partial to ants, beetles, berries and nectar. Evergreen and semi-evergreen forests, second growth and mangroves. **TN** Previously recognised as conspecific with Grey-headed (now Grey-faced) Woodpecker *P. canus.*

Great Slaty Woodpecker *Mulleripicus pulverulentus*   51cm

Rare and local resident in E. **ID** Giant slate-grey woodpecker with huge pale bill, long neck and long tail. Male has pinkish-red moustachial patch and pink on lower throat. **Voice** Loud goat-like bleating; very loud cackle in flight. **HH** Forages chiefly on trunks and major branches; usually in small parties, birds follow each other in flight with slow deliberate wingbeats. Has large home range. Evergreen and semi-evergreen forests with large trees, mainly in SE; records in NE presumably wanderers. **Threatened** Globally (VU).

Grey-capped Woodpecker *Picoides canicapillus*   14cm

Uncommon localised resident. **ID** Has grey crown (blackish on sides and towards nape), blackish stripe behind eye, and fulvous coloration to underparts with heavy black streaking. Also, tends to show diffuse blackish malar stripe and whitish throat. Male has crimson on nape. See Appendix 1 for differences from Indian Pygmy Woodpecker. **Voice** High, quickly repeated *tit-tit-errrrrr* rattle; also, weak, repeated *kik.* **HH** Often forages with mixed parties of insectivorous birds. Typically in tree canopy, its small size enables it to forage on outermost twigs. Often perches crosswise and hangs upside-down. Deciduous and mangrove forest, also (scarcer) in evergreen forest. **AN** Grey-capped Pygmy Woodpecker. **TN** Previously in genus *Dendrocopos.*

Fulvous-breasted Woodpecker *Dendrocopos macei*   18–19cm

Common and widespread resident. **ID** A medium-sized woodpecker with white-barred black mantle and wings, and lightly streaked dirty buff underparts. Male has red on crown, female black. **Voice** An explosive *tchick,* rapid *pik-pipipipipipipipi* and soft chattering *chik-a-chik-a-chit.* **HH** Often with mixed feeding flocks of other species. Forages chiefly on tree trunks and larger high branches. Village groves, open wooded country, tea estates, forest edge and deciduous forest.

Streak-throated Woodpecker

Streak-breasted Woodpecker

Black-naped Woodpecker

Great Slaty Woodpecker

Grey-capped Woodpecker

Fulvous-breasted Woodpecker

Common Kestrel *Falco tinnunculus* 32–35cm

Uncommon winter visitor. **ID** Long, rather broad tail; wingtips more rounded than on most falcons. Male has greyish head with diffuse dark moustachial stripe, rufous upperparts heavily marked with black, and grey tail with black subterminal band. Female and juvenile have rufous crown and nape streaked with black, diffuse and narrow dark moustachial stripe, rufous upperparts heavily marked with black, and dark barring on rufous tail. See Appendix 1 for comparison with Lesser Kestrel *F. naumanni.* **Voice** High-pitched, shrill *kee-kee-kee.* **HH** Usually found singly or in pairs. Hovers with rapidly beating wings and fanned tail, while scanning ground for prey. Open country.

Red-headed Falcon *Falco chicquera* 31–36cm

Scarce localised resident. **ID** Powerful falcon with pointed wings and longish tail. Flight usually fast and dashing. Adult has rufous crown, nape and narrow moustachial stripe, pale blue-grey upperparts with fine dark barring, white underparts finely barred with black, and grey tail with broad black subterminal band. In flight, blackish primaries contrast with rest of upperwing. Sexes alike, with female larger. Juvenile is similar but darker, with fine dark shaft streaking on crown, fine rufous fringes to upperparts, and underparts with fine rufous-brown barring. **Voice** Shrill *ki-ki-ki-ki-ki*, rasping *yak, yak, yak* and screaming *tiriri, tiriririeee.* **HH** Usually hunts cooperatively in pairs, one bird pursuing prey and the other cutting off its escape. Open country, including main rivers, usually with some large palms or other trees; also in urban areas, where it nests on pylons. **AN** Red-necked Falcon.

Amur Falcon *Falco amurensis* 28–31cm

Scarce passage migrant, mainly in May and also Oct/Nov. **ID** In all plumages has red to pale orange cere, eye-ring, legs and feet. Shape similar to Eurasian Hobby but with slightly more rounded wingtips and longer tail. Male dark grey, with rufous thighs and undertail-coverts, and white underwing-coverts. First-year male shows mixture of adult male and juvenile characters. Female has dark grey upperparts, short moustachial stripe, whitish underparts with some dark barring and spotting, and orange-buff thighs and undertail-coverts; uppertail barred; underwing white with strong dark barring and dark trailing edge. Juvenile is similar to female but has rufous-buff fringes to upperparts, rufous-buff streaking on crown and boldly streaked underparts. **Voice** Shrill, screaming *kew-kew-kew* when settling at roost may continue throughout night. **HH** Highly gregarious and crepuscular. Migrates during daytime in South Asia, typically in loose scattered flocks. Hunts by hawking insects and by hovering. Open country.

Eurasian Hobby *Falco subbuteo* 30–36cm

Scarce passage migrant and winter visitor. **ID** Slim, with long pointed wings and medium-length tail. Hunting flight swift and powerful, with stiff beats interspersed by short glides, acrobatic in pursuit of prey. Adult has broad black moustachial stripe, cream underparts with bold blackish streaking, and rufous thighs and undertail-coverts. Juvenile has dark brown upperparts with buffish fringes, pale buffish underparts, which are more heavily streaked, and lacks rufous thighs and undertail-coverts. **Voice** Calls *kew-kew-kew* or *ki ki-ki* at nest site. **HH** Markedly crepuscular. Often perches on isolated trees. Well-wooded areas; also, open country and cultivation in winter.

Oriental Hobby *Falco severus* 27–30cm

Rare presumed resident in Chittagong Hill Tracts. **ID** Similar to Eurasian Hobby in structure, flight action and appearance, although slightly stockier, with shorter tail. Slimmer in wings and body than Peregrine. Adult has complete blackish hood (lacking white cheeks of Eurasian), bluish-black upperparts and sides of breast (suggesting half-collar), and unmarked rufous underparts and underwing-coverts. Straight cut to black cheeks and absence of any barring on underparts help to distinguish from *peregrinator* subspecies of Peregrine. Juvenile has browner upperparts and heavily streaked rufous-buff underparts. **Voice** Rapid *ki-ki-ki-ki.* **HH** Habits very similar to Eurasian. Forested hills.

Common Kestrel

♀

♂

♂

♀

♂ imm

♀

♂

Red-headed Falcon

juv

ad

ad

ad

♂

♀

juv

♂ imm

Amur Falcon

ad

juv

juv

ad

Eurasian
Hobby

Oriental
Hobby

Laggar Falcon *Falco jugger*   43–46cm

Rare presumed resident but status uncertain. **ID** Large falcon, although smaller, slimmer-winged and less powerful than Saker Falcon (see Appendix 1). Adult has rufous crown, dark stripe through eye extending to nape, narrow but long and prominent dark moustachial stripe, brownish-grey to dark brown upperparts (can be greyer than illustrated), and rather uniform uppertail. Underparts and underwing-coverts vary, can be largely white or heavily streaked, but lower flanks and thighs usually wholly dark brown; typically shows dark panel across underwing-coverts. Juvenile similar to adult, but crown duller, moustachial is broader, and underparts very heavily streaked (almost entirely dark on belly, flanks and underwing-coverts), and has greyish bare parts; differs from juvenile Peregrine in paler crown, finer moustachial stripe, more heavily marked underparts and unbarred uppertail. **Voice** Usually silent; shrill *whi-ee-ee* seldom heard except in breeding season. **HH** Usually seen perched on a regularly used vantage point, such as treetop or post. Also circles high overhead. Hunts mainly by flying rapidly and low and seizing prey on ground. Open areas in NW, C and NE regions. **Threatened** Nationally (VU).

Peregrine Falcon *Falco peregrinus*   38–51cm

Scarce winter visitor, may be resident in Chittagong Hill Tracts in SE. **ID** Heavy-looking falcon with broad-based, pointed wings and short, broad-based tail. Flight strong, with stiff, shallow wingbeats and occasional short glides. *F. p. calidus*, a winter visitor that is most regularly recorded, has slate-grey upperparts, broad and clean-cut black moustachial stripe and whitish underparts with narrow blackish barring; juvenile *calidus* (not illustrated) has browner upperparts, heavily streaked underparts, broad moustachial stripe and barred uppertail. May show pale supercilium. Rare resident *F. p. peregrinator* has dark grey upperparts with more extensive black hood (and less pronounced moustachial stripe), and rufous underparts with dark barring on belly and thighs; juvenile *peregrinator* has darker brownish-black upperparts than adult, and paler underparts with heavy streaking. **Voice** That of *peregrinator* is unrecorded, except for *chir-r-r-r* uttered close to nest. Generally silent away from breeding areas. **HH** Bold falcon, highly skilful in flight. Pursues flying prey rapidly, finally rising above it and stooping with terrific force, wings almost closed. Coastal and large freshwater wetlands, particularly large rivers in winter, also occasionally in urban areas; *F. p. peregrinator* in part-forested hills.

Vernal Hanging-parrot *Loriculus vernalis*   14cm

Scarce local resident in E. **ID** Small (sparrow-sized), stocky green parrot with red rump and uppertail-coverts and red bill. Adult has yellowish-white iris. Male has turquoise throat patch, which is lacking or much reduced in female. Immature similar to female, but red rump and uppertail-coverts are mixed with green, and has brown iris. **Voice** Distinctive di- or trisyllabic rasping flight call *de-zeez-zeet*, occasionally given when at rest. **HH** Mainly keeps to tall treetops. Short, rapid wingbeats and undulating flight. Evergreen and semi-evergreen forest.

Grey-headed Parakeet *Psittacula finschii*   36cm

Rare resident or visitor in E. **ID** Ashy-grey head, yellowish-green wash to upperparts (pronounced on nape) and lilac-blue tail with yellowish-cream tip. Female similar to male but has darker green mantle and lacks maroon shoulder patch. Larger than female and immature Blossom-headed; has longer tail with larger yellowish-white tip, darker grey head with black chin-stripe, marked blue-green collar, and larger bill. Immature has brownish-green head, later becoming dull slate-grey, and lacks black chin-stripe and half-collar of adult. **Voice** Loud, shrill whistles: *sweet... sweet, swit* etc. **HH** Agile, fast-flying flocks. Semi-evergreen forest in Chittagong Hill Tracts. **Threatened** Nationally (VU).

Laggar Falcon

ad

juv

ad

ad

juv

ad
peregrinator

ad
peregrinator

juv
peregrinator

ad
calidus

ad
peregrinator

Peregrine Falcon

ad
calidus

juv
peregrinator

♂

♀

ɪᴍ

Grey-headed Parakeet

♂

imm

Vernal Hanging-parrot

Blossom-headed Parakeet *Psittacula roseata* 36cm

Local scarce resident in NE. **ID** Male has pale pink and lilac-blue on head, yellow upper mandible, uniform green neck and pale yellow tip to tail. Female has pale greyish-blue head, indistinct collar, maroon shoulder patch and pale yellow tail-tip; smaller body and daintier head and bill than Grey-headed, with lilac cast to paler grey head (lacking black chin-stripe and half-collar); yellow upper mandible. Juvenile has greenish head. **Voice** Shrill *tooi-tooi*, higher-pitched and less harsh than Grey-headed. **HH** Does not associate with people, unlike Rose-ringed. Roosts communally. In flight, weaves through trees with great agility. Evergreen and semi-evergreen forest, second growth and tea estates, attracted to fruiting figs in forest. One record in SE could be a vagrant or escape.

Plum-headed Parakeet *Psittacula cyanocephala* 36cm

Rare resident or visitor, mainly to far W. **ID** Male from Blossom-headed by plum-red and purplish-blue head, turquoise collar and white-tipped blue-green tail. Female similar to female Blossom-headed, has greyish head, and yellowish collar and upper breast. Juvenile has greenish head. **Voice** As Blossom-headed. **HH** Habits similar to Blossom-headed. Can be destructive to crops and orchards. Well-wooded areas. Birds recorded in Dhaka may have been escapes or releases from captivity.

Red-breasted Parakeet *Psittacula alexandri* 38cm

Locally common resident in E and C. **ID** Male has lilac-grey crown and ear-coverts (with variable pinkish wash), broad black chin-stripe, deep lilac-pink breast and belly, greenish-yellow lesser wing-coverts and yellowish-tipped blue-green tail. Female is similar, but has blue-green tinge to head, purer peach-pink breast and black upper mandible. Immature is duller, with green underparts and orange-red bill. **Voice** Short, sharp nasal *kaink*. **HH** Usually quiet while feeding in treetops. Has large communal roosts. Can raid crops. Sometimes in large flocks. The common parakeet of tea estates and second growth in hills; moves deeper into forests and keeps in smaller flocks in breeding season.

Alexandrine Parakeet *Psittacula eupatria* 53cm

Rare resident now most likely to be encountered in Dhaka city. **ID** From Rose-ringed Parakeet by combination of larger size, maroon shoulder patch, and massive bill. Deeper, more raucous call and slower and more laboured flight are additional pointers. Male has black chin-stripe joining pink and turquoise hind-collar, both of which are lacking on female and immature. Immature has less distinct maroon shoulder patch and shorter tail. **Voice** Loud guttural *keeak* or *kee-ah*, deeper and more raucous than Rose-ringed's call. **HH** Quite wary. Small parties climb about high in fruiting trees. Flocks raid orchards and crops. Flies with deliberate wingbeats, uttering a harsh, loud scream. Roosts communally. Originally deciduous sal forests, but colonies nesting in large trees now almost lost, one colony nesting in building cavities in central Dhaka.

Rose-ringed Parakeet *Psittacula krameri* 42cm

Common and widespread resident. **ID** From Alexandrine by smaller size, lack of maroon shoulder patch and smaller bill. Dark blue-green (rather than pale yellowish) dorsal aspect of tail is a further feature. Male has black chin-stripe joining pink hind-collar. Female lacks the chin-stripe and collar and is all green (with indistinct pale green collar). **Voice** Shrill, loud and variable *kee-ah*, higher-pitched and less guttural than Alexandrine. **HH** Flocks raid orchards and crops in large numbers. Can form enormous communal roosts, often with crows and mynas. Constantly screeches and squabbles. Often associated with habitation and cultivation; also open woodland and second growth.

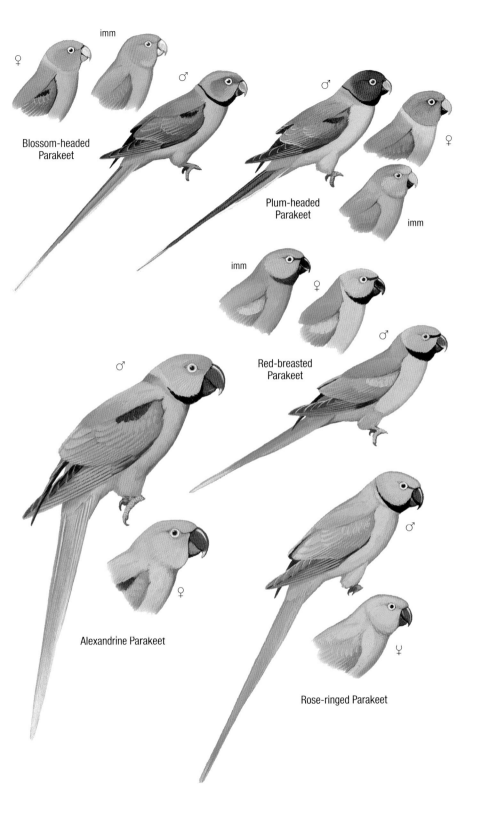

♀
imm
♂
Blossom-headed
Parakeet

♂
Plum-headed
Parakeet
♀
imm

imm
♀
Red-breasted
Parakeet
♂

♂
Alexandrine Parakeet
♀

♂
Rose-ringed Parakeet
♀

Blue-naped Pitta *Hydrornis nipalensis* 25cm

Local resident in NE and SE, including Chittagong Hill Tracts. **ID** Large pitta with fulvous sides of head and underparts, and uniform green upperparts. Male has glistening blue hindcrown and nape. Female has rufous-brown crown and smaller, greenish-blue patch on nape. Juvenile is mainly brown, streaked and spotted with buff, with brownish-white supercilium; wings and tail are brownish-green. **Voice** Powerful double whistle. **HH** Habits like Western Hooded. Broadleaved evergreen forest and moist shaded ravines, particularly with dense bamboo. **TN** Formerly in genus *Pitta*.

Blue Pitta *Hydrornis cyaneus* 23cm

Rare breeding visitor to NE (two records from Lawachara NP) and possibly in Chittagong Hill Tracts in SE. **ID** Large pitta with pinkish-red on hindcrown, black stripe through eye, black moustachial stripe, and bold black spotting and barring on underparts. In flight, shows small white wing patch. Male has blue upperparts and pale blue wash to underparts. Female has dark olive upperparts with variable blue tinge, and breast is washed with buff; occasionally lacks pinkish-red on hindcrown. Juvenile is mainly dark brown, streaked and spotted with rufous-buff, with prominent buff supercilium and dark eyestripe. **Voice** Song is liquid *pleoow-whit*, the first part falling, the second part sharp and short. **HH** Habits like Western Hooded. Broadleaved evergreen forest, moist ravines. **TN** Formerly in genus *Pitta*.

Indian Pitta *Pitta brachyura* 19cm

Local summer visitor, fairly common in sal forests in NW and C regions. **ID** Has bold black stripe through eye contrasting with white throat and supercilium, and buff lateral crown-stripes separated by black centre to crown. Underparts are buff, with reddish-pink lower belly and vent. Upperparts are green, with shining blue uppertail-coverts and forewing. In flight, shows small white wing patch. Juvenile is much duller with lateral crown-stripes scaled with black. **Voice** Sharp two-note whistle, second note descending, *pree-treer.* **HH** Habits like Western Hooded. Most easily seen shortly after arrival (Apr–May) when very vocal and active establishing territories. Deciduous forest with dense undergrowth.

Mangrove Pitta *Pitta megarhyncha* 20cm

Resident, restricted to mangroves within Sunderbans, where widespread. **ID** Similar to Indian Pitta, but larger with longer bill. Crown is uniform rufous-brown, with buffish sides which may show as indistinct supercilium. Lacks white patch below eye of Indian. Upperparts are darker green than Indian, wing-coverts and rump/uppertail-coverts are deeper blue, and white wing patch is larger. Juvenile is duller than adult with black scaling on crown. **Voice** A loud, slurred *tae-laew.* **HH** Mangroves.

Western Hooded Pitta *Pitta sordida* 19cm

Summer visitor, uncommon in NE and SE forests. **ID** Has largely black head with chestnut crown and nape, glistening blue forewing and uppertail-coverts, green breast and flanks, and scarlet belly and vent with black abdomen patch. In flight, shows larger white wing patch than Indian. Juvenile duller with black scaling on crown, white patch on median coverts, brownish chin and dirty white throat, and brownish underparts with dull pink belly and vent. **Voice** Song is loud, explosive double whistle, *wieuw-wieuw;* calls include *skyeew* contact call and bleating, whining note. **HH** Usually seen singly or in pairs. Keeps mainly to forest floor. Forages by flicking leaves and other vegetation and probing leaf litter and damp earth. Usually progresses on ground by long hopping bounds. Skulking and often most easily located by call. Sings and roosts in trees or bushes, sometimes high up. Evergreen forest, often near water. **AN** Hooded Pitta. **TN** Treated here as separate from extralimital Eastern Hooded Pitta *P. novaeguineae.*

Mangrove Whistler *Pachycephala cinerea* 17cm

Resident, restricted to Sunderbans. **ID** Rather drab, with thick black bill, grey-brown upperparts, greyish-white throat and breast merging into silvery-white of rest of underparts (some with whiter throat). **Voice** Variable phrase with 2–4 introductory notes and last note louder, shriller and more explosive. **HH** Mangroves.

♂ Blue-naped Pitta

♀

imm

♂ Blue Pitta

♀

imm

imm

Indian Pitta

ad

imm

Mangrove Pitta

ad

imm

Western Hooded Pitta

ad

Mangrove Whistler

ad

Long-tailed Broadbill *Psarisomus dalhousiae* 28cm

Local presumed resident, rare in Chittagong Hill Tracts and historically in NE. **ID** 'Dopey-looking' with big head, large eye, stout lime-green bill, long and thin tail, and upright stance. Mainly green with black cap, blue crown, and yellow 'ear' spot, throat and collar. Shows white patch at base of primaries in flight. Juvenile has green cap and lacks blue crown spot. **Voice** Loud, piercing *pieu-wieuw-wieuw-wieuw* call. **HH** Arboreal, keeps in forest canopy or middle storey, in flocks of up to 20. Very upright posture when perched. Unobtrusive and lethargic when not feeding. Evergreen and semi-evergreen forest.

Grey-browed Broadbill *Serilophus rubropygius* 18cm

Rare resident in NE and SE (Kaptai NP). **ID** Big head, crested appearance, large eye, stout bluish bill, upright stance, and sluggish movements give rise to distinctive jizz. Has blackish supercilium, yellow eye-ring, pale chestnut tertials and rump, complex white-and-blue pattern to black wings, and black tail. Shows white patch at base of primaries in flight. Female has broken white necklace. Juvenile similar to adult but with dark bill. **Voice** A *ki-uu*, like rusty hinge. **HH** Evergreen and semi-evergreen forest. **AN** Silver-breasted Broadbill. **TN** Often treated as conspecific with Silver-breasted Broadbill *S. lunatus*.

Maroon Oriole *Oriolus trailli* 27cm

Scarce winter visitor to E. **ID** Maroon rump, vent and tail in all plumages. Adult has blue-grey bill and pale yellow iris. Male has black head, breast and wings contrasting with glossy maroon body. Female similar but duller maroon mantle, whitish belly and flanks with diffuse maroon-grey streaking. Immature has uniform brown upperparts, including wings, whitish underparts streaked dark brown, and brown iris. Juvenile has orange-buff fringes to upperparts and tips to coverts. **Voice** Rich fluty *pi-io-io* song, nasal squawking call. **HH** More often found in tree canopy than Black-hooded. Often with mixed feeding parties of other species. Evergreen forest.

Black-hooded Oriole *Oriolus xanthornus* 25cm

Common and widespread resident. **ID** Adult male has black head contrasting with golden-yellow body, yellow edges to black tertials and secondaries, and mainly yellow tail. Adult female is similar but has olive-yellow mantle. Immature has dark bill, yellow forehead, yellow streaking on head, black streaking on white throat, diffuse black streaking on yellow breast, and duller wings with narrow yellowish edges to flight feathers. **Voice** Song is mixture of mellow, fluty and harsh notes; harsh, nasal *kwaak* call. **HH** Actively forages in trees, often in canopy but also at lower levels; not shy and often in villages and around homesteads and gardens. Frequently seen flying from tree to tree, sometimes with itinerant bands of insectivorous birds. All forests, well-wooded areas and villages.

Indian Golden Oriole *Oriolus kundoo* 25cm

Scarce but widely distributed summer visitor. **ID** Adult male has small black eye patch, golden-yellow head and body, largely black wings with yellow carpal patch and prominent tips to tertials/secondaries, and yellow-and-black tail. Adult female has yellowish-green upperparts, blackish streaking on whitish underparts (with variable yellow on sides of breast), brownish-olive wings, yellow rump, and brownish-olive tail with yellow corners. Both sexes lack black nape-band of Black-naped. Immature similar to adult female, but bill and eyes are dark; initially duller on upperparts and more heavily streaked on underparts. **Voice** Loud, fluty *weela-whee-oh* song; harsh, nasal *kaach* call. **HH** Usually hidden in leafy canopy. Flight is powerful and undulating with fast wingbeats. Open deciduous woodland. **TN** Previously considered a race of Eurasian Golden Oriole *O. oriolus*.

Black-naped Oriole *Oriolus chinensis* 27cm

Winter visitor and passage migrant to E and C. **ID** Male from Indian Golden by broad black mask extending as band over nape, larger and broader bill, and extensive yellow panels in wing. Female similar to male, although upperparts (including panel in wing) greener, and quite different in appearance to female Indian Golden. Immature initially has dark bill and eye, more uniformly olive wings and heavily streaked underparts; nape-band and eyestripe can be very diffuse. Diffuse nape-band, if apparent, and stouter bill are best features from immature Indian Golden. See Appendix 1 for differences from Slender-billed Oriole. **Voice** A cat- or jay-like squeal; song like Indian Golden. **HH** Evergreen and deciduous forest and well-wooded areas.

Grey-browed
Broadbill

Long-tailed Broadbill

imm

♂

♀

♂

♂

♀

imm

Maroon Oriole

Black-hooded
Oriole

imm

♀

imm

♂

Indian
Golden Oriole

♂

♀

♀

imm

Black-naped
Oriole

imm

White-bellied Erpornis *Erpornis zantholeuca* 11cm

Local resident in E. **ID** Crested, olive-yellow and white, with beady black eye. Undertail-coverts are bright yellow. Has pinkish bill and legs. Juvenile is duller, with brownish cast to upperparts. **Voice** Song is short, high-pitched, descending trill *si-i-i-i*. Calls include subdued, metallic *chit* and *cheaan* alarm. **HH** Often found singly, occasionally in small parties or with other small birds. Although lively, it is quiet and unobtrusive. Forages chiefly at middle levels, also in undergrowth and canopy. Evergreen and semi-evergreen forest. **TN** Formerly placed in genus *Yuhina* but recent taxonomic work has shown that it is not related and should be in its own monotypic genus.

Small Minivet *Pericrocotus cinnamomeus* 16cm

Widespread but uncommon resident. **ID** Small. Male has grey upperparts, dark grey throat, orange wing panel and orange underparts. Female has pale throat and underparts with orange wash; wing-panel is orange. Juvenile has upperparts scaled with buff. **Voice** Continuous high-pitched *swee-swee* etc. **HH** Active in small groups, foraging at middle levels and in canopy. More open wooded areas than those preferred by other minivets – village groves, tea estates, deciduous sal forest.

Scarlet Minivet *Pericrocotus flammeus* 22–23cm

Widespread forest resident. **ID** Largest minivet in Bangladesh. From rare Long-tailed Minivet *P. ethologus* (see Appendix 1) by isolated patch of colour on secondaries, red in male and yellow in female. Male is bright orange-red and black. Female is yellow or yellow-orange and black with extensive area of yellow on forehead and ear-coverts. **Voice** Piercing, loud *twee-twee-tweetywee-tweetyweetywee*. **HH** Usually flitting among upper foliage in trees, in small flocks or mixed feeding parties outside breeding season. Evergreen, deciduous and mangrove forest.

Ashy Minivet *Pericrocotus divaricatus* 20cm

Scarce winter visitor to forests in E and Modhupur NP. **ID** Grey and white, lacking any yellow or red in plumage. In flight, shows whitish wingbar, which is especially prominent on underwing. Male has black cap and hindcrown with white forehead (white does not extend prominently behind eye as in Brown-rumped and has narrow band of black across top of bill, which is usually not apparent in Brown-rumped). Upperparts cleaner grey with rump the same shade as mantle (paler and browner in Brown-rumped) and underparts strikingly white. Female has grey 'cap', with black lores and narrow white forehead and supercilium. Head pattern cleaner than in Brown-rumped with black band across top of bill usually apparent; also, underparts whiter and upperparts (including rump) uniform grey. Immature has little or no white on head, and has brownish-grey upperparts, white tips to tertials and greater coverts, and faint brownish scaling on neck-sides and breast. **Voice** Flight call is more rasping than that of other minivets, an unmelodious *tchue-de... tchue- dee-dee... tchue-dee-dee*. **HH** Habits as Rosy Minivet. Evergreen forest, usually in flocks with Rosy Minivet.

Brown-rumped Minivet *Pericrocotus cantonensis* 20cm

Scarce winter visitor to NE, Modhupur NP and rarely SE. **ID** Grey, dull brown and white, lacking any yellow or red in plumage. In flight, shows striking wingbar, especially prominent on underwing. Male from Ashy by grey (rather than black) hindcrown and nape and more extensive white on forehead (extending noticeably behind eye and lacking dark band across top of bill). Also upperparts may be tinged brown, breast and upper belly washed vinous-greyish-brownish, rump notably paler than rest of upperparts with brownish wash, and wing-patch (if present) pale buff. Upperparts can appear greyish and underparts whitish, and more similar to Ashy, although rump appears paler. Compared to male, female has less distinct head pattern, and is paler above with rump less sharply contrasting. Female from Ashy by paler rump, browner upperparts, more diffuse white forehead lacking dark band across top of bill, and underparts less clean. **Voice** Whirring trill. **HH** Habits as Rosy Minivet. Evergreen and deciduous forest. Usually in flocks with Rosy Minivet. **AN** Swinhoe's Minivet.

Rosy Minivet *Pericrocotus roseus* 20m

Fairly common winter visitor to E and C forests. **ID** Male has grey-brown upperparts, white throat, pinkish-red edges to tertials, and pinkish underparts and rump. Female from other female minivets by combination of grey to grey-brown upperparts, with indistinct greyish-white forehead and supercilium, dull and indistinct olive-yellow rump, whitish throat, and pale yellow underparts. Juvenile has upperparts scaled with yellow. **Voice** Whirring trill. **HH** Upright stance when perched. Arboreal; forages for insects in canopy and middle storey by flitting about foliage, sometimes hovering in front of a sprig or making short sallies. Usually in small parties. Broadleaved forests.

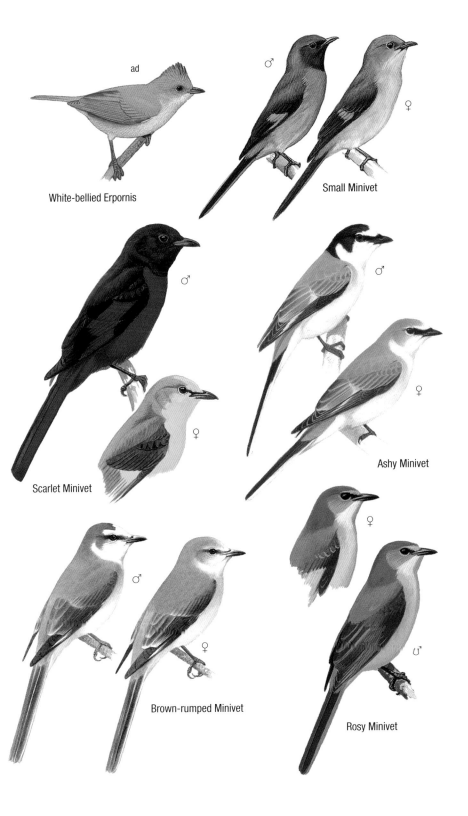

ad

White-bellied Erpornis

♂ ♀ Small Minivet

♂

♀ Scarlet Minivet

♂

♀ Ashy Minivet

♂ ♀ Brown-rumped Minivet

♀

♂ Rosy Minivet

Large Cuckooshrike *Coracina javensis* 30cm

Common and widespread resident. **ID** Male is grey on upperparts and underparts (latter without any barring) with blackish mask. Female has less pronounced dark mask and has barring on whitish belly and flanks. Both sexes have broad whitish fringes to wing feathers. Juvenile is heavily scaled with brown and white. **Voice** Song is rich fluty *pi-io-io*; also gives loud wheezy *jee-eet*. **HH** Found singly, in scattered pairs or loose flocks. Noisy. Usually perches in topmost branches of trees, although sometimes descends to bushes and ground to feed. Has distinctive habit of flicking each wing slightly, one after the other, when it alights. Open woodland, groves and trees in cultivation. **TN** Previously included in *C. macei* but this has now been split as extralimital Indian Cuckooshrike.

Black-winged Cuckooshrike *Lalage melaschistos* 24cm

Frequent winter visitor mainly to E. **ID** Male has dark slate-grey head and body, black wings, fine white tips to undertail-coverts and bold white tips to long tail. Female is similar, but paler grey, with wings not so contrastingly black, and has faint barring on belly and vent. Juvenile has head and body boldly scaled with white, with white tips to wing-coverts and tertials. **Voice** A descending monotonous *pity-to-be*. **HH** Generally singly or in pairs, often with minivets, drongos and other insectivorous birds. Gleans invertebrates from foliage, usually high in trees, sometimes in undergrowth. Has undulating flight. Forest, forest edges and groves. **TN** Formerly in genus *Coracina*.

Black-headed Cuckooshrike *Lalage melanoptera* 18cm

Locally uncommon resident but rare in NE. **ID** Male has dark slate-grey head, neck and upper breast, contrasting with pale grey mantle and rest of underparts; wings are darker grey than mantle, and have broad pale fringes to coverts and tertials. Female separated from female Black-winged by prominent supercilium, stronger dark grey-and-white barring on underparts, pale grey back and rump contrasting noticeably with shorter and squarer blackish tail, and broader white fringes to coverts and tertials. Juvenile upperparts barred with white. **Voice** Clear, mellow whistling notes followed by quick, repeated *pit-pit-pit*. **HH** Habits similar to Black-winged. Unobtrusive except for song from treetops. Open wooded areas, especially deciduous sal forest, and also mangrove edge, broadleaved forest edge, groves and second growth. **TN** Formerly in genus *Coracina*.

Ashy Woodswallow *Artamus fuscus* 19cm

Common resident. **ID** Adult has stout blue-grey bill, uniform slate-grey head, greyish-maroon mantle and pinkish-grey underparts. In flight, shows white-tipped tail and greyish-white band across uppertail-coverts. Juvenile has browner upperparts with buff fringes and paler grey throat with indistinct brownish barring. **Voice** Harsh *chek-chek-chek*. Song is drawn-out pleasant twittering, starting and finishing with harsh *chack* notes. **HH** In flocks of up to 30 birds. Spends much time hawking insect prey on the wing. Perches on dead branches near treetops, telegraph wires or other vantage points and makes frequent aerial sallies. Flies in wide circle with rapid wingbeats alternating with glides. Has distinctive habit of wagging its stubby tail when perched. Nest is shallow cup of fine fibres placed on horizontal branch. Open wooded country, agricultural land and villages.

Common Iora *Aegithina tiphia* 14cm

Widespread and common resident. **ID** Greenish upperparts, yellow underparts and prominent white wingbars. Male has blackish wings and tail, and may have some black markings on crown, nape and mantle in breeding plumage. Female has greenish-grey wings and greenish tail. Bill is stout and pointed, and has pale eye. **Voice** Song is long, drawn-out two-toned whistle; also, piping *tu-tu-tu-tu*. **HH** Singly or in pairs, each bird calling to the other frequently. Arboreal. Forages methodically among foliage of trees and bushes. Hops about branches and sometimes hangs upside-down. Any wooded areas from village groves to dense forest.

White-throated Fantail *Rhipidura albicollis* 19cm

Uncommon resident. **ID** Active sooty-black bird with broadly fanned white-tipped tail, narrow white supercilium and white throat, lack of spotting on wing-coverts and slate-grey underparts. Juvenile is browner, with body feathers, wing-coverts and tertials tipped with rufous; throat is dark. **Voice** Song is descending series of weak whistles, *tri... riri... riri... riri... riri*; squeaky *cheek* call. **HH** Active, foraging in middle storey of trees, often near main trunks. Breeds in forests but also in well-wooded village groves provided there is undergrowth; more numerous in coastal regions; more widely distributed in second growth and groves in winter.

Large Cuckooshrike

♀

♂

Black-winged Cuckooshrike

♂

♀

♂

♀

Black-headed Cuckooshrike

ad

Ashy Woodswallow

♂

Common Iora

♀

ad

White-throated Fantail

Bar-winged Flycatcher-shrike *Hemipus picatus* 15cm

Common forest resident. **ID** Dark cap contrasts with white sides of throat; has white wing patch and white rump. Male has blackish cap and brown mantle. Female has brown cap and brown mantle. **Voice** Continuous *tsit-ti-ti-ti-ti-ti* or *whiriri-whiriri-whiriri*; high-pitched trilling *sisisisisi* and insistent tit-like *chip*. **HH** Found in forest canopy, usually in pairs or small parties; often joins itinerant foraging flocks of mixed insectivorous species. Hunts insects among foliage and by making frequent aerial sallies like flycatcher. Mangroves (Sundarbans) and evergreen forest.

Large Woodshrike *Tephrodornis virgatus* 23cm

Resident, locally uncommon in E. **ID** From Common Woodshrike by larger size and bill, lack of pale supercilium, striking white lower back and rump, and uniform grey-brown tail. Female has poorly defined brown mask compared with male, with paler bill, darker eye, and brownish-grey crown and nape concolorous with mantle. Juvenile is scaled with buff and brown on upperparts, and tertials and tail feathers are diffusely barred and have buff fringes and dark subterminal crescents. **Voice** A musical *kew-kew-kew-kew*; also, harsh shrike-like calls. **HH** Quiet and unobtrusive. Seeks insects in foliage, on trunks and branches, frequently high in trees. Usually in small parties, and often joins with other insectivorous birds. Broadleaved forest, mostly evergreen; prefers moister habitats than Common. **TN** Formerly *T. gularis*.

Common Woodshrike *Tephrodornis pondicerianus* 18cm

Locally fairly common and widespread resident. **ID** From Large by smaller size, white supercilium above dark mask and dark brown tail with white sides. Iris is brown. Sexes similar. Juvenile has buffish-white supercilium, whitish spotting on crown and mantle, pale and dark fringes to tertials and tail feathers, and indistinct brown streaking on breast. **Voice** A plaintive whistling *weet-weet*, followed by quick, interrogative *whi-whi-whi-wheee* and accelerating trill, *pi-pi-i-i-i-i-i*. **HH** Edges and degraded areas of semi-evergreen forest, deciduous forest, tea estates and second growth.

Black-naped Monarch *Hypothymis azurea* 16cm

Common resident, more widespread in winter. **ID** Male almost entirely azure-blue, with black nape patch, black gorget across upper breast, beady black eye and black feathering at bill-base. Female has duller blue head, lacks black nape patch or gorget, and has blue-grey breast and grey-brown mantle, wings and tail. **Voice** Rasping, high-pitched *sweech-which* and ringing *pwee-pwee-pwee-pwee*. **HH** Arboreal from understorey to canopy. Flits actively in foliage. Captures insects by making aerial sorties from perch; also, hovers in front of leaves to disturb insects. Forests, wooded groves in winter.

Indian Paradise-flycatcher *Terpsiphone paradisi* 20cm (excluding tail streamers)

Chiefly local summer visitor to W and C. **ID** Male has black head and crest, with white or rufous upperparts and long tail-streamers. Intermediate birds occur, showing rufous and white in plumage. Female and immature are similar to rufous male but have shorter crest and short, square-ended tail. Juvenile is similar to female but shows indistinct pale centres and dark fringes to breast feathers. See Appendix 1 for account of Chinese Paradise-flycatcher *T. incei*. **Voice** Song is a slow warble, *peety-to-whit*, repeated quickly; calls and alarm include nasal *chechwe* and harsh *wee poor willie weep-poor willie*. **HH** Perches with upright stance, often high in trees, before darting out to catch flying insects. Active and graceful with undulating flight. Open forests, well-wooded areas, bamboo groves and orchards. **AN** Asian Paradise-flycatcher when Indian, Oriental and Chinese Paradise-flycatchers are treated as conspecific.

Oriental Paradise-flycatcher *Terpsiphone affinis* 20cm (excluding tail streamers)

Winter visitor and passage migrant in E. **ID** Differs from Indian by smaller bill and much shorter crest; rufous male and female have rufous rather than white vent. See Appendix 1 for account of Chinese Paradise-flycatcher. **Voice** Song less varied than Indian, a series of repeated upslurred notes. **HH** As Indian Paradise-flycatcher, also denser forest. **TN** Previously treated as conspecific with Indian and Chinese Paradise-flycatchers as Asian Paradise-flycatcher.

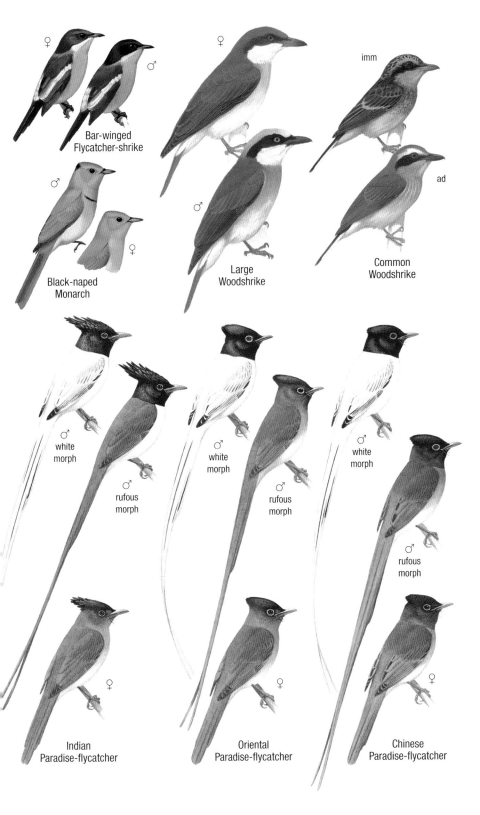

Bar-winged
Flycatcher-shrike

♀ ♂

Black-naped
Monarch

♂ ♀

♀

♂

Large
Woodshrike

imm

ad

Common
Woodshrike

♂
white
morph

♂
rufous
morph

♀

Indian
Paradise-flycatcher

♂
white
morph

♂
rufous
morph

♀

Oriental
Paradise-flycatcher

♂
white
morph

♂
rufous
morph

♀

Chinese
Paradise-flycatcher

Black Drongo *Dicrurus macrocercus* 28cm

Common, widespread resident. **ID** In winter from Ashy Drongo, which can appear dark and glossy, by blacker upperparts and shiny blue-black throat and breast, merging into black of rest of underparts; also usually shows white rictal spot and eye is duller. First-winter has whitish fringes to black underparts. Juvenile is uniform dark brown. **Voice** Harsh *ti-tiu* and *cheece-cheece-chichuk*; pairs duet during breeding season; good mimic. See Appendix 1 for comparison with White-bellied and Crow-billed Drongos. **HH** Often associates with grazing cattle. Crepuscular. Makes frequent sallies from vantage point to seize insects in mid-air or on ground. More open country than other drongos: open cultivation, around villages and suburbs of towns and cities.

Ashy Drongo *Dicrurus leucophaeus* 29cm

Fairly common winter visitor. **ID** Adult has dark grey underparts and slate-grey upperparts with blue-grey gloss; iris bright red. First-winter has brownish-grey underparts with indistinct pale fringes (underparts blacker and whitish fringes more distinct in Black). Juvenile as juvenile Black. See Appendix 1 for comparison with White-bellied Drongo. **Voice** Like Black but more varied and a good mimic; includes whistling *kil-ki-kil*. **HH** Usually uses bare branches near treetop as a vantage point. Very agile in pursuit of insects like other drongos. Forest and well-wooded areas.

Bronzed Drongo *Dicrurus aeneus* 24cm

Fairly common resident. **ID** Small, with flatter bill compared with Black and less deeply forked tail (which can be almost square-ended on moulting or juvenile birds). Note also habitat differences. Adult is strongly glossed metallic blue-green. Juvenile has brown underparts, and duller and less heavily spangled upperparts. **Voice** Loud, varied musical whistles and churrs. **HH** Usually singly or in pairs. A regular member of mixed feeding parties of insectivorous species. Shady areas in forest, in overgrown forest clearings and along forest paths. Evergreen, deciduous and mangrove forest.

Lesser Racquet-tailed Drongo *Dicrurus remifer* 25cm

Widespread, uncommon winter visitor. **ID** Tufted forehead without crest (giving rise to rectangular head shape with flattened crown), square-ended tail, and smaller size and bill than Greater Racquet-tailed; has smaller flattened racquets. As in Greater, tail-streamers and racquets can be missing or broken in adult and are absent in immature. **Voice** Loud, varied, musical whistling, screeching and churring with much mimicry. **HH** Found alone or in pairs; often associated with roving mixed flocks of smaller forest insectivores. Usually keeps to the leafy mid-level and canopy in forest clearings or forest edge and along streams. Evergreen forest and occasionally open wooded areas.

Hair-crested Drongo *Dicrurus hottentottus* 32cm

Quite widespread resident, fairly common. **ID** Broad tail with upward-twisted corners, and long downcurved bill. Adult has extensive spangling, shiny wings and hair-like crest. Juvenile is browner and lacks spangling; also lacks crest and has square-ended tail with less pronounced upward twist to outer feathers. **Voice** A loud *chi-wiii*, the first note stressed and the second rising, or sometimes *wiii* note given singly. **HH** Found alone or in small parties. Frequently joins flocks of insectivorous birds. Feeds mainly on flower nectar and also eats insects, which are located chiefly by searching flowers, leaves and tree trunks, although it also catches them on the wing. Evergreen, deciduous and mangrove forest; also open wooded areas. **TN** Some authorities consider extralimital Spangled Drongo *D. bracteatus* conspecific, when together known as Spangled Drongo.

Greater Racquet-tailed Drongo *Dicrurus paradiseus* 32cm

Locally frequent resident in E and SW. **ID** Adult from Lesser Racquet-tailed by larger size and less tidy appearance, larger bill, crested head, forked tail and longer, twisted tail-racquets. Tail-streamers and racquets can be missing or broken, and tail can appear almost square-ended when in moult (or in juvenile plumage). Juvenile initially lacks racquets, is less heavily glossed than adult and has much-reduced crest; first-winter plumage has white fringes to belly and vent. **Voice** Loud, varied musical whistling, screeching and churring, with much mimicry. **HH** Habits similar to other drongos, but more sociable, often in small groups and joins mixed foraging parties of laughingthrushes and other species. Frequently crepuscular. Forages in lower and middle storeys of forest. Evergreen and semi-evergreen forest, also mangroves in Sundarbans.

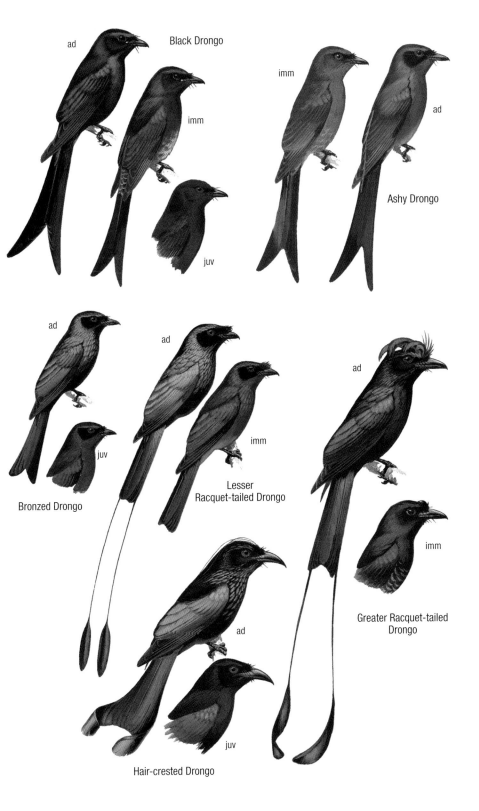

Black Drongo

ad

imm

juv

Ashy Drongo

imm

ad

Bronzed Drongo

ad

juv

Lesser
Racquet-tailed Drongo

ad

imm

Greater Racquet-tailed
Drongo

ad

imm

Hair-crested Drongo

ad

juv

Brown Shrike *Lanius cristatus* 18cm

Common and widespread winter visitor. **ID** Uniform rufous-brown upperparts including tail, prominent white supercilium, black mask, and paler underparts with warm buff flanks. Compared with vagrant Isabelline Shrike *L. isabellinus* (Appendix 1), has thicker bill, more graduated tail and lacks white patch at base of primaries (apparent in male Isabelline). Female has paler lores and faint dark scaling on breast and flanks compared with male. Has darker mask than female Isabelline with more prominent white supercilium. Juvenile has dark scaling to upperparts and underparts (mantle uniform in first-winter). **Voice** Rich varied chattering song; grating call. **HH** Watches for prey from top of bush, tree, post or telegraph wire. Swoops down to catch prey from the ground or in flight. Bushy areas within farmland, village groves, second scrub and forest edge.

Bay-backed Shrike *Lanius vittatus* 17cm

Rare, localised resident in far W, vagrant elsewhere. **ID** Adult has black forehead, pale grey crown and nape, deep maroon mantle, whitish rump and white patch at base of primaries. Juvenile is similar to juvenile Long-tailed (race *L. s. erythronotus*) and best told by smaller size and shorter tail, more uniform greyish/buffish base colour to upperparts, pale rump, and more intricately patterned wing-coverts and tertials (with buff fringes, dark subterminal crescents and central marks). First-year like washed-out version of adult; lacks black forehead. **Voice** Pleasant, rambling warbling song with much mimicry; harsh churring call. **HH** Habits like Brown. Open dry bushy areas, and bushes in cultivation in Chapai Nawabganj and Rajshahi Districts; winter vagrant to Dhaka and Chittagong.

Long-tailed Shrike *Lanius schach* 25cm

Common, widespread resident. **ID** Adult has black crown and nape, sandy-brown upperparts, flanks and sides to black tail, and small white patch on black primaries. Juvenile has dark scaling to upperparts and flanks; dark greater coverts and tertials fringed rufous. Resident race is *L. s. tricolor*, but vagrant western *L. s. erythronotus* has been recorded and intergrades with grey crown and mantle, rufous scapulars and upper back, and narrow black forehead may occur. **Voice** Pleasant subdued, rambling, warbling song with much mimicry; harsh grating call. **HH** Habits like Brown. Bushes in cultivation, open woodland and grasslands.

Grey-backed Shrike *Lanius tephronotus* 25cm

Fairly common, widespread winter visitor mainly to C and E regions. **ID** Adult has uniform grey upperparts, including crown, with rufous on uppertail-coverts. From *erythronotus* race of Long-tailed by cold grey coloration to upperparts, lack of, or very narrow, black forehead band, and absence of (or very indistinct) white patch at base of primaries. Juvenile has brown ear-coverts; cold grey upperparts (except for rufous-brown uppertail-coverts), with indistinct black subterminal crescents and buff fringes to feathers, and dark scaling on breast and flanks. Uniform cold grey base coloration to upperparts is best distinction from juvenile *erythronotus* race of Long-tailed. **Voice** Harsh grating call; mimics other birds. **HH** Habits like Brown. Bushes in cultivation, open scrub, second growth and forest edge.

Grey-headed Canary-flycatcher *Culicicapa ceylonensis* 13cm

Common winter visitor. **ID** Distinctive, with grey head and breast, greenish mantle and yellow belly, flanks and vent. Upright stance, crested appearance and flycatcher-like behaviour serve to distinguish it from similarly plumaged *Seicercus* and *Abroscopus* warblers. **Voice** Loud, high-pitched interrogative, repeated *chik…whichee-whichee* song; clear *kitwik…kitwik* and soft *pit…pit…pit* calls. **HH** Usually the most obvious member of mixed flocks of small insectivorous birds, perching on prominent branches from which it sallies out to catch flying insects. Forests, wooded areas and village groves.

Great Tit *Parus major* 14cm

Common, widespread resident. **ID** Black breast centre and line down belly, greyish mantle, greyish-white breast-sides and flanks, and white wingbar. Juvenile has yellowish-white cheeks and underparts, and yellowish-olive wash to mantle. **Voice** Extremely variable. Song includes loud, clear whistling *weeter-weeter-weeter*, *wreet-chee-chee*; calls include *tsee tsee tsee* and harsh churring. **HH** Often singly or in pairs; sometimes joins mixed roving parties outside breeding season. Forages in middle and especially lower level of open forests and well-wooded country. **TN** Form occurring in Bangladesh is sometimes treated as separate species, Cinereous Tit *P. cinereus*.

1st-winter

Brown Shrike

ad

ad

Bay-backed Shrike

juv

imm

ad
erythronotus

juv
erythronotus

Long-tailed Shrike

juv

juv
tricolor

Grey-backed
Shrike

ad
tricolor

ad

Grey-headed
Canary-flycatcher

ad

ad

juv

Great Tit

Rufous Treepie *Dendrocitta vagabunda* 46–50cm

Common and widespread resident. **ID** Adult has uniform slate-grey hood (extending to breast), rufous-brown mantle and scapulars, pale grey wing-coverts and tertials contrasting with black of rest of wing, fulvous-buff underparts, and black-tipped silver-grey tail. In flight, shows pale grey wing panel, whitish subterminal tail-band and rufous rump. Juvenile is similar to adult but has browner hood (less well demarcated from mantle), buffish wash to wing-coverts, and tail feathers have pale buffish tips. **Voice** Variety of harsh, metallic and mewing notes, the most distinctive being a loud, flute-like *ko-ki-la*, often mixed with harsh rattling cry. **HH** Usually in pairs or family parties, sometimes larger gatherings; frequent member of mixed foraging parties. Chiefly arboreal, keeping high in trees and climbing about trunks and branches with ease. Village groves, open wooded country and deciduous forest.

Grey Treepie *Dendrocitta formosae* 36–40cm

Uncommon resident in E. **ID** Dull-coloured treepie. From Rufous by blackish face contrasting with grey crown, nape and underparts, dull brown mantle, black wings with white patch at base of primaries, grey rump and rufous undertail-coverts. Juvenile is similar to adult but has narrower black forehead, dusky throat concolorous with upper breast, browner crown and nape, whitish belly and rufous tips to wing-coverts. **Voice** Wide variety of calls; often a loud, metallic, undulating *klok-kli-klok-kli-kli*. **HH** Habits similar to those of Rufous, though typically in flocks. Evergreen and semi-evergreen forest.

Common Green Magpie *Cissa chinensis* 37–39cm

Scarce resident in NE, SE and N edge of C. **ID** Mainly lurid green, with red bill and legs, black mask, rufous-chestnut wings, black-and-white-tipped tertials and secondaries, and long, graduated black-and-white-tipped green tail. In captivity, green coloration can bleach to pale blue, and chestnut wings can fade to olive-brown. Juvenile is similar to adult, but has dull yellow bill and legs, shorter crest and paler underparts. **Voice** Highly variable: harsh *chakakakakakakak* or *chkakak-wi*; high-pitched *wi-chi-chi, jao... wichitchit... wi-chi-chi, jao*, with shriller *jao* notes and complex high, shrill whistles combined with mimicry. **HH** In pairs or small parties, often with roaming mixed-species flocks. Forages in forest understorey or on ground under thick vegetation. Moist evergreen and semi-evergreen forest.

House Crow *Corvus splendens* 40cm

Abundant widespread resident. **ID** Two-toned appearance, with paler nape, neck and breast. Adult has gloss to black of plumage, and 'collar' is well defined (though it becomes paler with wear). Juvenile lacks gloss, and 'collar' is duskier and less well defined. **Voice** Main call is flat, dry *kaaa-kaaa*, weaker than Large-billed. **HH** Gregarious when feeding and roosting. Bold and cunning, omnivorous and an opportunistic feeder. Scavenges at rubbish dumps, in streets and on riverbanks. On ground, moves with confident strides and occasional sideways hops. Around habitation and cultivation, absent from forest and strangely absent from villages and tea estates in part of NE around Srimangal town.

Large-billed Crow *Corvus macrorhynchos* 46–49cm

Common and widespread resident. **ID** Lacks any contrast between head and neck/breast, unlike House Crow; bill is stouter with more pronounced curve to culmen. **Voice** Call a very nasal *nyark, nyark*, first rising then falling. **HH** Usually singly, in pairs or small groups, but roosts communally in large numbers. Inquisitive, bold and omnivorous. Frequently scavenges and eats carrion. Wide habitat range including forests. **TN** Form occurring in Bangladesh is recognised by some authorities as Eastern Jungle Crow *C. levaillantii*.

ad

ad

imm

imm

Grey Treepie

Rufous Treepie

Common Green
Magpie

ad

ad

House Crow

Large-billed Crow

ad

Ashy-crowned Sparrow-lark *Eremopterix griseus* 12cm

Rare local resident, not recorded in NE. **ID** Male has grey crown, and black throat, supercilium and underparts. Upperparts are fairly uniform sandy-grey. Female from other larks by combination of stout greyish bill, uniform head (lacking dark eyestripe), rather uniform upperparts (with almost unstreaked mantle and scapulars), indistinct and diffuse breast streaking, and blackish underwing-coverts (latter can be difficult to see in field). **Voice** Aerial display song comprises short flute-like *tweedle-deedle-deedle* as bird rises, and drawn-out whistle *wheeh* in descent. **HH** Open dry areas including cultivation.

Horsfield's Bushlark *Mirafra javanica* 14cm

Rare local resident in SE and central coast region; also, records from one site in SW. **ID** Stocky, stout-billed, broad-winged lark with slight crest and rufous panel in wing. Has comparatively uniform brownish-buff ear-coverts, weak and rather restricted spotting on upper breast, and whitish throat with brownish to rufous-buff breast-band. Tail has whitish outer feathers. Compared with Oriental Skylark, bill is distinctly shorter and stouter, and crest shorter. **Voice** Sweet and full song, with much mimicry, delivered from top of bush or in song flight. **HH** Habits like Bengal Bushlark, but has more varied song and distinctive display flight, towering high on winnowing or flickering wings. Also sings when perched. Virtually restricted to coastal grassland and cultivation near Chittagong. **TN** Sometimes split as Singing Bushlark *M. cantillans*.

Bengal Bushlark *Mirafra assamica* 15cm

Common, widespread resident. **ID** Has diffusely streaked brownish-grey upperparts, buffish supercilium and dirty rufous underparts (with paler throat and greyish flanks). Prominent rufous wing panel. **Voice** Song is a repeated series of thin, high-pitched disyllabic notes, usually delivered in prolonged song flight; call is a series of variable, thin, high-pitched short notes. **HH** Walks and runs on ground; strong and undulating flight like other larks. Fallow cultivation and short grassland.

Sand Lark *Alaudala raytal* 12cm

Locally common resident. **ID** Small, stocky lark with comparatively short tail, and rather rounded wings with distinctive, rather jerky and fluttering flight. Additional features from Eastern Short-toed Lark are finer bill, rather uniform sandy-grey upperparts (with streaking most prominent on crown), whitish underparts with fine sparse streaking across breast (lacking dark patch on breast-side), and prominent blackish sides to tail contrasting with white outer tail feathers and very pale rump and uppertail-coverts. Primaries extend beyond tertials on closed wing (primaries equal to length of tertials in Eastern Short-toed Lark). **Voice** Rolling, deep and guttural *prr... prr* call; aerial display song is a series of short, rapidly delivered and repeated undulating warbling notes. **HH** Runs about in zigzagging spurts. Sandbanks on islands (chars) and adjacent land alongside main rivers (Brahmaputra–Jamuna and Ganges–Padma). **TN** Previously in genus *Calandrella*.

Eastern Short-toed Lark *Calandrella dukhunensis* 14cm

Scarce passage migrant and winter visitor, not recorded in SW. **ID** Stouter bill than Sand Lark, with more prominent supercilium and eyestripe, and more obvious streaking on upperparts including prominent dark centres to median coverts. Dark breast-side patches often apparent, although breast may be streaked and more similar to Sand Lark. Tertials reach primary tips on closed wing (fall short of primaries in Sand Lark). Also has warm sandy-buff upperparts, variable rufous-buff breast-band, and rufous-buff ear-coverts and wash on flanks; coloration warmer than Sand Lark. See Appendix 1 for comparison with Hume's Lark. **Voice** Dry *tchirrup* or *chichirrup* flight call. **HH** Gregarious in winter and passage, running and flying about restlessly. Open, short grass and fallow cultivation. **TN** Until recently treated as conspecific with Greater Short-toed Lark *C. brachydactyla*.

Oriental Skylark *Alauda gulgula* 16cm

Fairly common and widespread breeding resident. **ID** Well-streaked upperparts and breast, with a small crest. Additional features which help identification from other larks are fine bill (compared with the bush larks), buffish-white outer tail feathers, and indistinct rufous wing-panel. **Voice** Song, usually delivered from high in sky, comprises bubbling warbles and shorter whistling notes with much variation; grating, throaty *bazz, bazz* call. **HH** Grassland and cultivation, particularly in more open wetlands and chars.

Ashy-crowned Sparrow-lark

♂

♀

Horsfield's Bushlark

ad

Bengal Bushlark

ad

Sand Lark

ad

Eastern Short-toed Lark

ad

Oriental Skylark

ad

Zitting Cisticola *Cisticola juncidis* 10cm

Common and widespread resident. **ID** Breeding adult has diffusely streaked grey-brown crown and often rather distinct buff or rufous rump. Non-breeding adult has longer tail, more heavily streaked upperparts and often less distinct rump; some birds have brighter rufous-buff coloration to upperparts but lack the more clearly defined rufous nape of Golden-headed Cisticola. First-winter has sulphur-yellow wash to underparts and less heavily streaked upperparts. See Golden-headed for other differences from that species. **Voice** Repeated *pip* uttered in distinctive display flight quite high above fields. **HH** Active and excitable, frequently flicking wings and cocking and spreading tail. Displays frequently when breeding; circles widely over its territory in 'bouncing' flight, beating its wings as it rises, drops a little and then rises again while singing. Paddy-fields and other crops, and dry and marshy grasslands.

Golden-headed Cisticola *Cisticola exilis* 10cm

Rare local resident in NE, SE and C regions. **ID** In all plumages from Zitting by blacker tail with narrow buffish or greyish tips (broader white tips in Zitting), unstreaked rufous nape and sides of neck, and rufous (rather than whitish) supercilium. Breeding male has unstreaked creamy-white crown and underparts, rufous-brown wash to nape and sides of neck, and unstreaked olive-grey rump. Female and non-breeding male have heavily streaked crown and mantle, and more closely resemble Zitting. Tail much longer in non-breeding plumage. First-winter has yellow wash to underparts. **Voice** Song comprises one or two jolly doubled notes introduced by a buzzy wheeze: *bzzeeee... joo-ee*; the wheeze is often repeated separately. **HH** Habits similar to Zitting's; display flight is faster, less jerky and ends in 'nose-dive' at high speed. Expanses of tall dense grassland with some bushes.

Rufescent Prinia *Prinia rufescens* 11cm

Scarce resident in E. **ID** Breeding adult has grey crown and ear-coverts, white supercilium and black bill. Non-breeding adult from Grey-breasted by larger and stouter bill (with pale base to lower mandible), more rufescent upperparts and tail, stronger rufous wash on flanks and nasal buzzing call. Juvenile has greyish-olive cast to crown, yellowish on underparts. **Voice** Song a rhythmic *chewp chewp chewp*. **HH** Tall grassland, grass under forest.

Grey-breasted Prinia *Prinia hodgsonii* 11cm

Fairly common resident except in SW. **ID** Adult breeding has grey 'cap' and upperparts, and variable greyish breast-band. Non-breeding adult has white supercilium and dark lores, olive-brown upperparts with rufescent cast, and white to greyish underparts. **Voice** Rhythmic undulating *tirr-irr-irr-irr* song; laughing, high-pitched *hee-hee-hee-hee* call. **HH** Gregarious in winter. Always on move; creeps about in bushes and runs about mouse-like. Deciduous forest, bushes at forest edges, scrub and second growth.

Graceful Prinia *Prinia gracilis* 11cm

Local resident in NW, SW and C. **ID** Small prinia with streaked grey-brown upperparts, white underparts and cross-barred tail. **Voice** Fast, rhythmic, wheezy warbling song *ze-r witze-r wit*; nasal, buzzing *bzreep* call. **HH** Often perches in open. Feeds actively in bushes, in grass and on ground. Tall to medium-height *Saccharum* grassland and bushes near water in chars of Ganges–Padma and Brahmaputra–Jamuna rivers.

Yellow-bellied Prinia *Prinia flaviventris* 13cm

Local resident in Sundarbans, NE and SE. **ID** Adult has fine white supercilium (sometimes lacking), white throat and breast, slate-grey forecrown and ear-coverts, dark olive-green upperparts, yellowish belly and vent. Juvenile has yellowish olive-brown upperparts (no blue-grey on head) and uniform pale yellow underparts. **Voice** Sharp *chirp* followed by a five-note trill song. **HH** Very active, foraging in tall grass and occasionally clambering to top of grass stems to look around; sometimes feeds on ground. Tall grass, reeds and scrub along river valleys.

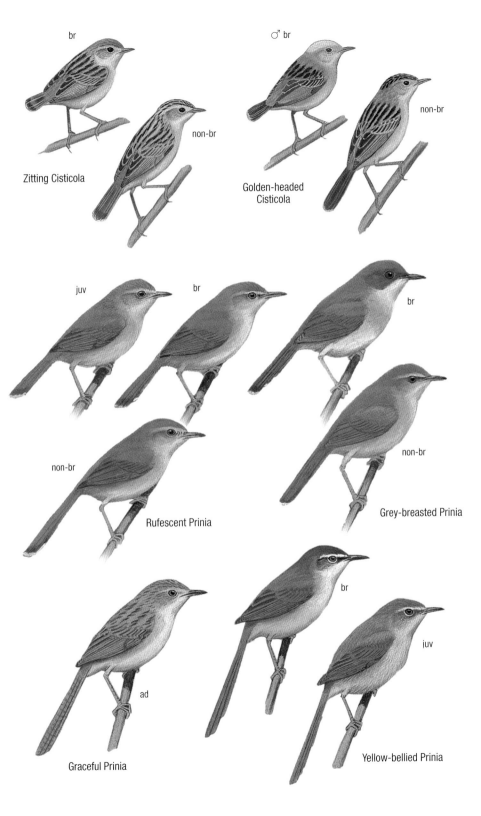

br

Zitting Cisticola

non-br

♂ br

Golden-headed
Cisticola

non-br

juv

br

non-br

Rufescent Prinia

br

non-br

Grey-breasted Prinia

ad

Graceful Prinia

br

juv

Yellow-bellied Prinia

Ashy Prinia *Prinia socialis* 13cm

Rare presumed resident in NW and C, but with single isolated records in NE and SE. **ID** Has white supercilium (sometimes lacking), slate-grey crown and ear-coverts, red eye, slate-grey mantle, orange-buff wash on underparts and prominent black subterminal marks (and whitish tips) to tail feathers. Juvenile has greenish upperparts and buffish-yellow underparts. **Voice** Song is wheezy *jimmy-jimmy-jimmy.* **HH** Very lively and quite tame. Forages low in grasses and bushes, and on ground, staying close to cover. Tall grass and scrub.

Plain Prinia *Prinia inornata* 13cm

Fairly common and widespread resident. **ID** Breeding adult has black bill, grey-brown upperparts and whitish underparts, with largely white outermost rectrices. Non-breeding adult has longer tail, pale base to lower mandible, warm brown upperparts, more rufescent wings and tail, buff tips and dark subterminal marks on rectrices, and warm buff wash to underparts. Juvenile is more rufescent. **Voice** Song is rapid, wheezy trill, *tlick tlick tlick.* **HH** When perched, tail is often held cocked and slightly fanned, and frequently jerked and side-switched. Feeds actively in low vegetation. Normal flight is weak and jerky and over only short distances. Reeds, grassland, edges of cultivation, scrub and forest edge.

Swamp Grass-babbler *Laticilla cinerascens* 14cm

Very rare, probably extirpated, resident of NW, C and NE. **ID** Has cold olive-grey coloration to streaked upperparts, uniform greyish-white underparts, with grey flanks and undertail-coverts. Narrow eye-ring and lores are whitish. Considerably larger than similarly coloured Graceful Prinia, with broader tail (lacking white tips and dark subterminal spots). **Voice** Song is elaborate fast and varied warble. **HH** Formerly tall grassland and swamp-thickets of Brahmaputra and haors. **AN** Swamp Prinia. **TN** Previously included in *Prinia* and sometimes treated as conspecific with Rufous-vented Grass-babbler *L. burnesii.* **Threatened** Globally (EN).

Common Tailorbird *Orthotomus sutorius* 13cm

Abundant and widespread resident. **ID** Has long (slightly downcurved) pale bill, rufous forehead and forecrown, greenish upperparts and dull whitish or buffish underparts. Breeding male has elongated central tail feathers. Juvenile lacks rufous on forecrown. **Voice** Song is loud *pitchik-pitchik-pitchik.* **HH** Confiding. Hunts actively, usually low in vegetation and on ground; occasionally climbs quite high in trees. Tail constantly cocked over back. Flies weakly and over short distances. Bushes in gardens, cultivation edges and forest edge.

Dark-necked Tailorbird *Orthotomus atrogularis* 13cm

Scarce resident in E and C. **ID** From Common Tailorbird by rufous extending onto hindcrown, yellow bend of wing, yellow flanks and undertail-coverts, brighter green upperparts and different call. Male has blackish lower throat and sides of breast (although Common, when calling or singing, reveals dark bases to feathers of neck). Juvenile lacks rufous on forecrown. **Voice** Song is rather high, agitated, repeated *pirrah.* **HH** Dense scrub and evergreen and semi-evergreen forest.

Thick-billed Warbler *Arundinax aedon* 19cm

Scarce winter visitor. **ID** From large *Acrocephalus* species by short, stout bill (lacking dark tip), rounded head (with crown feathers often raised) and 'plain-faced' appearance (lacks prominent supercilium and any suggestion of darker eyestripe). Tail appears long and graduated, and wings short. In fresh plumage, upperparts have rufous suffusion, especially to fringes of remiges, rump and uppertail-coverts (not apparent on Clamorous Reed-warbler), and is warmer buff on breast and flanks. **Voice** Gives a hard *shak,* tongue-clicking *tuc* and loud chattering calls. **HH** Skulks in dense bushes, usually moving about with clumsy, heavy hops. Tall grass, scrub, reeds and bushes at edges of forest and cultivation, generally in drier habitats than Clamorous. **TN** Formerly placed in *Acrocephalus* but now recognised as monospecific genus *Arundinax* (with this name pre-dating *Phragamaticola* which is sometimes used).

ad

ad

Ashy Prinia

non-br

br

Plain Prinia

ad

Swamp Grass-babbler

♀

♂

Common Tailorbird

♂

Dark-necked Tailorbird

♀

ad

Thick-billed Warbler

Black-browed Reed-warbler *Acrocephalus bistrigiceps* 13cm

Local winter visitor to NE. **ID** From Paddyfield Warbler by broader and more clear-cut supercilium and more pronounced blackish lateral crown-stripes; also has shorter tail and longer primary projection beyond tertials, plus dark grey (rather than pale brown) legs and feet. Rufescent above, with warm buff sides of breast and flanks in fresh plumage (upperparts olive-brown in worn plumage). **Voice** Call is soft, repeated *chuk*. **HH** Tall wet reed-swamp thickets in haors.

Blyth's Reed-warbler *Acrocephalus dumetorum* 14cm

Common and widespread winter visitor. **ID** Compared with Paddyfield has longer bill (usually lacking well-defined dark tip), olive-brown to olive-grey upperparts and uniform wings. Supercilium is comparatively indistinct and barely apparent behind eye. Has noticeable olive cast to upperparts and edges of remiges in fresh plumage, but in first-winter upperparts can have slight rufous cast. See Appendix 1 for comparison with Large-billed Reed-warbler. **Voice** Fairly soft *chek* or grating *chek-tchr* call. **HH** Typically hops and creeps about within bushes, also higher in trees and lower in ground cover. Frequently flicks, raises and fans tail. Bushes and trees at forest edge and in cultivation, swamp-thickets and gardens.

Paddyfield Warbler *Acrocephalus agricola* 13cm

Uncommon visitor, mainly to NE haors. **ID** Compared with Blyth's Reed-warbler has more prominent white supercilium behind eye (often with diffuse dark upper edge), more pronounced dark eyestripe, shorter bill usually with well-defined dark tip, and typically shows dark centres and paler edges to tertials (uniform on Blyth's). Rufous cast to mantle and rump in fresh plumage, when Blyth's is more olive on upperparts, and flanks are more strongly washed with buff. Worn upperparts greyer or sandier but retains rufous cast to rump (absent on Blyth's). Legs and feet yellowish-brown to pinkish-brown (dark grey in Blyth's). **Voice** Soft *dzak* call. **HH** Skulking and lively, adept at climbing up and down plant stems. Frequently flicks and cocks tail and raises crown feathers. Usually flies only short distances low over reeds, with tail slightly spread and depressed. Reeds and swamp-thickets.

Oriental Reed-warbler *Acrocephalus orientalis* 18cm

Scarce winter visitor, mainly to NE. **ID** Almost indistinguishable, except in hand, from Clamorous Reed-warbler. Smaller, with shorter and squarer tail, than Clamorous. Often has streaking on sides of neck and breast, and well-defined whitish tips to outer rectrices. Streaking on underparts can, however, be lacking or not visible in the field, while a few Clamorous can also show faint streaking. **Voice** Calls are similar to those of Clamorous. **HH** Reeds and swamp-thickets.

Clamorous Reed-warbler *Acrocephalus stentoreus* 19cm

Uncommon winter visitor. **ID** Large *Acrocephalus* with long bill, short primary projection, whitish supercilium and unstreaked underparts. Lacks white at tip of tail (which helps to distinguish from Oriental Reed-warbler). In fresh plumage, has olive cast to upperparts and variable buff wash to breast and flanks. **Voice** Loud, repeated *karra-karra-karet-karet* song; loud and deep, hard *tak* or soft *karrk* call. **HH** Often forages in bushes. Movements slower and less agile than smaller *Acrocephalus*. Reeds and swamp-thickets around wetlands. **TN** The subspecies present – *brunnescens* – is sometimes treated as separate species, Indian Reed-warbler.

Black-browed Reed-warbler

ad

ad fresh

Blyth's Reed-warbler

ad worn

ad fresh

Paddyfield Warbler

ad worn

ad worn

ad fresh

Oriental Reed-warbler

Clamorous Reed-warbler

ad worn

ad fresh

Pallas's Grasshopper-warbler *Locustella certhiola*   13.5cm

Uncommon localised winter visitor to E and SW. **ID** Larger and more robust than rarer Common Grasshopper-warbler (see Appendix 1). Has more prominent supercilium contrasting with greyer crown, more heavily streaked mantle with rufous tinge, rufous rump and uppertail-coverts, cleaner and narrower fringes to tertials (often whiter tips), rufous olive-brown breast-sides and flanks and unstreaked undertail-coverts. In flight, shows dark, white-tipped tail, which contrasts with rufous uppertail-coverts. Juvenile has yellowish wash to underparts, light breast-spotting, less distinct supercilium, and more olive-brown and less heavily streaked crown and mantle. **Voice** A sharp metallic *pit* and hard, drawn-out, descending rattle *trrrrrrrrrr*. **HH** Creeps among dense vegetation in reeds, tall grass and swamp-thickets. **AN** Rusty-rumped Warbler.

Baikal Grasshopper-warbler *Locustella davidi*   13cm

Uncommon and local winter visitor to NE haors. **ID** From Spotted Grasshopper-warbler by broader white edges to undertail-coverts, relatively short tail compared with undertail-coverts, paler brown upperparts, brownish wash on sides of neck and across breast (lacking grey breast-band of Spotted), typically weaker and browner spotting on throat and breast, whitish supercilium, and pale base to lower mandible (bill all dark in Spotted). **Voice** Calls include a hard, raspy, irregularly spaced *tshuk*. **HH** Very skulking, creeping among stems of dense wetland vegetation. **AN** Baikal Bush-warbler, David's Bush-warbler. **TN** Formerly placed in genus *Bradypterus*.

Spotted Grasshopper-warbler *Locustella thoracica*   13cm

Uncommon and local winter visitor to NE haors. **ID** Has spotting on throat and breast (sometimes indistinct), greyish supercilium, dark olive-brown upperparts with rufescent cast, grey ear-coverts and breast, olive-brown flanks, and boldly patterned undertail-coverts with narrow white edges. Juvenile lacks grey on underparts and has cloudy olive-brown spotting and faint olive-yellow wash to underparts. See account for Baikal Grasshopper-warbler for differences from that species. **Voice** Repeated, rhythmic *trick-i-di* song; calls include long, drawn-out, harsh *tzee-eenk*. **HH** As Baikal Grasshopper-warbler. **AN** Spotted Bush-warbler. **TN** Formerly placed in genus *Bradypterus*.

Striated Grassbird *Megalurus palustris*   22–28cm

Locally common resident, particularly in E. **ID** Babbler-like, with prominently streaked upperparts, whitish supercilium, rufous crown, whitish underparts with fine brown streaking on breast and flanks, and long, graduated tail. In fresh plumage, supercilium and underparts are strongly washed with pale yellow and streaking on underparts is partially obscured. **Voice** Song is clear, drawn-out whistle, ending in short explosive *wheeechoo*. **HH** Often perches conspicuously on top of reeds or bushes, jerking tail and flicking wings. In breeding season, males sing from exposed perches and in short gliding flight. Tall wet grassland and reed-swamp wetlands.

Bristled Grassbird *Chaetornis striata*   20cm

Very local breeding visitor to NW, C and NE. **ID** From Striated by stout bill, less distinct supercilium (barely apparent behind eye), and shorter (broader) tail with pale shafts, buffish-white tips and with more prominent dark cross-barring (undertail appears blackish with broad whitish tips, although these can be lost through wear). In fresh plumage, underparts are washed with buff and feathers of upperparts and wings have broad buff fringes. Upperparts become greyer-brown and underparts whiter when worn. Male has black bill; mainly yellowish in female. **Voice** Song is disyllabic *trew-treuw* repeated at 2–3-second intervals. **HH** When breeding, males perform song flight, rising 20m or more, circling widely for *c.* 10 minutes or longer before descending; also sings from exposed perch. Favours tall, wet grassland, including *Saccharum spontaneum* in chars of main rivers and haors. **Threatened** Globally (VU) and nationally (EN).

Indian Grass-babbler *Graminicola bengalensis*   18cm

Rare and very local resident in NE, formerly in C. **ID** Dark upperparts with rufous streaking on crown and lower mantle, and white or buff streaking on nape and upper mantle. Largely rufous rump and wings. Tail blackish, broadly tipped white. Underparts white with rufous-buff breast-sides and flanks. **Voice** Song a subdued, high *er-wi-wi-wi-*, *you-wuoo*, *yu-wuoo*, followed by series of harsh notes and ending with wheezy sounds. **HH** Outside breeding season, skulks in grassland. When breeding, males are very noisy, singing from reed-tops and soaring into air. Relatively undisturbed dense, wet grasslands in haors. **AN** Rufous-rumped Grassbird. **TN** Now placed in family Pellorneidae (ground babblers). **Threatened** Nationally (EN).

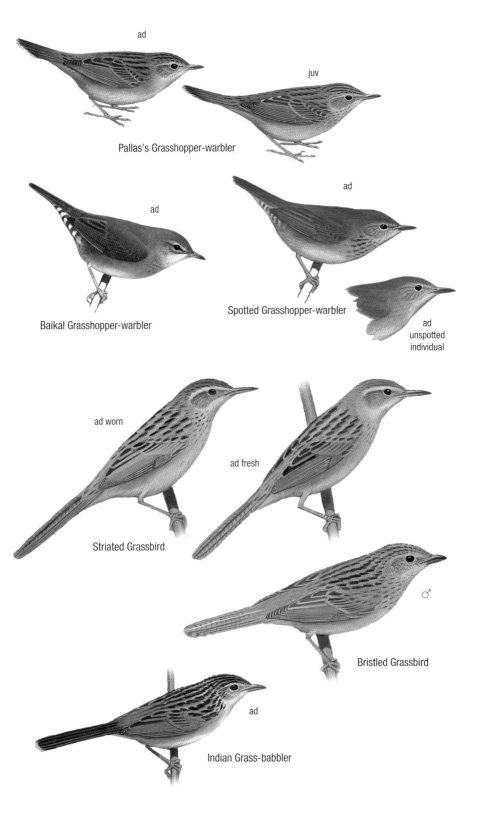

ad

juv

Pallas's Grasshopper-warbler

ad

Baikal Grasshopper-warbler

ad

Spotted Grasshopper-warbler

ad
unspotted
individual

ad worn

ad fresh

Striated Grassbird

♂

Bristled Grassbird

ad

Indian Grass-babbler

Asian House Martin *Delichon dasypus* 12cm

Rare winter visitor to E. **ID** Larger than Nepal House Martin, with whitish throat and vent, and forked tail (although tail appears almost square-ended when spread). Has pale dusky grey-brown wash to underparts and undertail-coverts. Juvenile has browner upperparts, stronger dusky wash to underparts, broad white tips to tertials and squarer-shaped tail. **Voice** Flight call is scratchy *priit –priit.* **HH** Similar to other hirundines; most records over large wetlands and nearby open fields.

Nepal House Martin *Delichon nipalense* 13cm

Scarce, localised resident in Chittagong Hill Tracts. **ID** Smaller and more compact than Asian House Martin, with almost square-ended tail. Underwing-coverts and undertail-coverts are black, contrasting sharply with white underparts. Extent of black on throat varies but variably mottled blackish throat gives rise to dark-headed appearance. Juvenile is duller and browner on upperparts and has buffish wash to underparts. **Voice** Call a high-pitched *chi-i.* **HH** Forages on the wing, continually flying with somewhat fluttering action, often quite high. Recorded over forests, river valleys, hills and around villages.

Streak-throated Swallow *Petrochelidon fluvicola* 11cm

Rare but apparently increasing visitor to W. **ID** A small, compact swallow with slight fork to long broad tail and weak, fluttering flight. Adult separated from other swallows by combination of lightly streaked chestnut crown and nape, dirty off-white underparts (with brown streaking on chin, throat and breast), narrow white streaks on mantle, and brownish rump. Juvenile has duller, browner crown, and brown-toned mantle and wings, with buff fringes (most obvious on scapulars and tertials). From juvenile Plain Martin by larger size, chestnut cast to crown, blue gloss to mantle and heavily streaked throat. Streaked throat is best distinction from juvenile Wire-tailed Swallow (see Appendix 1). **Voice** A twittering *chirp* and sharp *trr trr* in flight. **HH** Typical hirundine behaviour, see Red-rumped Swallow. Usually seen over lakes and rivers. **TN** Formerly placed in genus *Hirundo.*

Red-rumped Swallow *Cecropis daurica* 16–17cm

Uncommon winter visitor. **ID** From Barn Swallow by rufous-orange sides of neck (creating collar and dark-capped appearance), rufous-orange rump, finely streaked buffish-white underparts and black undertail-coverts. Compared with Barn, is bulkier and has slower and more buoyant flight, gliding strongly and for prolonged periods. Juvenile has duller upperparts, paler neck-sides (and collar) and rump, buff tips to tertials and shorter tail-streamers. 'Striated Swallow' *C. d. striolata* has been recorded in NE: this form is larger, with boldly streaked underparts and poorly defined rufous collar. **Voice** Distinctive *treep* call; twittering song. **HH** Typical hirundine behaviour. Catches tiny invertebrates on the wing in swift, agile, sustained flight. Perches readily on telegraph wires and other exposed perches. Often feeds in flocks with other hirundine and swift species. Open country. **TN** Formerly placed in genus *Hirundo.* 'Striated Swallow' is treated here as a subspecies of Red-rumped Swallow, but some authorities afford it species status, *C. striolata.*

Barn Swallow *Hirundo rustica* 18cm

Common winter visitor; some present throughout year, particularly in NW and NE. **ID** Adult has bright red forehead and throat, blue-black breast-band and upperparts and long tail-streamers. Underparts vary from white to rufous. Immature has duller orange forehead and throat, browner and less well-defined breast-band, duller upperparts and shorter tail-streamers. **Voice** Varied twittering song; call a clear *vit vit,* louder *vheet vheet* when alarmed. **HH** Swift and agile flight with frequent banks and turns. Gregarious outside breeding season, when often congregates on telegraph wires and in reed-swamps to roost. Cultivation, towns, cities, wetlands and open country.

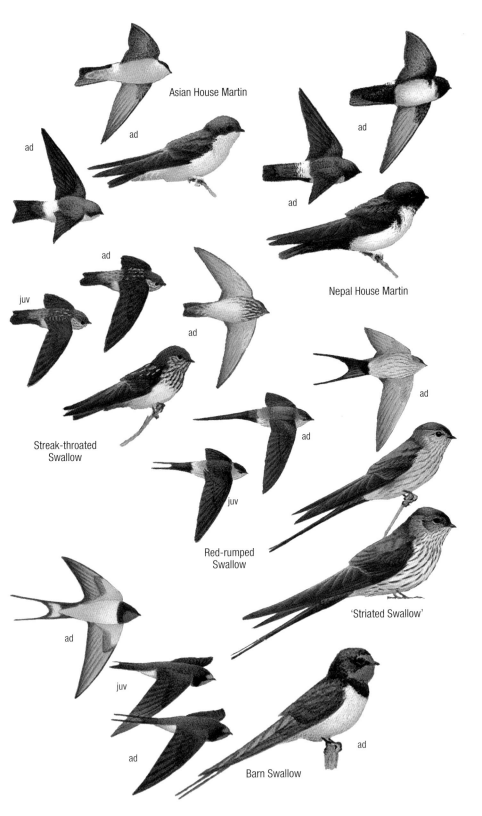

Asian House Martin

ad

ad

ad

ad

ad

Nepal House Martin

juv

ad

ad

Streak-throated
Swallow

ad

ad

juv

Red-rumped
Swallow

'Striated Swallow'

ad

juv

ad

ad

Barn Swallow

Asian Plain Martin *Riparia chinensis* 12cm

Uncommon resident, particularly along main rivers. **ID** Pale brownish-grey throat and breast, merging into dingy white rest of underparts. On some, throat is paler than breast, and may show suggestion of breast-band. Underwing darker than on Collared Sand Martin, flight weaker and more fluttering, and has shallower indent to tail. See Appendix 1 for differences from Pale Sand Martin. **Voice** Weak, high-pitched twittering song; rasping *chrrr* call. **HH** Usually hawks insects over water in swift, agile, sustained flight, sometimes high in air. Nests in sandy banks of rivers and rarely found far from these sites. **AN** Plain Martin, Brown-throated Martin. **TN** Previously treated as conspecific with African Plain Martin *R. paludicola*.

Collared Sand Martin *Riparia riparia* 13cm

Uncommon winter visitor. **ID** Adult from Asian Plain by white throat and half-collar, and by brown breast-band which is generally well defined against whitish underparts. In addition, appears stockier and more purposeful in flight, and has more prominently forked tail. See Appendix 1 for differences from Pale Sand Martin. **Voice** Rasping call. **HH** Habits like Asian Plain. Around large waterbodies.

White-throated Bulbul *Alophoixus flaveolus* 22cm

Local resident in E. **ID** Adult striking with prominent brownish crest and puffed-out white throat. Also has whitish lores and grey ear-coverts (streaked with white), yellow breast and belly, and olive-green upperparts with rufous-brown cast to wings and tail. Juvenile has browner upperparts, more rufescent wings and tail, and brownish wash to underparts. **Voice** A chacking *chi-chack chi-chack chi-chack* and a nasal *cheer.* **HH** Noisy; heard more often than seen. Creeps and clambers about bushes and lower forest storey in flocks; sometimes ascends to canopy. Often puffs out throat feathers. Undergrowth in dense evergreen and semi-evergreen forest.

Olive Bulbul *Iole virescens* 19cm

Local resident in E. **ID** A rather nondescript, olive bulbul, with slight crest. Likely to be confused only with rare Flavescent Bulbul (see Appendix 1), but is smaller and has proportionately shorter tail and longer bill, which is brownish-horn with paler lower mandible. Also lacks supercilium, has unstreaked pale olive-yellow underparts, pale rufous vent and rufous-brown tail. Lores and ear-coverts are distinctly paler than crown, giving rise to capped appearance. Legs and feet are brownish-pink (black on Flavescent). **Voice** A disyllabic musical *whe-ic.* **HH** Evergreen forest and second growth. **TN** The form found in Bangladesh has recently been proposed as separate species 'Cachar Bulbul' *I. cacharensis.*

Ashy Bulbul *Hemixos flavala* 20cm

Uncommon resident in E. **ID** A distinctive, crested bulbul, with black mask, tawny ear-coverts, grey upperparts and greyish-brown tail, olive-yellow wing-panel, white throat and pale grey breast merging into whitish belly. Juvenile is similar to adult, but has shorter crest, browner upperparts and duller wing panel. **Voice** A loud ringing call of four or five notes, the second or third being highest in pitch and the last one or two descending. **HH** In pairs in breeding season, otherwise in noisy parties. Arboreal; forages in middle and upper storeys. Feeds on berries, nectar and insects, the last sometimes caught in aerial sallies. Evergreen and semi-evergreen forest.

Asian Plain Martin

ad

ad

ad

Collared Sand Martin

ad

ad

ad

juv

White-throated Bulbul

ad

Olive Bulbul

ad

ad

ad

Ashy Bulbul

Black Bulbul *Hypsipetes leucocephalus* 25cm

Recorded in NE and SE, where it may be regular and resident in Chittagong Hill Tracts. **ID** A slate-grey to blackish bulbul, with slight crest. Has bright red bill, legs and feet, pale fringes to undertail-coverts and shallow fork to tail (at times recalling a drongo). Juvenile lacks crest; has whitish throat, grey breast-band, brownish cast to upperparts and brownish bill, legs and feet. A white-headed form, possibly *H. l. leucocephalus*, with all white head, has been recorded in NE. *H. l. nigrescens* could occur and is dull black compared with paler slate-grey *H. l. psaroides* found in Bangladesh. **Voice** Song is monotonously repeated series of three or four rising and falling notes; also screeching and mewing notes. Large flocks give continuous shrill, nasal, chattering babble. **HH** Mainly broadleaved forest and plantations.

Black-crested Bulbul *Pycnonotus flaviventris* 22cm

Locally frequent resident in NE, SE and C. **ID** A black-headed bulbul with olive-green upperparts and yellow underparts. Has erect black crest, uniform olive-green wings, uniform olive-brown tail and yellow iris. Juvenile has dull black head and shorter crest, and paler yellow underparts. **Voice** Song is sweet, musical and lacks harsh notes. **HH** Rather quiet and retiring for a bulbul. Found singly, in pairs or sometimes in groups in lower forest storey and in bushes. Evergreen, semi-evergreen and deciduous forest. **AN** *P. melanicterus* when not recognised as species separate from other forms of Black-crested Bulbul in the Indian Subcontinent.

Red-whiskered Bulbul *Pycnonotus jocosus* 20cm

Locally common and widespread resident. **ID** Striking with glossy black crown and erect crest, red patch behind eye, white patch on lower ear-coverts bordered below by black moustachial stripe, white underparts with complete or broken breast-band, and red vent. Lacks white rump (see Red-vented Bulbul). Juvenile lacks red 'whiskers'; crest and nape are lighter brown compared with adult, and vent is paler rufous-orange. **Voice** Calls include lively *pettigrew* or *kick-pettigrew*. **HH** Active, noisy and confiding. Often perches conspicuously on tops of small bushes. Sometimes in flocks with other bulbuls outside breeding season. Open forest, second growth and tea estates, rarer in village groves.

Red-vented Bulbul *Pycnonotus cafer* 20cm

Common and widespread resident. **ID** Short-crested black head with brown cheeks, white rump, white-tipped black tail and red vent. White rump helps separate from other bulbuls in flight. Juvenile has browner head, rufous edges to flight feathers, buffish cast to white rump and lacks white tips to tail. **Voice** Cheery *be-care-ful* or *be quick-quick*; alarm call is sharp repetitive *peep*. **HH** Bold, tame and quarrelsome. In pairs or small, loose flocks according to season. Feeds mainly on fruits and berries, also insects caught in aerial sallies. Open deciduous forest, second growth, open bushy wooded areas and trees around habitation in villages and towns.

Black-headed Bulbul *Brachypodius atriceps* 18cm

Uncommon resident in E. **ID** A crestless, olive-green and yellow bulbul with black head. From Black-crested by lack of crest, striking yellow panel on wing, and broad black subterminal and yellow terminal bands to tail. Iris is striking pale blue. Male is brighter olive above and yellower below than female. Juvenile darker and duller; chin and throat more olive-green, forehead dark olive; upperparts darker olive, primaries browner, contrasting less with yellow wing panel. **Voice** Song a hesitant series of short, tuneless whistles; call a ringing metallic *chewp*. **HH** Arboreal, from treetops down to undergrowth, in pairs or family parties. Evergreen and semi-evergreen forest. **TN** Formerly in genus *Pycnonotus*.

Black Bulbul

juv

ad

ad

Red-whiskered Bulbul

ad

juv

juv

Black-crested Bulbul

ad

juv

juv

ad

Red-vented Bulbul

Black-headed
Bulbul

Yellow-browed Warbler *Phylloscopus inornatus* 10–11cm

Common winter visitor. **ID** In fresh plumage, has greenish-olive upperparts, yellowish-white supercilium, two broad yellowish-white wingbars and whitish underparts often with touch of yellow. In worn plumage, supercilium, wingbars and underparts are whiter, and upperparts become greyer. Has pale base to lower mandible, and legs are paler flesh-brown or greyish-brown. Call is very distinctive. Compared with Greenish Warbler, smaller, and has broader wingbars and pale tips to tertials. See Appendix 1 for differences from very similar Hume's Leaf-warbler, and also from Lemon-rumped Leaf-warbler. **Voice** Loud, rising *che-wiest* call. **HH** Hunts actively in trees and bushes, favouring sunny edges of forest clearings. Village groves, open woodland and forest.

Dusky Warbler *Phylloscopus fuscatus* 11cm

Common and widespread winter visitor. **ID** Whitish underparts, often with buff on sides of breast and flanks, and dark brown to paler greyish-brown upperparts. Prominent supercilium and stronger dark eyestripe than Siberian Chiffchaff, lacks olive-green edges to wing feathers and yellow at bend of wing, has paler legs and pale base to lower mandible. *P. f. weigoldi* could occur, which is darker above and duskier below, with touch of yellow on underparts. See Appendix 1 for comparison with Smoky Warbler. **Voice** Hard *chack chack* call is diagnostic. **HH** Secretive, skulking in dense low cover, often on or close to ground, and sometimes in low to mid branches of wetland trees. Reed-swamp, thickets and swamp forest near water.

Siberian Chiffchaff *Phylloscopus tristis* 11cm

Scarce winter visitor. **ID** Whitish or buffish supercilium, and greyish to brownish upperparts with olive-green cast to rump, wings and tail. Underparts whitish, with buffish or greyish on sides of breast and flanks. Less stocky than Dusky Warbler. Blackish bill and legs, less prominent supercilium (with obvious whitish crescent below eye) and absence of wingbar help separate from Greenish Warbler. **Voice** Calls include plaintive *peu*. **HH** Forages actively at all levels, from tall treetops down to bushes, undergrowth and sometimes on ground. Feeds chiefly by gleaning but will also hover and make short aerial sallies. Open woodland, second forest growth and bushes. **TN** Often treated as conspecific with Common Chiffchaff *P. collybita*.

Tickell's Leaf-warbler *Phylloscopus affinis* 11cm

Uncommon winter visitor. **ID** Greenish-brown upperparts, greenish edges to wing feathers and bright lemon-yellow underparts. Supercilium is similar in coloration to throat, and dark eyestripe contrasts with yellowish ear-coverts. Worn birds may lack greenish cast to upperparts, and have paler yellow supercilium and underparts. **Voice** Short *chip...whi-whi-whi-whi* call; *chit* call, not as hard as Dusky. **HH** Bushes and small trees at forest edge, and particularly swamp forests and thickets in NE.

Green-crowned Warbler *Phylloscopus burkii* 11cm

Fairly common winter visitor to NE and C. **ID** Has yellow eye-ring, yellowish-green face and green crown. Compared with Whistler's, sides of crown are blacker, has narrower eye-ring (broken at rear) and usually lacks wingbar. See Whistler's for further information. **Voice** Song has variety of quite rich phrases, e.g. *weet-weeta-weeta-weet* interspersed with short trills. **HH** Forages actively in forest understorey and second growth, gleaning and by making frequent aerial sallies, often in mixed feeding flocks. Evergreen and deciduous forest, village bamboo groves. **TN** Formerly placed in genus *Seicercus*.

Grey-crowned Warbler *Phylloscopus tephrocephalus* 11–12cm

Scarce winter visitor mainly to NE. **ID** Much like Green-crowned, but crown and supercilium greyer (less olive). Sides of crown are blacker than in Whistler's, lacks prominent wingbar and has narrower eye-ring (broken at rear). **Voice** Song easily told from similar *Phylloscopus* species by inclusion of tremolos and trills. **HH** Understorey of evergreen and semi-evergreen forest with bamboos. **TN** Formerly placed in genus *Seicercus*.

Whistler's Warbler *Phylloscopus whistleri* 11–12cm

Scarce winter visitor. **ID** Very similar to Green-crowned; dark sides of crown are not as black and are diffuse on forehead, and yellow eye-ring is broader at rear. Generally, upperparts are duller greyish-green, underparts are duller yellow, and wingbar is usually more distinct. Shows more white in outer tail feathers; there is much white on basal half of outer web of outermost tail feathers (generally lacking white in Green-crowned). **Voice** Song is simple *witchu-witchu* and is best means of separation from Green-crowned. **HH** Behaviour as Green-crowned. Understorey of evergreen and deciduous forest, and village bamboo groves. **TN** Formerly placed in genus *Seicercus*.

ad fresh

Dusky Warbler

Yellow-browed Warbler

ad fresh

ad fresh

ad worn

Siberian Chiffchaff

ad fresh

ad

Tickell's Leaf-warbler

Green-crowned Warbler

ad

ad

Grey-crowned Warbler

Whistler's Warbler

Greenish Warbler *Phylloscopus trochiloides*   9.5–10.5cm

Common winter visitor. **ID** Variable in appearance in the Indian Subcontinent, but only the nominate form is recorded with certainty in Bangladesh. Has oily-green upperparts (with darker crown), mottled ear-coverts, dusky underparts with diffuse oily yellow wash and mainly dark bill (with orange at base of lower mandible); often shows trace of second (median-covert) wingbar. Larger size, narrow wing-bar(s) on otherwise uniform wing and different call are best features from Yellow-browed Warbler. Uniform crown (lacking pale crown-stripe) is best feature from Western Crowned and Blyth's Leaf-warblers. Very similar in appearance to Large-billed Leaf-warbler and Green Warbler (see Appendix 1), and best told by call. **Voice** Has *chis-weet* call and *chis-weet, chis-weet* song. **HH** Forages from canopy down to low bushes in all forest types, open woodland and village groves.

Yellow-vented Warbler *Phylloscopus cantator*   10cm

Uncommon winter visitor to E and C. **ID** From Blyth's Leaf-warbler by yellow throat, upper breast and undertail-coverts contrasting with white lower breast and belly. In addition, is smaller, has brighter yellow supercilium and crown-stripe contrasting with darker lateral crown-stripes, and has brighter yellowish-green upperparts. **Voice** Song consists of several single notes on same pitch and ends in two slurred notes: *seep, seep, seep to-you*, with the accent on *you*; double call is softer than those of other *Phylloscopus* with accent on second note. **HH** Often in mixed itinerant flocks in winter. Forages actively in middle and lower storey. Has habit of spreading tail and flicking it upwards when calling. Evergreen and semi-evergreen forest.

Blyth's Leaf-warbler *Phylloscopus reguloides*   11cm

Locally common in winter. **ID** Very similar to Western Crowned. Head pattern tends to be more striking, with yellower supercilium and crown-stripe, and darker lateral crown-stripes (which can be almost black). Underparts generally have distinct yellowish wash, and upperparts are darker and purer green. Wingbars are more prominent (being broader, and often divided by dark panel across greater coverts). **Voice** Constantly repeated *kee-kew-i* call; *ch-ti-ch-ti-chi-ti-ch- ti-chee* trilling song. **HH** Unlike other leaf-warblers has habit of clinging upside-down to trunks like nuthatch. Mainly evergreen and semi-evergreen forests.

Western Crowned Leaf-warbler *Phylloscopus occipitalis*   11cm

Rare winter visitor to E and C. **ID** Very similar to Blyth's but appears larger and more elongated, with larger, longer-looking bill. Upperparts are generally duller greyish-green, and whitish underparts are strongly suffused with grey. Median- and greater-covert wingbars are less prominent and head pattern tends to be less striking (supercilium and crown-stripe are duller and contrast less with dusky-olive sides to crown which may be darker towards nape). Some birds are very similar to Blyth's in appearance. **Voice** Constantly repeated *chit-weei* call. **HH** All levels of vegetation. Forests.

Grey-bellied Tesia *Tesia cyaniventer*   9cm

Scarce winter visitor to E. **ID** From rare Slaty-bellied (see Appendix 1) by paler grey underparts, becoming almost whitish on throat and centre of belly, and by concolorous olive-green crown and mantle, with brighter lime-green supercilium. Also has more prominent black stripe behind eye, and stouter bill with dark tip and yellow basal two-thirds to lower mandible. Juvenile has dark olive-brown upperparts and olive cast to grey of underparts; like adult, has brighter green supercilium and dark eyestripe. **Voice** A series of strong, clear notes, ending in short, decending flourish. Loud and rattling *trrrrrrk* call. **HH** Skulking, very active and always on the move. Tangled undergrowth and ferns in thick evergreen and semi-evergreen forest, often near small streams.

Yellow-bellied Warbler *Abroscopus superciliaris*   9cm

Scarce local resident in E. **ID** White supercilium, dark crown and eyestripe, yellowish-olive upperparts, white throat and yellow rest of underparts. Separated from all *Phylloscopus* warblers by combination of rather long bill, brownish-grey crown, white throat and yellow rest of underparts, and by fairly narrow tail, lacking prominent undertail-coverts, giving distinctive profile. Tail appears pale brownish-buff from below. **Voice** Song a halting ditty of six notes, ascending at the end; *chrrt chrrt chrrt* call. **HH** Found singly, in pairs or in mixed, roving flocks of insectivorous species. Very active, feeds chiefly by making short aerial sallies in undergrowth and middle levels of evergreen forest with stands of bamboo.

ad fresh

Yellow-vented Warbler

Greenish Warbler

ad fresh

ad fresh

Blyth's Leaf-warbler

ad fresh

Western Crowned Leaf-warbler

ad

juv

Grey-bellied Tesia

ad

Yellow-bellied Warbler

Yellow-eyed Babbler *Chrysomma sinense*   18cm

Local resident in NE. **ID** Long-tailed babbler with rounded head and stout, dark bill (recalling giant prinia in shape). Most distinctive features are yellow iris and broad orange orbital ring, white lores and supercilium, chestnut upperparts, striking white throat and breast merging into richer buff lower belly and flanks, and yellow legs and feet. Juvenile has browner bill, dark eye and duller eye-ring. **Voice** Song a variable, rapid twittering trill *tri-rit-ri-ri-ri-ri*, ending in a two-note *toway-twoh*; call is loud, plaintive *teeuw-teeuw-teeuw*, repeated 2–4 times. **HH** In parties of 12 or more outside breeding season. Skulking, often foraging close to ground. Often calls briefly from conspicuous perch before diving into cover. Has laboured, jerky flight. Tall moist grassland; also thickets, bamboo and reeds, often near water. **Threatened** Nationally (VU).

Black-breasted Parrotbill *Paradoxornis flavirostris*   19cm

Former resident in NE, not recorded since 19th century. **ID** Medium-sized parrotbill with rufous-brown head and olive-brown upperparts, black patch on ear-coverts and huge yellow bill. From Spot-breasted Parrotbill by black breast patch and solid black chin (with black barring on white throat and malar area), rufous-buff (rather than pale buff) underparts, darker rufous-brown crown and nape, even stouter bill and different call. **Voice** A striking whistled *phew-phew-phew-phuit*, ascending in pitch and volume. **HH** Reedbeds and tall grass. **Threatened** Globally (VU).

Spot-breasted Parrotbill *Paradoxornis guttaticollis*   19cm

Former resident in NE, not recorded since 19th century. **ID** Medium-sized parrotbill with rufous head and upperparts, black patch on ear-coverts and large yellow bill. From Black-breasted by black arrowhead-shaped spotting on buffish-white throat and breast (lacking bold black breast patch), pale buff (rather than rufous-buff) underparts, brighter rufous crown and nape, less stout bill and different call. **Voice** Typical territorial calls consist of 3–7 loud, even-pitched staccato notes, *whit-whit-whit-whit-whit-whit-whit* or more plaintive series. **HH** Grass and scrub, also bushes and bamboo.

Striated Yuhina *Yuhina castaniceps*   13cm

Rare presumed resident in Chittagong Hill Tracts. **ID** Subspecies occurring in Bangladesh *S. c. castaniceps* has rufous crown, which is finely scaled with pale buff and is concolorous with ear-coverts. Also shows very short white supercilium, white streaking on greyish-olive mantle, uniform greyish-white underparts and white tips to graduated tail. **Voice** A loud cheeping or twittering *chir-chit... chir-chit*; also, short, nasal, descending *chrrr*. **HH** Second growth and evergreen forest with thick, bushy understorey. **TN** Sometimes placed in genus *Staphida*.

Indian White-eye *Zosterops palpebrosus*   10cm

Common and widespread resident. **ID** Distinctive, with prominent white eye-ring, black bill and lores, green to yellowish-green upperparts, bright yellow throat and vent, and whitish rest of underparts with variable greyish wash. **Voice** Plaintive *cheer* or *prree-u* call; tinkling jingle song. **HH** Outside breeding season, in flocks of up to 50 birds, which continually utter plaintive contact calls. Favours flowering shrubs and trees. Forages actively and often hangs upside-down. Forest and wooded areas. **AN** Oriental White-eye.

Yellow-eyed Babbler

ad

Black-breasted Parrotbill

ad

Spot-breasted Parrotbill

ad

Striated Yuhina

ad

Indian White-eye

ad

White-browed Scimitar-babbler *Pomatorhinus schisticeps*　　22cm

Local resident in E. **ID** Striking white supercilium contrasting with black ear-coverts and downcurved yellow bill. Clean white centre to breast and belly, and chestnut sides to breast with variable white streaking. Yellow bill and slate-grey crown are best features from Red-billed Scimitar-babbler (see Appendix 1). Head pattern is best feature from Large Scimitar-babbler. **Voice** Single note followed by trilled hoot or evenly spaced three-note call that varies from a hoot to a whistle. **HH** Forages chiefly in thick forest undergrowth and dense scrub, favours bamboo thickets in evergreen and semi-evergreen forest. Single sighting from Sundarbans (SW) suggests that Indian Scimitar-babbler *P. horsfieldii* might occur but this requires confirmation.

Large Scimitar-babbler *Erythrogenys hypoleucos*　　28cm

Rare resident in E. **ID** From other scimitar-babblers by combination of larger size, lack of white supercilium, stouter, straighter, dull-coloured bill, dark eye, grey ear-coverts, variable rufous mottling on supercilium and sides of neck, white streaking on grey sides of breast, dark olive-brown upperparts with slight rufous tinge, and stout grey legs and feet. **Voice** Loud variable piping notes, usually three per phrase. **HH** Undergrowth in evergreen and semi-evergreen forest. **TN** Formerly in genus *Pomatorhinus*.

Grey-throated Babbler *Stachyris nigriceps*　　12cm

Local resident in E. **ID** Has blackish crown with white streaking, black lateral crown-stripe and greyish supercilium, greyish submoustachial stripe, blackish throat, and buff underparts. **Voice** High-pitched, tinkling trill, prefaced with single note and brief pause. **HH** Usually keeps well under cover. Often in itinerant, mixed foraging parties outside breeding season. Undergrowth and bamboo thickets in evergreen and semi-evergreen forests.

Chestnut-capped Babbler *Timalia pileata*　　17cm

Local resident in E and formerly in C. **ID** A stocky, thick-necked babbler with bright chestnut cap, reddish iris, white forehead and supercilium, and thick black bill and black lores (resulting in masked appearance). Has white throat and breast (streaked finely with black), bordered with slate-grey on sides of neck, buffish-olive flanks and vent, and buff belly. Tail is faintly barred. **Voice** Fast, high-pitched, descending trill, varying in length; *tzt* contact note and *pic-pic-pic* alarm call. **HH** In small parties outside breeding season. Usually keeps out of sight in thick cover although may sun itself for a few seconds or sing in the open. Typically clambers up and down grass stems, systematically searching for insects. Dense scrub with tall grasses in degraded forest and tea estates.

Pin-striped Tit-babbler *Mixornis gularis*　　11cm

Common resident except in NW. **ID** Rather scruffy-looking babbler with rufous-brown cap, yellow eyes, pale yellow lores and supercilium, olive mantle, brown wings and tail, and pale yellow underparts with fine black streaking on throat and breast. **Voice** Loud, monotonous *chuk-chuk-chuk* repeated up to 15 times; *bizz-chir-chur* alarm call. **HH** Particularly noisy babbler, readily located by distinctive call. Creeps and clambers about unobtrusively, searching for insects. In pairs or parties, often with other species, depending on season. Bushes, creepers, tangles and undergrowth in all forest types. **AN** Striped Tit-babbler. **TN** Formerly in genus *Macronus*.

Rufous-fronted Babbler *Cyanoderma rufifrons*　　12cm

Scarce resident in E. **ID** Rufous cap, white throat (faintly streaked with black), fine grey supercilium and greyish lores, and buffish breast and flanks, becoming paler on belly. Juvenile as adult but has more rufous fringes to wings and tail, paler underparts and paler rufous cap. **Voice** *Pee-pi-pi-pi-pi-pi-pi.* **HH** Thick undergrowth in evergreen forest including second growth. **TN** Formerly placed in genera *Stachyris* or *Stachyridopsis*.

White-browed Scimitar-babbler

ad

ad

Large Scimitar-babbler

Chestnut-capped Babbler

ad

Grey-throated Babbler

ad

Pin-striped Tit-babbler

ad

Rufous-fronted Babbler

ad

White-hooded Babbler *Gampsorhynchus rufulus* 23cm

Locally rare in Chittagong Hill Tracts. **ID** Bull-headed, long-tailed babbler. Adult has white head and underparts (with buff wash on flanks), contrasting with rufous-brown upperparts and tail (latter tipped pale buff); iris and bill are strikingly pale. Has variable white on wing-coverts, but this is usually obscured by feathers of mantle. Juvenile has rufous-orange crown and ear-coverts, rufous-brown upperparts, and whitish throat becoming buff on underparts. **Voice** Usual call a harsh, stuttering rattle or cackle; contact calls soft and quiet *wit*, *wet* and *wyee* notes. **HH** Gregarious and often with other species. Forages in thick undergrowth and bamboo in evergreen forest and associated second growth. **Threatened** Nationally (EN).

Puff-throated Babbler *Pellorneum ruficeps* 15cm

Locally common resident in E and C. **ID** Comparatively long-tailed species. Has rufous crown, prominent buff supercilium, white throat (often puffed out), and bold brown spotting/streaking on breast and sides of neck. **Voice** Halting, impulsive song, *sweee ti-ti-hwee hwee hwee ti swee-u*, rambles up and down scale; calls include plaintive whistled *ne-menue* and throaty churrs. **HH** Very skulking, staying on or near ground. Runs over forest floor or makes long hops when foraging. Found in pairs or family parties. Bushy undergrowth in forests, dense scrub and second growth.

Marsh Babbler *Pellorneum palustre* 15cm

Rare presumed resident in NE; no records since 1980s so possibly extirpated. **ID** Marsh-dwelling babbler with white throat, grey supercilium to back of eye, white eye-ring (not depicted) and bold brown streaking on breast and flanks. Superficially resembles Puff-throated, but smaller, lacks rufous crown and buff supercilium, and has rufous-buff wash to sides of throat, breast and flanks. **Voice** Loud *chi-chew* call. **HH** Normal habitat is reedbeds and tall grassland, but Bangladesh records include wet forest scrub. **Threatened** Globally (VU).

Spot-throated Babbler *Pellorneum albiventre* 14cm

Rare presumed resident in NE; no records since 1980s so possibly extirpated. **ID** From Buff-breasted by white throat with faint arrowhead-shaped grey spotting (which may be absent). Otherwise similar in coloration to Buff-breasted. **Voice** Rich, thrush-like song. **HH** Scrub and second growth associated with forest.

Buff-breasted Babbler *Trichastoma tickelli* 15cm

Scarce, declining resident in E. **ID** A rather plain, well-proportioned babbler with dark olive-brown upperparts with slight rufescent cast, and mainly buff underparts except for white belly. Compared with Abbott's, more neatly proportioned, with smaller head and proportionately longer tail, smaller bill, buff lores, throat and breast (some with quite noticeable brown streaking), and duller buff to olive-brown flanks and darker olive-brown upperparts. Also lacks rufous coloration to uppertail-coverts and sides of tail. Confusable with the more similar-sized Spot-throated but has faint whitish shaft streaking on more rufescent crown, buff throat and square-ended tail. **Voice** Song is loud, sharp *wi-twee*. **HH** Undergrowth in evergreen and semi-evergreen forest and bamboo thickets. **TN** Formerly in genus *Pellorneum*. **Threatened** Nationally (EN).

ad

White-hooded Babbler

juv

ad

Puff-throated Babbler

ad

Marsh Babbler

ad

Spot-throated Babbler

ad

Buff-breasted Babbler

Abbott's Babbler *Malacocincla abbotti* 17cm

Locally common resident, absent from NW. **ID** 'Top heavy', with large bill and short tail. Has unspotted white throat and breast, grey lores and supercilium, rufous uppertail-coverts and tail, and rufous-buff flanks and vent. **Voice** Song is 3–4 whistled notes with last note highest; sometimes duets with mate, giving one or two *peep* notes. **HH** Singly or in pairs. Skulking and keeps mainly to undergrowth and ground. Reluctant to fly; if disturbed, hides in cover. Dense tangles and thickets in forest, including forest edge along stream banks; also, dense cover in village groves, including betel gardens in SC and E.

Brown-cheeked Fulvetta *Alcippe poioicephala* 15cm

Uncommon local resident in E. **ID** Large, rather nondescript fulvetta, lacking any pattern to head. Has greyish crown and nape, grey-brown to olive-brown mantle, brown to rufous-brown wings and tail, and greyish-white to buff underparts, with paler throat and centre of breast. **Voice** Attractive, bustling, whistled song, *tiew-teuw-tu-tee-tiu-tiu-wheet* repeated. **HH** Undergrowth in evergreen and semi-evergreen forest, particularly in bamboo stands.

Nepal Fulvetta *Alcippe nipalensis* 12cm

Uncommon resident in E. **ID** A distinctive fulvetta with grey head and blackish lateral crown-stripes, prominent white eye-ring, olive-brown upperparts and whitish underparts with buff flanks. **Voice** Short buzzes and metallic *chit* notes; also, short fast trill of varying speed. **HH** Found in energetic parties, often with other small babblers, continually calling as they move about. Forages mainly in bushes, undergrowth and small trees, sometimes on ground. Dense undergrowth in evergreen and semi-evergreen forest.

Striated Babbler *Argya earlei* 21cm

Locally common resident. **ID** Has dark streaking to upperparts, cross-barred tail, and streaked or mottled appearance to fulvous throat and breast. Forehead and lores have greyish cast and moustachial whitish. Also, blue-grey legs and feet (varying to olive-brown) and golden-yellow iris. **Voice** Song is loud, repeated series of *tiew-tiew-tiew-tiew* calls, interspersed with *quip-quip-quip* calls from other group members. **HH** In noisy flocks, individuals following each other. Clambers about vegetation when foraging. Tall grass, reedbeds and thickets, especially in damper habitats and wetlands. **TN** Formerly in genus *Turdoides*.

Common Babbler *Argya caudata* 19.5cm

Rare presumed resident discovered in 21st century in far W. **ID** From Striated by unstreaked whitish throat and unstreaked centre to breast (with streaking on underparts restricted to breast-sides). Also, slightly smaller, colder whitish or greyish-buff underparts, yellowish legs and feet, and darker, more orange-brown iris. **Voice** A series of pleasant, rapid fluty whistles, and louder, more drawn-out *pieuu-u-u pie-u-u pi-e-u-u.* **HH** Dry scrub. **TN** Formerly in genus *Turdoides*.

Jungle Babbler *Turdoides striata* 25cm

Common resident in W and C. **ID** Greyish, often rather scruffy-looking babbler with yellowish bill. Has mottled appearance to throat and upper breast and pale streaking on mantle and underparts. **Voice** A harsh *ke-ke-ke*, frequently becoming chorus of excited, discordant squeaking and chattering. **HH** Gregarious, noisy and excitable, birds calling to each other almost continuously. Mainly feeds on ground, hopping about and busily turning over leaves. Has characteristic habit of fluffing-out rump feathers, and drooping wings and tail. Village groves and gardens, open wooded areas and deciduous forest.

Abbott's Babbler

Brown-cheeked Fulvetta

Nepal Fulvetta

Striated Babbler

Common Babbler

Jungle Babbler

Lesser Necklaced Laughingthrush *Garrulax monileger* 27cm

Local resident in E. **ID** Very similar to Greater Necklaced, with white supercilium, black necklace, rufous-orange flanks, greyish-white panel on wing and bold white tips to largely blackish outer tail feathers. Smaller than Greater with finer, dark bill, yellow eye with dark eye-ring, dark lores, and brownish (rather than slate-grey) legs and feet. Lower black border of ear-coverts does not extend to bill and white throat is bordered with rufous-orange adjacent to necklace. Has narrower necklace thinner and often almost obscured by rufous at centre (although necklace can be uniformly broad on some birds), and white of breast-sides extends as crescent below black necklace. Also has olive-brown (not dark grey) primary coverts, which are concolorous with rest of coverts. Juvenile similar, but with dusky necklace. **Voice** Loud, mellow *tee-too-ka-kew-kew-kew* song. **HH** When not breeding, often in mixed feeding parties with Greater Necklaced and other larger insectivorous birds; forages among leaf litter and also in mid-levels of trees among dead leaves. Evergreen and semi-evergreen forests.

White-crested Laughingthrush *Garrulax leucolophus* 28cm

Scarce resident restricted to SE. **ID** Large, with white crest and black mask. Also, contrasting white throat and upper breast, grey nape, chestnut mantle and band across lower breast, and dark olive-brown wings and tail. **Voice** Frequent bursts of cackling laughter. **HH** Usually in small noisy groups in dense undergrowth of evergreen forest.

Greater Necklaced Laughingthrush *Garrulax pectoralis* 29cm

Locally common resident in E and C. **ID** Larger than Lesser Necklaced, with stouter (paler-based) bill, dark eye and yellow eye-ring, complete black moustachial stripe (bordering either black or white, or streaked black-and-white ear-coverts), uniform buff or white throat without two-toned appearance, blackish primary coverts which contrast with mantle and wings, broader necklace which is clearly defined at centre, and slate-grey (rather than brownish) legs and feet. As Lesser Necklaced, shows broad white tip to blackish outer tail in flight. **Voice** Loud *what-what-who-who* song. **HH** Habits similar to scarcer Lesser Necklaced; evergreen, semi-evergreen and sal forests.

Rufous-necked Laughingthrush *Garrulax ruficollis* 23cm

Uncommon resident in E, rarely in C. **ID** Small, mainly olive-brown laughingthrush, with prominent rufous patch on sides of neck, and black face and throat. Also has grey crown and nape, and rufous vent and centre of lower belly. Tail is uniform brownish-black. Juvenile is duller, with browner crown. **Voice** An incredibly varied and vocal songster. Vocalisations include shrill whistles that run up the scale, scolding whistles, descending trills, chittering babbles and hoarse squawks. **HH** Skulks on ground and in undergrowth or low bushes. Tea estates, scrub and second growth in hills and also in wetlands.

Rusty-fronted Barwing *Actinodura egertoni* 23cm

Rare former resident in NE; no record since 19th century. **ID** Has prominent barring across wings. The only barwing recorded in Bangladesh. Distinguishing features include rufous 'front' to head, stout yellowish bill, uniform grey crown and nape merging into olive-brown mantle (lacking prominent streaking), and diffusely barred rufous-brown tail. Juvenile has crown and nape rufous-brown and concolorous with mantle, making rufous forehead less prominent. **Voice** Three-note whistle *ti-ti-ta*, the first note accentuated, the last lower; also feeble *cheep*. **HH** Forages by clambering about among bushes and undergrowth, sometimes in canopy. Dense thickets in humid, broadleaved evergreen forest.

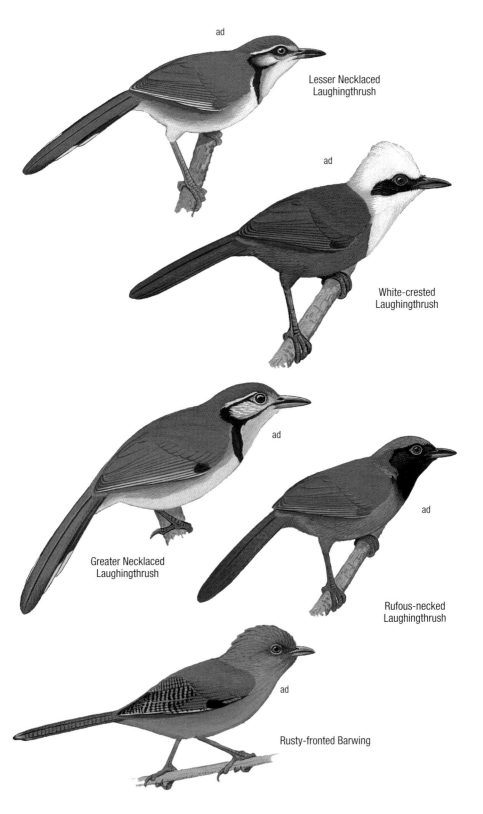

Lesser Necklaced
Laughingthrush

White-crested
Laughingthrush

Greater Necklaced
Laughingthrush

Rufous-necked
Laughingthrush

Rusty-fronted Barwing

Chestnut-bellied Nuthatch *Sitta cinnamoventris* 12cm

Rare resident in Chittagong Hill Tracts. **ID** Male always shows striking white cheek patch contrasting with dark chestnut underparts. Female is similar, but underparts are paler and cheek patch less striking. In both sexes, scalloping on undertail-coverts is white (grey in Indian Nuthatch, see Appendix 1), and crown/nape are concolourous with mantle. **Voice** Song is single pure whistle every few seconds; calls include high mouse-like *seet*, lower and more squeaky *vit* and full *chup*. **HH** Forages on tree trunks and branches, often head down. Evergreen forests. **TN** Sometimes merged with Indian Nuthatch, when treated as *S. castanea*.

Velvet-fronted Nuthatch *Sitta frontalis* 10cm

Locally fairly common resident in E and SW and rarer in C. **ID** Striking, with violet-blue upperparts, black forehead, black-tipped red bill, startling yellow iris and eye-ring, and lilac suffusion to ear-coverts and underparts. Male has black eyestripe extending behind eye (absent in female) with stronger lilac suffusion on underparts (more cinnamon, less lilac in female). Juvenile has blackish bill and duller and greyer upperparts; underparts lack lilac suffusion and are washed with orange-buff. **Voice** Song is fast, hard rattle. **HH** Forages on tree branches and trunks from canopy down to undergrowth but not on ground. All forest types, rarer in sal forest.

Common Starling *Sturnus vulgaris* 21cm

Rare winter visitor to E and C. **ID** Adult breeding is metallic green and purple with yellow bill. Adult non-breeding has dark bill; upperparts are heavily spangled with buff, wing feathers have broad buff fringes and underparts are boldly spotted with white. Juvenile is entirely dusky brown, with whiter throat, and buff fringes to wing-coverts and flight feathers. **Voice** Song is varied combination of chirps, twitters, clicks and drawn-out whistles, with much mimicry; flocking birds give slurred *scree-scree*. **HH** Sometimes in small flocks, forages mainly on ground with confident, waddling walk; often stalks around grazing livestock. Feeds chiefly on invertebrates; also eats fruits and berries. Cultivation and damp grassland.

Asian Pied Starling *Gracupica contra* 23cm

Abundant, widespread resident. **ID** Black and white, with white cheek patch and scapular line. Has orange orbital skin and base to large, pointed yellowish bill. In flight, white uppertail-coverts contrast with black tail. Juvenile has black of plumage replaced by brown; white cheeks are washed with brown and less distinct, and breast-band is not clearly defined. **Voice** Assortment of high-pitched musical, liquid notes. **HH** Mainly terrestrial. Seeks invertebrates by digging in damp ground; also eats cereal grain, fruit and flower nectar. Usually in small parties outside breeding season and forms noisy communal roosts. Cultivation, damp grassland and habitation. **AN** Pied Myna. **TN** Formerly placed in genus *Sturnus*.

Brahminy Starling *Sturnia pagodarum* 21cm

Scarce resident in NW and C. **ID** Myna-like profile. In flight, shows white sides and tip to dark tail, and uniform wings without white wing patch. Adult has black crest, and rufous-orange sides of head and underparts. Has yellowish bill with blue base, and blue or yellow skin behind eye. Juvenile lacks crest, but has grey-brown cap, paler orange-buff underparts, duller bill and eye patch. **Voice** Song is short, gurgling, drawn-out cry followed by bubbling yodel. **HH** Open wooded areas, including villages.

Chestnut-tailed Starling *Sturnia malabarica* 20cm

Common and widespread resident. **ID** Adult has grey head and upperparts, with whitish forehead and throat, and whitish lanceolate feathers across crown, nape, and sides of neck and breast. Underparts are rufous (variable in extent), and tail is mainly chestnut with grey central feathers. Bill is yellow with bluish base and eye is whitish. Female is more uniformly pale grey and underparts are paler rufous-buff. Juvenile has pale sandy-grey upperparts and greyish-white underparts. **Voice** Sharp disyllabic metallic note and mild tremulous whistle. **HH** Chiefly arboreal, also in flowering bushes and sometimes on ground. Forages by hopping about branches for nectar, berries, figs and insects. Usually feeds in flocks, often with other starlings and mynas, while constantly chattering and squabbling. Open wooded areas, groves, villages and towns.

Velvet-fronted
Nuthatch

♂   ♀

Chestnut-bellied
Nuthatch

♂   ♀

non-br

br

juv

Common
Starling

juv

ad

Asian Pied
Starling

juv

juv

ad

Brahminy Starling

♂

Chestnut-tailed
Starling

Common Myna *Acridotheres tristis* 25cm

Abundant, widespread resident. **ID** Brownish myna with yellow orbital skin, white wing patch and white tail-tip. Adult has glossy black on head and breast merging into maroon-brown of rest of body. Juvenile is duller, with brownish-black head and paler brown throat and breast. **Voice** Song is disjointed, noisy and tuneless, with gurgling and whistling, and much repetition; distinctive alarm call, a harsh *chake-chake*. **HH** Usually in pairs or small parties. Bold, tame and pugnacious. Scavenges in built-up areas; flocks follow grazing cattle or feed in cultivation. Omnivorous diet. Forms large noisy communal roosts. Habitation, cultivation.

Bank Myna *Acridotheres ginginianus* 23cm

Local resident in W and C. **ID** From Common by smaller size, bluish-grey coloration, small frontal crest, orange-red orbital patch, orange-yellow bill, red eye, orange-buff patch at base of primaries and on underwing-coverts, and orange-buff tip to tail. Has capped rather than hooded appearance of Common. Juvenile is duller and browner than adult, with buffish-white wing patch and rufous-buff tips to tail. **Voice** Similar to Common's, but not as loud and strident. **HH** Habits similar to Common's, but more gregarious. Feeds chiefly on ground; also regularly rides on backs of grazing animals. Cultivation, moist grassland. Mainly around rivers with sandy banks and bridges, providing nest sites for colonies.

Jungle Myna *Acridotheres fuscus* 23cm

Common, widespread resident. **ID** Adult resembles Bank Myna, but has more prominent frontal crest, white patch at base of primaries, white tip to tail and lacks bare orbital skin. Eye is pale. Black of crown and ear-coverts merges into grey or grey-brown of upperparts (with less distinct 'cap' than on Bank). Bill is orange, with dark blue base to lower mandible. Juvenile is browner, with darker brown head; has pale shafts on ear-coverts, pale mottling on throat, all-yellow bill and frontal crest is much reduced. **Voice** Song similar to Common. **HH** In pairs, family parties or flocks, according to season. Less bold than Common, not as commensal with people and less of a scavenger. Strides jauntily on ground when foraging; also feeds in trees and bushes. Cultivation near well-wooded areas, open forest, villages and edges of habitation.

Great Myna *Acridotheres grandis* 25cm

Rare, very local resident in Chittagong region; also one site in NE. **ID** Similar to Jungle Myna, but has uniform blackish-grey upperparts (showing little contrast between crown and mantle, and rump and tail), and uniform dark grey underparts (including belly and flanks), strongly contrasting with white undertail-coverts. Further, has more prominent frontal crest, all-yellow bill and reddish to orange-brown iris. Juvenile is browner and lacks prominent frontal crest; throat is diffusely mottled with white on some birds. Brown belly with diffuse brownish-white fringes and broad whitish tips to brown undertail-coverts are best distinctions from juvenile Jungle. **Voice** Song very similar to Common. **HH** Habits very similar to Jungle. Cultivation, grassland. **TN** Formerly treated as conspecific with White-vented Myna *A. cinereus.*

Common Hill Myna *Gracula religiosa* 25–29cm

Uncommon resident in E. **ID** Large myna with yellow wattles and large orange to yellow bill. Plumage is entirely 'black' except for prominent white wing patches. Adult has purple-and-green gloss to plumage and bright orange bill. Juvenile has duller yellowish-orange bill, paler yellow wattles and less gloss to plumage, with non-glossy brownish-black underparts. **Voice** Extremely varied, loud piercing whistles, screeches, croaks and wheezes. **HH** Active and noisy. Feeds in trees, mainly on fruits and berries, also flower buds, nectar and insects, sometimes in fruiting bushes. Evergreen and semi-evergreen forest.

Asian Glossy Starling *Aplonis panayensis* 20cm

Scarce resident in SE, one record in NE. **ID** Adult glossy greenish-black, with bright red eye and stout black bill. Juvenile has blackish-brown upperparts with variable greenish gloss and buffy-white underparts which are heavily streaked with blackish-brown (also with variable greenish gloss); eye is yellowish-white. **Voice** Call consists of sharp ringing whistles, *tseu...tseu* etc. **HH** Chiefly arboreal, often feeding with other starlings in flowering and fruiting trees. Coconut groves, forest edge and clearings.

Common Myna

Bank Myna

ad

juv

juv

Jungle Myna

ad

Great Myna

ad

juv

Common Hill Myna

ad

juv

ad

Asian Glossy Starling

Scaly Thrush *Zoothera dauma* 26–27cm

Rare winter visitor to C and NE. **ID** Boldly scaled upperparts and underparts with golden-olive panels on wing, pale face, large black eye and dark patch on ear-coverts. In flight, shows whitish panels on underwing. **Voice** Song is a slow, broken *chirrup...chwee...chwee...weep...chirrol...chup*; grating *tsshhh* call. **HH** Shy and retiring. Forages among leaf litter. Forest and well-wooded areas.

Orange-headed Thrush *Geokichla citrina* 21cm

Uncommon resident. **ID** Adult has orange head and underparts, and white shoulder patch; male has blue-grey mantle, female has olive-brown wash to mantle. Juvenile has buffish-orange streaking on upperparts and mottled breast. Shows white banding on underwing in flight. May show diffuse dark vertical bar at rear of ear-coverts. **Voice** Rich, sweet, variable song. **HH** Mainly forages on the ground, often comes onto forest paths at dusk. Moist places in forests and well-wooded groves. **TN** Formerly in genus *Zoothera*.

Black-breasted Thrush *Turdus dissimilis* 22cm

Scarce winter visitor, mainly to NE; also C and recorded in Sundarbans. **ID** Male has black head and breast, grey upperparts, orange lower breast and flanks, and white belly and vent. Female has dark olive-grey upperparts, plain face, prominent dark (streaked) malar stripe and dark spotting across olive-grey upper breast, whitish throat (with variable dark streaking) and submoustachial stripe, and orange lower breast and flanks. **Voice** Song is sweet and mellow, typical of the genus, consisting of 3–8 notes per phrase; calls are resounding *tup-tup...tup-tup-tup-tup-tup* etc. and thin *see*. **HH** Forest and nearby wooded areas.

Tickell's Thrush *Turdus unicolor* 21cm

Rare winter visitor. **ID** Small, compact thrush with rather plain face, small yellowish or pale brown bill and pale legs. Male pale bluish-grey, with whitish belly and vent. Has yellow bill and fine yellow eye-ring. First-winter male is similar but with pale throat and submoustachial stripe and dark malar stripe, and often has spotting on breast. **Voice** Song has weak, monotonous, repeated disyllabic or trisyllabic phrases, e.g. *chilliyah-chilliyah*, *tirlee-tirleechelia-chelia*; soft *juk-juk* call. **HH** Most records from deciduous forest, also coastal wooded areas and NE forest.

Black-throated Thrush *Turdus atrogularis* 26.5cm

Rare winter visitor mainly to C and E. **ID** Adult male has black supercilium, throat and breast (with narrow white fringes in fresh plumage), grey upperparts and whitish underparts. Female is similar to male, but typically has white or buffish throat and black-streaked malar stripe, and black gorget of spotting across breast. First-winter has fine white supercilium, white tips to greater coverts and pale-fringed tertials. First-winter male resembles adult female. First-winter female is less heavily marked and has finely streaked breast and flanks. See Appendix 1 for differences from Rufous-throated Thrush. **Voice** Calls include a shrill rattle in alarm and high-pitched squeaky contact notes. **HH** Forages on ground in open wooded areas and groves. **TN** Formerly treated as conspecific with Rufous-throated Thrush under the name Dark-throated Thrush *T. ruficollis*.

Orange-headed Thrush

♂

♀

Scaly Thrush

ad

♂

♀

Black-breasted Thrush

♂
1st-winter

♂

♀

Tickell's Thrush

♀
1st-winter

♂

♀

Black-throated Thrush

Oriental Magpie-robin *Copsychus saularis* 20cm

Widespread common resident, the national bird of Bangladesh. **ID** In all plumages, has white wing patch and white sides to long tail. Male has glossy blue-black head, upperparts and breast. Female has bluish-grey head, upperparts and breast. Juvenile has indistinct orange-buff spotting on upperparts, and orange-buff wash and diffuse dark scaling on throat and breast. **Voice** Spirited, clear and varied whistling song; plaintive *swee-ee* or *swee-swee*; alarm is harsh *chr-r*. **HH** Confiding and conspicuous. Partly crepuscular. Tail usually held cocked and frequently lowered and fanned, then closed and jerked up while wings are dropped and flicked. Gardens, groves and open woodland.

White-rumped Shama *Kittacincla malabarica* 25cm

Locally common resident, rare in W. **ID** Long, graduated dark tail with white sides and rump. Male has glossy blue-black upperparts and breast, rufous-orange underparts. Female duller, with brownish-grey upperparts; tail shorter and squarer. Juvenile has orange-buff spotting on upperparts, and orange-buff throat and breast with fine scaling. **Voice** Song comprises rich melodious phrases; call is musical *chir-chur* and *chur-chi-churr*; alarm is harsh scolding. **HH** Keeps close to ground in undergrowth and low trees in evergreen and deciduous forest. **TN** Formerly in genus *Copsychus*.

Lesser Shortwing *Brachypteryx leucophris* 13cm

Rare winter visitor, possibly resident in E. **ID** Small and short-tailed, with long pinkish legs. Male pale slaty-blue with white throat and belly. Female has rufous-brown upperparts, white throat and belly, and rufous-brown wash and diffuse scaling on breast and flanks. Both sexes have fine white supercilium, which is often obscured. First-year male similar to female but greyer above and on breast. **Voice** Brief melodious warbling song, accelerating into rapid jumble; calls include hard *tock-tock* and plaintive whistle. **HH** Very skulking, usually on ground. Thick undergrowth in moist evergreen forest.

Bluethroat *Cyanecula svecica* 15cm

Widespread, locally common winter visitor, more frequent in haors and riverine chars. **ID** Has prominent white supercilium and rufous tail-sides in all plumages. Male has variable blue, black and rufous pattern to throat and breast (obscured by whitish fringes in fresh plumage). Female has black submoustachial stripe and band of black spotting across breast; older females can have breast-bands of blue and rufous. **Voice** Deep *chack* or *chack-chack* calls. **HH** Secretive and terrestrial. Often cocks and fans tail. Scrub, reeds, tall grass and cultivation; often near water. **TN** Formerly in genus *Luscinia*.

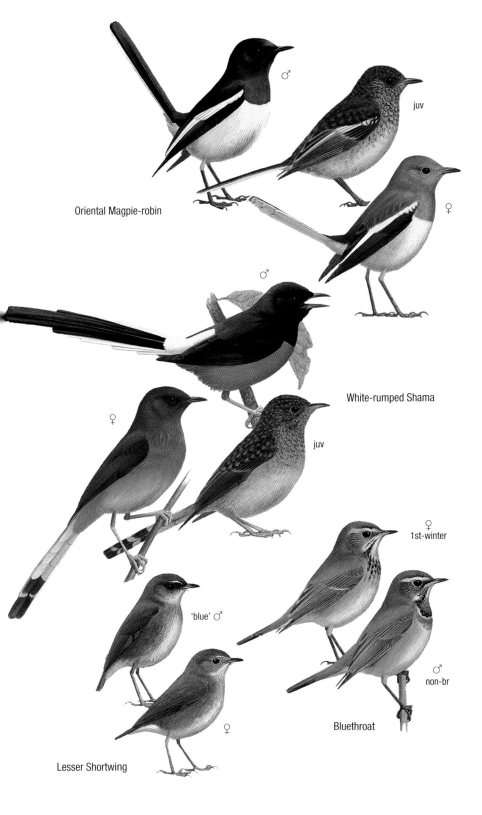

Oriental Magpie-robin

juv

♂

♀

White-rumped Shama

♂

juv

♀

Lesser Shortwing

'blue' ♂

♀

Bluethroat

♀
1st-winter

♂
non-br

Indian Blue Robin *Larvivora brunnea* 15cm

Scarce passage migrant, rare in winter. **ID** Male has slate-blue upperparts, broad white supercilium, (usually) black ear-coverts, and rufous-orange underparts with whitish centre to belly and vent. Female has olive-brown upperparts, and orange-buff to brownish-buff underparts with striking white throat, belly and vent; lacks prominent supercilium. First-year male variable; some have buffish supercilium, dull blue upperparts, and dull orange breast and flanks. Has long, pale legs and feet (compare with Firethroat). Short tail is frequently bobbed and fanned. **Voice** Song comprises three or four piercing whistles followed by rapid, tumbling notes *tit-tit-titwit-tichu-tichu-chuchu-cheeeh*; hard *tek-tek-tek* call. **HH** Secretive; usually on ground in forest, woodland and dense scrub. **TN** Formerly in genus *Luscinia*.

Firethroat *Calliope pectardens* 15cm

Rare winter visitor to NE; also recorded in Sundarbans. **ID** Male has flame-orange throat and breast bordered at sides with black, white patch on sides of neck and white sides to base of blackish tail. Non-breeding male has upperparts and tail as breeding male, but underparts are similar to female. First-winter male has olive-brown upperparts, with slaty-blue back and wing-coverts; tail as adult male, and underparts as adult female. Female has olive-brown upperparts, with rufous cast to uppertail-coverts, and buffish underparts; tail is brown. Very similar to female Indian Blue Robin, but with orange-buff vent and undertail-coverts, and dark legs and feet. **Voice** Call is deep *tock tock*. Song is masterful, lengthy and varied with sweet, repeated musical notes interspersed with harsher discordant notes. **HH** Dense wetland thickets and swamp forest, usually near ditch or water channel. **TN** Formerly in genus *Luscinia*.

Siberian Rubythroat *Calliope calliope* 14cm

Scarce winter visitor mainly to E and SW. **ID** Male lacks black breast-band and white sides and tip to tail of Chinese Rubythroat, and has olive-brown upperparts. Female separated from Chinese by olive-brown upperparts, olive-buff wash to breast and flanks, lack of white tip to tail, and pale brown or pinkish legs. First-winter as adult (i.e. male has red throat), with retained juvenile buff tips to greater coverts and tertials. **Voice** Calls include loud, clear double whistle *ee-uh* and hard *schak*. Long, pleasant scratchy warbling song sometimes heard before spring migration. **HH** Shy; skulks in bushes and thick undergrowth in second growth, wetlands and tea estates. **TN** Formerly in genus *Luscinia*.

Chinese Rubythroat *Calliope tschebaiewi* 14cm

Local winter visitor to NE. **ID** Male from Siberian by black breast (fringed white when fresh), white moustachial stripe, greyer upperparts, and blackish tail with white sides and tip. Female from Siberian by grey-brown upperparts, grey breast and flanks contrasting with belly, white tail-tip, black legs. First-winter as adult female, but buff tips to greater coverts and tertials. Male White-tailed Rubythroat (could occur as vagrant) lacks white moustachial stripe. **Voice** Song is long series of rising and falling warbling trills and twitters; call is harsh *ke*. **HH** Habits similar to Siberian but more secretive. Dense wetland thickets, scrub and reed-swamp. **TN** Formerly in genus *Luscinia* and treated as conspecific with White-tailed Rubythroat *C. pectoralis*.

Indian Blue Robin

Firethroat

Siberian Rubythroat

Chinese Rubythroat

White-tailed Blue Robin *Myiomela leucura* 18cm

Scarce winter visitor to NE and C; one breeding record from NE. **ID** White patches on tail in all plumages (visible as tail is slowly dipped and spread). Male blue-black, with glistening blue forehead and shoulders; has concealed white patch on side of neck. Female olive-brown, with whitish lower throat. First-year male similar to female but with blue on uppertail-coverts. **Voice** Song is a loud, clear, rather hurried jangling of seven or eight notes; calls include thin one- or two-note whistle and low *tuc*. **HH** Usually remains under cover. Undergrowth in dense, moist evergreen and semi-evergreen forest, rarely in swamp-thickets. **AN** White-tailed Robin.

Black-backed Forktail *Enicurus immaculatus* 23cm

Local and uncommon resident in E. **ID** The long black-and-white tail is often seen first when bird is flushed from forest streams. Has uniform black crown and mantle, white forehead, black throat and white underparts. Legs and feet are strikingly pink. Juvenile has shorter tail, lacks white forehead and supercilium, has brownish-black upperparts and dark scaling on white breast. See Appendix 1 for differences from other vagrant forktails. **Voice** Hollow *huu* call, followed by shrill *zeee*. **HH** Keeps to banks and beds of shallow, shaded forest streams, occasionally perching in adjacent undergrowth in evergreen forest.

Blue Whistling-thrush *Myophonus caeruleus* 33cm

Scarce winter visitor to E and C; probably nested in 1980s, at sea level near Cox's Bazar. **ID** Adult is dark blue-black, with head and body spangled with glistening silvery-blue. Forehead, shoulders and fringes to wings and tail are brighter blue. Has stout yellow bill. Juvenile is browner and lacks blue spangling. Wings and tail are duller blue than those of adult. **Voice** Melodic, rambling, whistling song; calls include shrill rasping *tzet* and shrill *kree*. **HH** Forests and wooded areas, particularly near streams and rivers.

Plumbeous Water-redstart *Phoenicurus fuliginosus* 12cm

Rare winter visitor to NE. **ID** Stocky and short-tailed; constantly flicks open tail while moving it up and down. Male slaty-blue, with rufous-chestnut tail. Female and first-year male have black-and-white tail and white spotting on grey underparts. **Voice** Song is rapidly repeated, insect-like *streeee-treee-tree-treeeh*; strident *peet-peet* alarm call. **HH** Flits restlessly between rocks in streams, frequently making flycatcher-like aerial sallies after insects. Waterfalls and adjacent stretches of hill streams. **TN** Often placed in genus *Rhyacornis*.

White-capped Water-redstart *Phoenicurus leucocephalus* 19cm

Rare winter visitor to NE; also recorded in NW and SE. **ID** Adult has white cap and rufous tail with broad black terminal band. **Voice** Song is weak, drawn-out, undulating whistle, *tieu-yieu-yieu-yieu*; far-carrying, upward-inflected *tseeittseeit* call. **HH** Flies from stone to stone in streams. Has distinctive habit of pumping and fanning tail, sometimes tilting it right over its back, often accompanied by deep curtsy. Waterfalls and adjacent stretches of hill streams. **AN** White-capped Redstart. **TN** Often placed in monspecific genus *Chaimarrornis*.

White-tailed Blue Robin

♂

♀

♀ juv

Black-backed Forktail

ad

Blue Whistling-thrush

juv

ad

Plumbeous Water-redstart

♂

♀

White-capped
Water-redstart

ad

Black Redstart *Phoenicurus ochruros*   15cm

Scarce winter visitor. **ID** Rufous-orange rump and outer tail feathers, unmarked wings. Male has blackish upperparts, black breast and rufous underparts. Female and first-year male are almost entirely dusky brown with rufous-orange wash on lower flanks and belly. **Voice** Song is scratchy trill, followed by short wheezy jingle; calls include short *tsip*, scolding *tucc-tucc* and rapid rattle. **HH** Vibrates tail when perched. Perches on bushes and often feeds on ground in open wooded and bushy areas.

Blue Rock-thrush *Monticola solitarius*   20cm

Uncommon winter visitor. **ID** Male indigo-blue, obscured by pale fringes in non-breeding and (especially) first-winter plumages. Female has bluish cast to slaty-brown upperparts, and buff scaling on underparts. *M. s. pandoo* is regular visitor, but male of rare *M. s. philippensis* has chestnut belly. **Voice** Short and repetitive song with fluty phrases, often with long pauses; calls include *chak, veeht veeht* and high *tsee*. **HH** Typically perches very upright on building or similar vantage point, frequently wagging tail up and down. Seeks insects mainly on ground, progressing by long hops, sometimes catching them in air. Among old buildings in open areas, and also along streams.

Grey Bushchat *Saxicola ferreus*   15cm

Rare winter visitor to NE, with single records from C and SE. **ID** Male has white supercilium and dark mask; upperparts grey to almost black, depending on extent of wear; underparts whitish with grey breast and flanks. Female has buff supercilium contrasting with dark brown ear-coverts, and rufous rump and tail-sides. First-winter as fresh-plumaged adult. **Voice** Song is brief and repeated, starting with two or three emphatic notes and ending with trill: *tree-toooh tu-treeeh t-t-t-t-tuhr*; calls include *zee-chunk*, and sharp *tak-tak*. **HH** Habits similar to Common Stonechat. Second growth, forest edge and scrub-covered hillsides.

Pied Bushchat *Saxicola caprata*   12.5–13cm

Locally uncommon in E. **ID** Male is entirely black except for white rump and patch on wing; duller due to rufous fringes on body in non-breeding and first-winter plumages. Female has dark brown upperparts and rufous-brown underparts, with rufous-orange rump. **Voice** Brisk, whistling *chip-chepee-chewee chu* song; calls include plaintive *chep chep-hee* or *chek chek trweet*. **HH** Habits very similar to Common Stonechat. Territorial throughout year. Bushes and tall grass, including tea estates.

White-tailed Stonechat *Saxicola leucurus*   12.5–13cm

Rare, local resident in chars of Ganges and Brahmaputra rivers. **ID** Male very similar to Common, but inner webs of all but central tail feathers are largely white; shows much white in tail in flight. Female has greyer upperparts, with diffuse streaking, and paler grey-brown tail. **Voice** Song is series of short phrases comprising rapid, squeaky, scratchy, wheezy and creaky notes that may fall and rise; alarm call is *peep-chaaa*. **HH** Habits very similar to Common. Tall grassland.

Common Stonechat *Saxicola torquatus*   12.5–13cm

Common winter visitor. **ID** Male has black head, white patch on neck, orange breast, and whitish rump (features obscured in fresh plumage). Female has streaked upperparts and orange on breast and rump. Tail darker than in female White-tailed. For differences from White-throated Bushchat, see Appendix 1. **Voice** Variable, rather melancholy warbling song; main calls are *hweet* and hard *tsak* which are often run together. **HH** Perches prominently on bush, tall plant or post; alert stance. Frequently flicks wings and jerks tail up and down while fanning it. Flies or hops to ground, picks up prey and returns to perch; also catches insects in aerial sallies. In scrub, wetland reed-swamp and bushes bordering cultivation. **TN** Siberian Stonechat *S. maurus* (the form found in Bangladesh) is often treated as separate species, but the relationships of the various forms are still unclear.

♂
♂ *philippensis*

♂ br
pandoo

Blue Rock-thrush

♂
♀

Black Redstart

♂ 1st-winter
pandoo

♀
pandoo

♂

♀

juv

♂ br

Pied Bushchat

♀

Grey Bushchat

♂ br

♀

♂
1st-winter

♂ br

♀

♂ non-br

White-tailed Stonechat

Common
Stonechat

Dark-sided Flycatcher *Muscicapa sibirica* 14cm

Rare passage migrant, mainly in NE. **ID** From Asian Brown Flycatcher by small dark bill and longer primary projection (exposed primaries are equal to or distinctly longer than tertials). Also, darker sooty-brown upperparts, breast and flanks more heavily marked, white crescent on neck-side and narrow white line down centre of belly. **Voice** Thin, high-pitched, repetitive phrases followed by quiet, melodious trills and whistles. **HH** Characteristically perches upright on wire or other favoured prominent vantage point, flying out to catch prey and returning to same perch. Open wooded areas.

Brown-breasted Flycatcher *Muscicapa muttui* 14cm

Rare passage migrant, mainly to E and C. **ID** Compared with Asian Brown, has larger bill with entirely pale lower mandible, pale legs and feet, rufous-buff edges to greater coverts and tertials, rufescent tone to rump and tail, more pronounced brown or grey-brown breast-band and warmer brownish-buff flanks. **Voice** A pleasant, feeble song; call a thin *sit*. **HH** Behaviour as Asian Brown. Open forest, groves and wooded areas.

Asian Brown Flycatcher *Muscicapa dauurica* 13cm

Rare passage migrant, mainly to E and C. **ID** Grey-brown with short tail, large head and huge eye with prominent eye-ring. From Dark-sided by larger bill with more extensive orange base to lower mandible, shorter primary projection and paler underparts (with light grey-brown wash to breast and flanks). **Voice** Song comprises short trills interspersed with two- or three-note whistling phrases, louder than Dark-sided; weak trilling *sit-it-it-it* call. **HH** Partly crepuscular. Perches in lower tree branches and makes aerial sallies. Open forest, groves and wooded areas.

Rufous-bellied Niltava *Niltava sundara* 18cm

Rare winter visitor to NE, also SE. **ID** Male has dark blue upperparts and orange underparts, with brilliant blue crown, neck patch, shoulder patch and rump. Female has well-defined oval-shaped white throat patch. Also has small blue patch on side of neck (often difficult to see). **Voice** Includes raspy *z-i-i-i-f-cha-chuk*, perhaps song, hard *tic*, thin *see* and low, soft *cha...cha*. **HH** Usually perches quietly in bush, or on branch low down in forest, occasionally darting out or dropping to ground to catch insects. Undergrowth in evergreen forest and second growth.

Verditer Flycatcher *Eumyias thalassinus* 16cm

Common winter visitor. **ID** Male is entirely greenish-blue, with brighter forehead and throat, and black lores. Not really confusable with any other flycatcher. Female is similar, but duller and greyer, and has dusky lores. Female confusable with male (but not female) Pale Blue-flycatcher, but has shorter bill, turquoise-blue upperparts and uniform greyish turquoise-blue underparts (lacking contrast between breast and belly). **Voice** Song comprises series of rapid, undulating, strident notes, gradually descending scale. **HH** Conspicuous and confiding. Hunts mainly by sallying forth from exposed perch. Open forests; forest canopy, clearings and edges; and wooded areas.

Pale Blue-flycatcher *Cyornis unicolor* 18cm

Rare winter visitor to E. **ID** Male from Verditer by longer bill and pale blue coloration (lacking greenish cast), with distinctly greyer belly. Has shining blue forecrown and dusky lores. Female is very different from Verditer; best separated by combination of large size, brownish-grey upperparts, uniform greyish underparts (lacking paler throat, with greyish-white centre of belly and dark buff undertail-coverts) and rufous-brown uppertail-coverts and tail. **Voice** Rich, melodious thrush-like song, unlike other *Cyornis* flycatchers with descending sequences, *chi, chuchichu-chuchichu-chucchi*, usually ending with harsh *chizz*. **HH** Usually frequents middle and upper storeys of forest, sometimes near ground. Pursues insects like typical flycatcher, but usually moves from perch to perch. Evergreen and semi-evergreen forest.

ad

Dark-sided Flycatcher

ad

Brown-breasted Flycatcher

♂

ad fresh

Asian Brown Flycatcher

ad worn

♀

Rufous-bellied Niltava

♀

♂

Verditer Flycatcher

♂

♀

Pale Blue-flycatcher

Pale-chinned Flycatcher *Cyornis poliogenys* 14cm

Locally uncommon resident in E and Sundarbans. **ID** Similar to females of other *Cyornis* flycatchers, and best separated by greyish crown and ear-coverts, prominent eye-ring, well-defined cream throat, and creamy-orange breast and flanks, with colour graduating into belly. Juvenile has diffuse buff spotting and black scaling on upperparts, and buff underparts with brown scaling on breast and flanks. **Voice** Song is high-pitched series of 4–11 notes, slightly rising and falling, sometimes interspersed with harsh *tchut-tchut* notes; repeated *tik* call. **HH** Hawks insects as typical flycatcher. Forages in bushes and undergrowth, sometimes on ground, where it resembles chat. Evergreen, semi-evergreen and mangrove forest. **AN** Pale-chinned Blue-flycatcher.

Hill Blue-flycatcher *Cyornis banyumas* 14cm

Rare winter visitor in NE and probable resident in Chittagong Hill Tracts. **ID** Male has dark blue upperparts with shining blue forehead and supercilium, and orange wash to throat and breast. From Blue-throated by orange throat. Female from female Blue-throated by orange of breast extending onto flanks; from Pale-chinned by pale orange throat concolourous with breast. See Appendix 1 for comparison with Large and Tickell's Blue-flycatchers. **Voice** Sweet, melancholy, warbling song; calls include scolding *trrt-trrt-trrt*. **HH** Dense evergreen forest.

Blue-throated Blue-flycatcher *Cyornis rubeculoides* 14cm

Scarce winter visitor to E, C and SW. **ID** Male from male Hill Blue-flycatcher by blue throat (some with orange wedge) and well-defined white belly and flanks. Female has narrow and poorly defined creamy-orange throat, and orange breast well demarcated from white belly. **Voice** Short, sweet song, with trilling notes; calls include *click click* and *chr-r chr-r* alarm. **HH** Frequents middle storey of forest. Sallies for insects but does not use regular perch. Open forest, wooded areas and groves.

Slaty-blue Flycatcher *Ficedula tricolor* 13cm

Local winter visitor to NE; also one record in SE. **ID** Small, slim, long-tailed flycatcher which typically feeds close to (or on) the ground, with tail cocked. Male has dark blue upperparts with brighter blue forehead, blue-black sides of head and breast contrasting with white throat, and blue-black tail with white patches at base. Female has fulvous underparts, warm brown upperparts, and rufous uppertail-coverts and tail. First-year male as female. **Voice** Song a series of high-pitched whistles, the final a lower, rapid trill; calls include rapid *tic-tic*. **HH** Undergrowth and dense vegetation in wetland swamp-thickets; also along streams in evergreen forest.

Snowy-browed Flycatcher *Ficedula hyperythra* 11cm

Rare winter visitor, mainly to NE; also C and NW. **ID** Small, compact, short-tailed flycatcher, with large head, typically found close to ground. Male has short, broad white supercilium, dark slaty-blue upperparts with rufous-brown wings, rufous-orange throat and breast, and white patches at base of tail (which can be difficult to see). Female by combination of small size and compact shape, dark olive-brown upperparts, orange-buff supercilium and eye-ring, and faint rufous panel on wing. Has pinkish legs and feet. **Voice** Quiet, high-pitched wheezy song *tsit-sit-si-sii*; thin, repeated *sip* call. **HH** Evergreen and deciduous forest with dense undergrowth; favours bamboo.

Little Pied Flycatcher *Ficedula westermanni* 11cm

Uncommon winter visitor to NE, Chittagong Hill Tracts and C. **ID** Small, compact, bull-headed flycatcher, with small dark bill. Male is strikingly black and white, including long and broad white supercilium. Female has brownish-grey upperparts (tinged with warm brown on forehead, lores and around eye), and whitish underparts with brownish-grey wash to breast and flanks. **Voice** Song is series of thin, high-pitched notes followed by a rattle, *pi-pi-pi-pi-churr-r-r-r-r-r*; mellow *tweet* call. **HH** Forests and open woodland.

Red-throated Flycatcher *Ficedula albicilla* 11–12cm

Common and widespread winter visitor. **ID** Always has white sides to long blackish tail, which is frequently held cocked. Male has orange throat bordered below by grey breast-band. Female and first-winter plumages have cold grey-brown upperparts, and underparts are whitish with greyish wash across breast. Bill is mainly dark. See Appendix 1 for differences from Red-breasted Flycatcher. **Voice** Calls include frequent buzzing *drrrrrrt*. **HH** Open forest, wooded areas and scrub at cultivation edge. **AN** Taiga Flycatcher. **TN** Formerly treated as conspecific with Red-breasted Flycatcher *F. parva*.

Pale-chinned Flycatcher

ad

juv

Hill Blue-flycatcher

♂

♀

Blue-throated Blue-flycatcher

♂

♀

♂

♀

Slaty-blue Flycatcher

Snowy-browed Flycatcher

♂

♀

Red-throated Flycatcher

♂

♀

♂

♀

ttle Pied Flycatcher

PLATE 95: FAIRY-BLUEBIRD, LEAFBIRDS AND SPIDERHUNTERS

Asian Fairy-bluebird *Irena puella* 25cm

Uncommon resident in E. **ID** Male has glistening violet-blue upperparts and black underparts. Female and first-year male entirely dull blue-green, with dusky lores and blackish flight feathers. Both sexes have striking red eye. Juvenile is entirely dull brown. **Voice** Liquid *tulipwae-waet-oo* and shorter liquid notes. **HH** Forages actively for fruit and nectar, usually in canopy. Evergreen and semi-evergreen forest.

Golden-fronted Leafbird *Chloropsis aurifrons* 19cm

Locally common resident. **ID** Adult is mainly green with golden-orange forehead (dull on some birds, especially females), purplish-blue throat, and broad golden-yellow collar (especially pronounced across breast). Lacks orange on belly of Orange-bellied. Juvenile is all green, with diffuse yellowish patch on forecrown and touch of blue in moustachial. See Appendix 1 for differences from Jerdon's Leafbird. **Voice** Song a cheery series of rising and falling liquid chirps, bulbul-like in tone. Wide variety of harsh and whistling notes; also mimics other birds. **HH** Found singly, in pairs or family parties according to season. Arboreal, typically inhabiting thick foliage in canopy. Searches leaves for insects; also feeds on berries and nectar. Forest and second growth.

Orange-bellied Leafbird *Chloropsis hardwickii* 20cm

Local resident in Chittagong Hill Tracts; rare winter visitor to NE. **ID** Male is striking with black of throat extending onto breast, orange belly and vent, large blue moustachial stripe, and purplish-blue flight feathers and tail. Female is largely green, with orange centre of belly and vent, and blue moustachial stripe. Juvenile is all green with blue moustachial and usually with touch of orange on underparts. **Voice** Wide variety of harsh and whistling notes; also mimics other birds. **HH** Habits similar to Golden-fronted. Evergreen forest and second growth.

Blue-winged Leafbird *Chloropsis moluccensis* 20cm

Scarce resident in E. **ID** From other leafbirds, in all plumages, by blue panel in wing and blue sides to tail. Also lacks well-defined golden-orange forehead of Golden-fronted. Male has small violet-blue moustachial stripe, black mask with diffuse yellow border, golden cast to crown and nape, and bright blue shoulder patch. Female is almost entirely green, with golden cast to crown and nape; throat is pale bluish-green, with brighter turquoise moustachial. Juvenile has green head, with slight suggestion of turquoise moustachial. See Appendix 1 for comparison with Jerdon's Leafbird, which has been recorded in far W of country. **Voice** Varied whistles, chuckles and rattles, with much mimicry; difficult to separate from other leafbirds. **HH** Evergreen and semi-evergreen forest.

Little Spiderhunter *Arachnothera longirostra* 16cm

Locally common resident in E. **ID** Has strikingly long, downcurved bill, unstreaked olive-green upperparts, whitish throat and breast merging into pale yellow of rest of underparts, whitish crescents above and below eye, and dark moustachial stripe. **Voice** Rapidly repeated *wit-wit-wit-wit....* song; abrasive *itch* call. **HH** Found singly or in pairs. Inhabits the lower storey of forest. Restless and noisy. Especially fond of wild banana flower nectar. Often clings upside-down while probing blossoms for nectar. Evergreen and semi-evergreen forest.

Streaked Spiderhunter *Arachnothera magna* 19cm

Locally uncommon resident in Chittagong Hill Tracts; rare winter visitor elsewhere in E. **ID** From Little by larger size, bold streaking on dark olive-green upperparts and bold streaking on yellowish-white underparts. Has orange legs and feet. **Voice** Strident chattering song; sharp *chirirrik* or *chirik chirik* call. **HH** Usually found singly or in pairs, often with itinerant foraging parties. Forages in upper storey. Fast-moving, flies strongly and swiftly from one tree to next. Very fond of nectar of wild bananas. Evergreen and semi-evergreen forest.

Golden-fronted Leafbird

Asian Fairy-bluebird

Orange-bellied
Leafbird

Blue-winged
Leafbird

Streaked Spiderhunter

Little Spiderhunter

Yellow-vented Flowerpecker *Dicaeum chrysorrheum* 10cm

Uncommon resident in E. **ID** From other flowerpeckers by blackish streaking on white or yellowish-white underparts and orange-yellow vent. Also has blackish malar stripe, curved black bill, whitish supercilium, red eye, bright olive-green upperparts with contrasting blackish primaries, and blackish tail. Juvenile has duller upperparts, and greyish-white underparts with paler yellow vent and less prominent streaking. **Voice** Short distinctive *dzeep*. **HH** Arboreal. Like other flowerpeckers very active, continually flying about restlessly, and twisting and turning in different attitudes when perched, while calling frequently. Evergreen and semi-evergreen forest and forest edge. Strongly prefers mistletoe berries.

Thick-billed Flowerpecker *Dicaeum agile* 10cm

Rare resident in E. **ID** From other flowerpeckers by combination of stout bluish-grey bill, indistinct dark malar stripe, lightly streaked breast, comparatively long and broad, fairly dark tail with white tip (which can be indistinct), and orange-red iris. Juvenile has pinkish bill. **Voice** A *tchup-tchup* call, not as hard as *chick* call of Pale-billed. **HH** Has distinctive habit of jerking its tail from side to side as it feeds or hops about. Arboreal and highly active, feeding mainly on peepul and banyan *Ficus*, also on mistletoe berries. Evergreen and semi-evergreen forest.

Orange-bellied Flowerpecker *Dicaeum trigonostigma* 9cm

Uncommon resident in Sundarbans and SE. **ID** Male is blue-grey and orange. Female from other female flowerpeckers by combination of olive-grey throat and upper breast, yellowish-orange underparts, greenish upperparts (with variable blue-grey cast to crown and mantle), blue-grey wing-panel and dull orange rump contrasting with dark tail. Juvenile is similar to female, but duller, with olive-brown upperparts, olive-yellow rump, and olive-yellow centre to belly and undertail-coverts; bill is orange with dark tip. Juvenile male has blue edges to rectrices. **Voice** Song is sharp, high-pitched, metallic *ptit.ptit.ptit.ptit.ptit.tsi*; also high, thin, rapid *psee-psee-psee-psee-psee-psee*, each note with an upward inflection. Call is harsh *dzit*. **HH** Mainly mangrove forest; also edges of evergreen forest.

Pale-billed Flowerpecker *Dicaeum erythrorhynchos* 8cm

Common resident. **ID** From Plain Flowerpecker by pinkish bill. Very plain with greyish-olive upperparts with slight greenish cast, and pale greyish underparts with variable yellowish-buff wash. **Voice** Hurried chittering song is lower-pitched than Thick-billed; sharp *chik chik chik* call. **HH** Remains in pairs in breeding season; otherwise in small parties. Constantly on the move, all the while calling. Flits about actively, usually in canopy. Has strong bounding and dipping flight. Feeds chiefly on mistletoe berries. Open forest, wooded areas and village groves.

Plain Flowerpecker *Dicaeum minullum* 8.5cm

Uncommon local resident in E. **ID** From Pale-billed by fine dark bill (with paler base to lower mandible). Has olive-green upperparts and edges to flight feathers and dusky greyish-olive underparts with yellow on throat and belly, which are further differences from Pale-billed; juvenile has browner upperparts and greyish-white underparts. **Voice** Repeated staccato *tzik*, more piercing than Pale-billed call. Distinctive *tzierr* is repeated as song. **HH** Extremely active and restless, usually in pairs or small parties. Very fond of mistletoe berries; also takes nectar, insects and spiders. Edges and clearings of evergreen and semi-evergreeen forests. **TN** Often considered conspecific with Nilgiri and Andaman Flowerpeckers as Plain Flowerpecker *D. concolor*.

Scarlet-backed Flowerpecker *Dicaeum cruentatum* 9cm

Common resident in E and Sundarbans, local in C. **ID** Male has scarlet upperparts and black 'sides' to whitish underparts. Female has scarlet rump contrasting with blackish tail, olive-brown upperparts and faint buffish wash to whitish underparts. Juvenile as female but lacks scarlet rump (usually with touch of orange); has bright orange-red bill and fine whitish supercilium. **Voice** Thin, repeated *tissit, tissit....* song; hard, metallic *tip..tip..tip* etc. call. **HH** Typical flowerpecker habits, see Pale-billed. Evergreen, semi-evergreen and mangrove forest, locally in deciduous forest.

Yellow-vented Flowerpecker

Thick-billed Flowerpecker

Orange-bellied Flowerpecker

Pale-billed Flowerpecker

Plain Flowerpecker

Scarlet-backed Flowerpecker

Ruby-cheeked Sunbird *Chalcoparia singalensis* 11cm

Common resident in E, C and Sundarbans. **ID** Shorter, straighter bill than other sunbirds, with rufous-orange throat and yellow underparts. Male has metallic green upperparts and 'ruby' cheeks. Female lacks 'ruby' cheeks and has dull olive-green upperparts. Juvenile is entirely yellow below. **Voice** Disyllabic *wee-eeast* with rising inflection. Song is rapid, high-pitched *switi-ti-chi-chu...tusi-tit...swit-swit...switi-ti-chi-chu...switi-ti-chi-chu.* **HH** Small, active sunbird, continually flitting about on low branches and in mid-storey. Unlike other sunbirds, associates with small insectivorous bird parties in winter. Evergreen, deciduous and mangrove forest. **TN** Formerly placed in genus *Anthreptes.*

Purple-rumped Sunbird *Leptocoma zeylonica* 10cm

Common resident. **ID** Male has narrow maroon breast-band, maroon head-sides and mantle, metallic green shoulder patch, and yellow lower breast and belly with distinctive greyish-white flanks. Does not have eclipse plumage. Female has greyish-white throat, yellow breast, whitish flanks, olive rump and rufous-brown wing-panel. Juvenile is uniform yellow below; rufous-brown on wings distinguishes it from Purple. **HH** Cultivation and second growth. **TN** Formerly placed in genus *Nectarinia.*

Maroon-bellied Sunbird *Leptocoma brasiliana* 10cm

Locally common resident in E. **ID** Small, short-billed sunbird. Male has purple throat, black mask and mantle, maroon breast and belly, and blue rump. Does not have eclipse plumage. Female has dull yellowish underparts with more olive throat, and olive-green upperparts; face is plain with fine eye-ring. **HH** Evergreen and semi-evergreen forest. **TN** Formerly placed in genus *Nectarinia.* Also formerly treated as conspecific with Purple-throated Sunbird *L. sperata.*

Purple Sunbird *Cinnyris asiaticus* 10cm

Common resident. **ID** Male is metallic blue-green and purple, becoming blacker on belly and vent. Female has uniform yellowish underparts, with faint supercilium and darker mask (whiter below in worn plumage). Eclipse male as female but has broad blackish stripe down centre of throat and breast, metallic blue wing-coverts and glossy black wings and tail. Juvenile is brighter yellow than female on entire underparts. **Voice** Pleasant descending *swee-swee-swee swit zizi-zizi* song; buzzing *zit* and high-pitched upward-inflected and slightly wheezy *swee* or *che-wee.* **HH** Typical sunbird, but probably more insectivorous than other species. Sometimes makes short aerial sallies like flycatcher. Male sings from bare treetop while jerking its body from side to side and raising his wings to reveal bright orange pectoral tufts. Deciduous forest, open wooded areas and gardens. **TN** Formerly placed in genus *Nectarinia.*

Crimson Sunbird *Aethopyga siparaja* 11cm

Locally common resident. **ID** Male has crimson mantle, scarlet throat and breast, and yellowish-olive belly. Female has yellowish-olive underparts; lacks yellow rump or prominent white on tail. Immature male as female but with red throat and breast. **Voice** Rapid, tripping song of 3–6 sharp, clear notes, *tsip-it-sip-it-sit.* **HH** Especially fond of red flowers. Forages mainly in undergrowth and mid-storey. Evergreen, deciduous and mangrove forest.

Ruby-cheeked Sunbird

♂

♀

Purple-rumped Sunbird

♂

juv

♀

♀

Maroon-bellied Sunbird

♂

♂ eclipse

Purple Sunbird

♀

♂ br

♀

♂

Crimson Sunbird

Black-breasted Weaver *Ploceus benghalensis*   14cm

Uncommon and local resident in NW and C, wintering also in NE. **ID** Breeding male has yellow crown and black breast. Throat may be black or white, and some variants have white ear-coverts and throat. In non-breeding male, female and juvenile plumages, breast-band is blotchy or restricted to small patches at sides, and may show indistinct, diffuse streaking on lower breast and flanks. In these plumages, has yellow supercilium (often white behind eye), distinct yellow patch on side of neck and yellow submoustachial stripe (with black malar); similar to Streaked Weaver, except crown, nape and ear-coverts are more uniform; rump also indistinctly streaked and, like nape, contrasts with heavily streaked mantle. **Voice** Soft, barely audible *tsi tse tsisik tsisik tsik tsik tsik* song; soft *chit-chit* call. **HH** Habits very similar to Baya. Nest similar to Baya but placed in tall grasses with top dome of nest interwoven into a number of standing grass or reed stems, and the vertical entrance tube is shorter. Tall, moist, seasonally flooded grassland and reedy marshes along main rivers.

Streaked Weaver *Ploceus manyar*   14cm

Rare and local resident previously in SW, C and SE; few recent records in NE. **ID** Breeding male has yellow crown, dark brown head-sides and throat, and heavily streaked breast and flanks. Other plumages typically show boldly streaked underparts. However, may be only lightly streaked on underparts, especially juvenile, when best told from Baya Weaver by combination of yellow supercilium and neck patch, heavily streaked crown, dark or heavily streaked ear-coverts, and pronounced dark malar and moustachial stripes. When streaking is absent on underparts, streaked crown, nape and rump are best features from Black-breasted. **Voice** Song is soft, continuous trill, *see-see-see-see-see*, ending in *o-chee*; loud *chirt chirt* call. **HH** Habits very similar to Baya. In winter, chiefly feeds on flowering heads of *Phragmites*, bulrushes and *Saccharum* grasses by clinging to upright stems. Reedy marshes.

Baya Weaver *Ploceus philippinus*   15cm

Locally common resident. **ID** Breeding male of nominate race occurring in W has yellow crown, dark brown ear-coverts and throat, unstreaked yellow breast, and yellow streaking on mantle and scapulars. Breeding male *burmanicus* (the predominant race in Bangladesh) has greyer face, buff or pale grey throat, and buff breast. Non-breeding male, female and juvenile usually have unstreaked buffish underparts; streaking can be as prominent as on poorly marked Streaked, but has less distinct and buffish supercilium, lacks yellow neck patch, and lacks pronounced dark moustachial and malar stripes. Non-breeding male, female and juvenile *burmanicus* are more rufous-buff on supercilium and underparts. Probably much intergradation between these two races in Bangladesh. **Voice** Song is soft *chit chit chit* followed by drawn-out wheezy whistle, *chec-ee-ee*; *chit-chit-chit* call. **HH** Highly gregarious throughout year. Feeds extensively on ripening cereal crops and insect crop pests. Forages by hopping on ground and picking seeds from tops of upright grass stems. Roosts communally all year. Colonial breeder, building retort-shaped nest with long vertical entrance tube, suspended from branch usually of an isolated tree, typically a palm. Open country near water with scattered bushes and tall trees and cultivation.

House Sparrow *Passer domesticus*   15cm

Abundant, widespread resident. **ID** Breeding male has grey crown with chestnut sides and nape, and black throat and upper breast; duller in non-breeding plumage when head pattern and black throat/breast partly obscured by pale fringes. Female has pale buff supercilium, dark brown streaking on buffish mantle and unstreaked greyish-white underparts (faintly washed with buff on flanks). Juvenile as female, with broader buff-brown fringes to upperparts; juvenile male has greyish chin. **Voice** Monotonous *chirrup* call; *chur-r-r-it-it* alarm; song comprises long series of *chirrup*, *cheep* and *churp* notes. **HH** Markedly gregarious outside breeding season, forming large communal roosts. Forages mainly by hopping on ground, also perches on ripening cereal heads and pecks out seeds. Villages, towns and cities; also cultivation in winter.

Eurasian Tree Sparrow *Passer montanus*   14cm

Rare and local resident in NW and NE. **ID** Adult has dull chestnut crown, black spot on whitish ear-coverts, small black throat patch not extending onto breast, and white collar separating chestnut nape from brown-streaked mantle. Sexes are alike. Juvenile is similar to adult but has paler chestnut crown and diffuse black patches on ear-coverts and throat. **Voice** Monotonous *chip chip* call, harder than similar call of House; song is a running-together of this note, interspersed with *tsweep* or similar calls. **HH** Habits very similar to House. Suburbs and fields at edges of towns and villages.

♂ br

♂ br

♂ br

Streaked Weaver

♂ non-br

Black-breasted Weaver

♀

♂ non-br

♂ br
philippinus

♀
philippinus

♀

♂ br
burmanicus

♂ non-br
philippinus

♀
burmanicus

Baya Weaver

♂

♀

ad

House Sparrow

Eurasian Tree Sparrow

Red Avadavat *Amandava amandava* 10cm

Rare and local resident, mainly along main rivers and in NE, in 19th century near Dhaka and in 1940s in SW. **ID** Breeding male is mainly red with irregular white spotting. Non-breeding male and female have grey-brown upperparts and buffish-white underparts; best told by red bill, red rump, and white tips to wing-coverts and tertials. Juvenile lacks red in plumage; buff wingbars and tertial fringes, pink bill-base, and pink legs and feet help separate from juveniles of other munia species. **Voice** Weak, high-pitched warbling song; calls include a thin *teei* and a variety of high-pitched chirps and squeaks. **HH** Habits similar to Scaly-breasted Munia. Tall grassland, reedbeds and wet scrub.

Indian Silverbill *Euodice malabarica* 11–11.5cm

Uncommon resident. **ID** Male has fawn-brown upperparts, whitish face and underparts with barred flanks, long and pointed black tail, and white rump and uppertail-coverts. Female is duller with plainer face, and flanks are less barred. Juvenile lacks barring on flanks, has dark mottling on rump, and tail is shorter and more rounded. **Voice** A *tchrip!* or *tchreep!* contact call; repeated *chir-rup!* flight call; song is series of short, abrupt trills. **HH** Habits similar to Scaly-breasted Munia. Outside breeding season, roosts communally in old nests of weavers or their own. Cultivation, grassland and scrub. **TN** Formerly placed in genus *Lonchura*.

White-rumped Munia *Lonchura striata* 10–11cm

Uncommon resident in E and C. **ID** A dark breast, streaked upperparts and white rump are best features. Furthermore, has rufous-brown on side of head/neck, rufous-brown to whitish fringes on dark brown breast and faint brownish streaking to greyish-buff belly; juvenile barred brown-buff on throat and breast. **Voice** Song is rising and falling series of twittering notes; calls include twittering *tr-tr-tr*, *prrit* and *brrt*. **HH** Habits similar to Scaly-breasted but usually in smaller groups. Clearings and edges of evergreen and semi-evergreen forest, including nearby scrub.

Scaly-breasted Munia *Lonchura punctulata* 10.7–12cm

Common and widespread resident. **ID** Adult has chestnut-brown face, throat and upper breast, whitish underparts boldly scaled with black, and olive-yellow to rufous-orange on uppertail-coverts and edges of tail. Juvenile has uniform brown upperparts and buff to rufous-buff underparts, with whitish belly (probably indistinguishable from juvenile Tricoloured). **Voice** Typical song is series of *klik-klik-klik* or *tit-tit-tit* notes followed by short series of whistles and churrs, and ending with longer *weeee*; contact calls include repeated *tit-ti tit-ti* and loud *kit-teee kit-teee*. **HH** Forages for grass and bamboo seeds on ground or by clinging to stems and pulling seeds directly from seed heads. Gregarious outside breeding season and roosts communally. If disturbed, a flock typically flies up together in close-knit pack and moves off with fast, whirring wingbeats to nearby cover. Flight of individuals is undulating, but the flock maintains a fairly direct course. Often roosts in nests, either those used previously for breeding or specially built for roosting. Open second forest, bushes, village groves and cultivation.

Tricoloured Munia *Lonchura malacca* 11.5cm

Scarce resident, mainly in C; origins uncertain and may have been introduced. **ID** Adult has black head and upper breast, rufous-brown upperparts, and black belly centre and undertail-coverts. Lower breast and flanks are white. Juvenile has uniform brown upperparts and buff to whitish underparts. **Voice** Songs include thin, evenly rising, high, whining nasal whistle; calls include nasal, slightly downturned, abrupt squeaky-toy *nyek, nyek...*. **HH** Habits similar to Scaly-breasted, usually with Chestnut Munia. Marshes, tall grassland and cultivation. Normal range is peninsular India; birds in Bangladesh may have become established naturally or be escapes from captivity. **AN** Black-headed Munia.

Chestnut Munia *Lonchura atricapilla* 11cm

Uncommon and local resident. **ID** Similar to Tricoloured but with chestnut lower breast and flanks. Juvenile is warmer buff on underparts than Tricoloured. **Voice** Varied, short, squeaky, nasal flight calls. **HH** Habits similar to Scaly-breasted but keeps to dense wetland vegetation. Tall, wet grassland, marshes and nearby cultivation, mainly along large rivers and in haors. **TN** Sometimes treated as conspecific with Tricoloured Munia.

♂ br

juv

Red Avadavat

♀

juv

ad

Indian Silverbill

ad

White-rumped Munia

ad

juv

Scaly-breasted Munia

ad

juv

Tricoloured Munia

ad

Chestnut Munia

Olive-backed Pipit *Anthus hodgsoni*   15cm

Common winter visitor. **ID** Greenish-olive cast to upperparts and edges to wing feathers. Has prominent supercilium (buffish in front of eye and white behind), stronger dark eyestripe and distinct whitish spot and blackish patch on rear ear-coverts (lacking on Tree Pipit). Worn upperparts become more greyish-olive, and breast loses warm buff wash. *A. h. yunnanensis*, the commoner of the two subspecies that occurs, is much less heavily streaked on upperparts than the nominate form. See Appendix 1 for differences from Tree Pipit. **Voice** Weak *see* flight call, fainter than Tree. **HH** Open wooded groves, forest edge.

Red-throated Pipit *Anthus cervinus*   15cm

Rare winter visitor and passage migrant to E. **ID** Adult has reddish throat and upper breast, which tend to be paler on female and on autumn/winter birds (lacking in first-winter birds). In all plumages has heavily streaked upperparts, often with pale 'braces', heavily streaked rump, well-defined and broad white wingbars, strongly contrasting blackish centres and whitish fringes to tertials, pronounced dark malar patch, and boldly streaked breast and (especially) flanks. Very different call, browner coloration to upperparts and absence of olive in wing help separate from non-breeding Rosy. **Voice** Typical call is long drawn-out *seeeeee*, but can be abbreviated. **HH** Large wetlands including coasts, marshes, wet grassland and stubble.

Rosy Pipit *Anthus roseatus*   15cm

Local winter visitor, mainly to NE haors; rarer in SE, C, and NW. **ID** Has prominent supercilium, dark lores and malar stripe, boldly streaked upperparts, olive cast to mantle and olive to olive-green edges to wing feathers. In non-breeding plumage, underparts are heavily streaked. Breeding adult has mauve-pink wash to underparts with irregular black spotting on breast and flanks; pinkish to buff supercilium is very prominent, and has whitish eye-ring. Female is less pink below with heavier breast streaking than male. **Voice** Weak *seep-seep* call. **HH** Wet grass and muddy borders of large wetlands.

Richard's Pipit *Anthus richardi*   17cm

Locally fairly common, quite widespread winter visitor. **ID** Large size, upright stance, larger bill, longer legs and hindclaw, and call are best features from otherwise similar Paddyfield Pipit. **Voice** Distinctive loud, explosive *schreep* call; also, shorter *chup.* **HH** On ground, progresses with swift runs combined with strutting walk on strong legs. When flushed, typically gains height and distance with deep undulations. May hover above ground, fluttering with dangling legs. Moist grassland and cultivation.

Paddyfield Pipit *Anthus rufulus*   15cm

Common and widespread resident. **ID** Smaller than Richard's, with different call; when flushed, has comparatively weak flight. Juvenile/first-winter Tawny Pipit (see Appendix 1) can appear very similar, but lores of Paddyfield often pale (though can be dark), and shows warm ginger-buff wash across breast and on flanks (underparts more uniform cream-white on Tawny); flight call differs. **Voice** Calls include weak *chup-chup-chup* or *chip-chip-chip*; song given perched or in aerial display is repetitive *chip-chip-chip.* **HH** Forages by running rapidly on ground, wagging its tail. Less powerful gait than Richard's. When flushed, usually flies short distance. Edges and banks of fields, short grassland, ploughed and fallow fields, and stubbles.

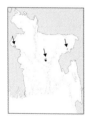

Blyth's Pipit *Anthus godlewskii*   16.5cm

Rare passage migrant in C, NW and NE; probably under-recorded. **ID** Compared with Richard's, very subtle differences are slightly smaller size and more compact appearance, shorter tail, shorter hindclaw, shorter and more pointed bill, shorter legs, and call. Distinctive shape of centres to adult buff-fringed median coverts (square-shaped black centres with broad pale buff tips; centres diffuse and more triangular in Richard's). Best distinctions from Paddyfield Pipit are call, larger size and pattern of adult median coverts (Paddyfield as Richard's Pipit). **Voice** Diagnostic powerful, wheezy *spzeeu* call; also, mellow *chup* or *chep* resembling Paddyfield or Tawny Pipit (see Appendix 1). **HH** Gait resembles Tawny; lacks Richard's strutting walk; flight similar to Richard's but without fluttering pause before landing. Marshes, grassland and cultivation.

fresh
hodgsoni

Olive-backed Pipit

worn
hodgsoni

♂ br

yunnanensis

Red-throated Pipit

1st-winter

♀ br

ad

juv

♂ br

non-br

Richard's Pipit

Rosy Pipit

juv

1st-winter

Paddyfield Pipit

ad

Blyth's Pipit

ad

Western Yellow Wagtail *Motacilla flava* 18cm

Locally fairly common winter visitor, also passage migrant; may be scarcer as status needs review due to probable confusion with Eastern. **ID** Breeding male has olive-green upperparts and yellow underparts, with considerable subspecies variation. Female in breeding plumage usually shows some features of breeding male. First-winter birds typically have brownish-olive upperparts, whitish underparts with variable yellowish wash, and buff or whitish median- and greater-covert wingbars and fringes to tertials. Some first-winters, however, have greyish upperparts and whitish underparts (lacking any yellow), and can closely resemble first-winter Citrine Wagtail. Best told by: narrower white supercilium which does not extend around ear-coverts; grey forehead concolorous with crown; dark lores resulting in complete dark eyestripe; pale base to lower mandible; and narrower white wingbars. Juvenile has dark malar stripe and band across breast. Two subspecies have been recorded (*beema* and *thunbergi*), and others may occur. *M. f. beema* has pale bluish-grey head and white supercilium. *M. f. thunbergi* has dark slate-grey crown with darker ear-coverts. *M. f. leucocephala* has whitish head. *M. f. feldegg* (including *melanogrisea*) has black head. *M. f. lutea* has yellowish head. *M. f. 'superciliaris'* is an intergrade, probably between *beema* and *feldegg*, and looks like the latter but with white supercilium. **Voice** Typical call is loud disyllabic *tswee-ip*, usually quite distinct from Citrine, although *feldegg* at least has a harsher *tsreep* call, more closely resembling Citrine. **HH** Shallow water, marshlands, damp grasslands. **AN** Yellow Wagtail. **TN** Eastern Yellow Wagtail is now treated as a separate species due primarily to remarkable genetic differences.

Eastern Yellow Wagtail *Motacilla tschutschensis* 18cm

Winter visitor and passage migrant confirmed from NE and coast (S), but likely to be more widespread due to confusion with (and challenge of separation from) Western. **ID** The racial identity of birds occurring in Bangladesh is far from clear and several are possible. Male breeding *M. t. plexa* and *M. t. macronyx* are not safely distinguishable using plumage from *M. flava thunbergi*. Male breeding *M. t. tschutschensis*, *M. t. simillima* and *M. t. angarensis* have white supercilium, grey crown and grey to blackish ear-coverts, but may not be separable from *M. flava beema*. Male breeding *M. t. taivana* is more distinctive with olive crown and back, yellow supercilium and dark ear-coverts. Female, non-breeding and immature birds are not safely identifiable in the field on plumage at present. Best distinguished from Western by call, supported by biometrics (hindclaw is longer in Eastern but there is some overlap) and genetic analysis. **Voice** Flight call is harsher than Western Yellow Wagtail and is much like Citrine. **HH** As Western Yellow Wagtail. **TN** Formerly considered conspecific with Western Yellow Wagtail (as Yellow Wagtail *M. flava*).

Citrine Wagtail *Motacilla citreola* 19cm

Locally fairly common winter visitor. **ID** Breeding male separable from Western and Eastern Yellow by yellow head and underparts, black (*M. c. calcarata*) or grey (nominate) mantle and broad white wingbars. Breeding female and non-breeding adult best told from Western and Eastern Yellow by broad yellow supercilium which surrounds ear-coverts to join yellow of throat, broad white wingbars (although narrower when worn), and greyish crown and mantle. Juvenile lacks any yellow, and has brownish crown, ear-coverts and mantle, buffish supercilium (with dark upper edge) and surround to ear-coverts, and buffish-white underparts with gorget of black spotting across breast. First-winter has grey upperparts and is similar to some first-winter Western and Eastern Yellow, and best told by: broader white supercilium, which usually surrounds ear-coverts; pale brown forehead; pale lores; all-dark bill; broader white wingbars; and white undertail-coverts. Separable from White Wagtail by absence of black breast-band and different call. By early Nov, first-winter Citrine has yellowish supercilium, ear-covert surround and throat. **Voice** Harsh *brrzzreep* call, more buzzing than Western Yellow. **HH** Wetlands, including haors, coast, wet fields, marshes and riverbanks.

juv
beema

♀
beema

Western Yellow Wagtail

♂ br
beema

grey-and-white
1st-winter
beema

♂
leucocephala

1st-winter
lutea

♂
lutea

1st-winter
thunbergi

♂
feldegg

♂
'superciliaris'

♂
thunbergi

♂
tschutschensis

♂
macronyx

♂
taivana

Eastern Yellow Wagtail

Citrine Wagtail

♂ br
calcarata

1st-winter
(late)
calcarata

juv
calcarata

1st-winter
(early)
calcarata

♂ br
citreola

♀
calcarata

Forest Wagtail *Dendronanthus indicus* 18cm

Regular passage migrant to E, C and SW; rare in winter. **ID** A forest-dwelling wagtail; from other wagtails by combination of broad yellowish-white median- and greater-covert wingbars and white patch on secondaries, double black breast-band (lower band broken in centre of breast), olive upperparts, white supercilium and whitish underparts. Sexes alike. **Voice** Strident metallic *pink* or *dzink-dzzt* call. Repetitive intense Great Tit-like see-sawing song. **HH** Characteristic habit of swaying its tail and hind part of body from side to side with rather deliberate motion, instead of wagging tail up and down like other wagtails. Glades and paths in forest and wooded groves.

Grey Wagtail *Motacilla cinerea* 19cm

Uncommon winter visitor. **ID** Much longer-tailed than other wagtails (especially noticeable in flight). In all plumages has white supercilium, grey upperparts and yellow vent and undertail-coverts. In flight, shows narrow white wingbar and yellow rump. At rest, shows whitish fringes to tertials but otherwise wings are blackish, lacking broad fringes to coverts present on Western Yellow and Citrine Wagtails. Breeding male has black throat, with rest of underparts yellow. Breeding female lacks well-defined black bib but may show black mottling on chin/throat. Non-breeding adult and first-winter plumages have white throat and pale yellowish to buffish-white underparts, with yellow vent. Juvenile much as non-breeding, but with brownish cast to upperparts, buffish supercilium and dark mottling on breast-sides. **Voice** Sharp *stit* or *zee-fit* call; song comprises series of call-like notes. **HH** Mainly along smaller rivers and streams, rarer at coast.

White-browed Wagtail *Motacilla maderaspatensis* 21cm

Uncommon resident, rarer in E and SW. **ID** Very large wagtail. Combination of black mantle and black head with white supercilium separates it from all subspecies of White. Sexes are similar; shows no variation in non-breeding plumage. First-winter is similar, but with greyer crown and mantle. Juvenile has brownish-grey head, mantle and breast, with white supercilium. **Voice** Distinctive, loud *chiz-zat* call; song comprises clear, high-pitched jumble of loud, pleasant whistling notes. **HH** Riverbanks, ponds and lakes.

White Wagtail *Motacilla alba* 19cm

Widespread and common winter visitor. **ID** Much variation in adults in breeding plumage (sexes are similar) with the following six subspecies occurring in the country: *M. a. alboides* has black head, mantle and breast, with white forehead and face patch. Upperparts of female are variably mixed with grey. *M. a. personata* similar to *alboides* but has grey mantle. *M. a. alba* has grey mantle, white forehead and face, and black hindcrown/nape, throat and breast. *M. a. leucopsis* has black mantle and back, with head pattern as *alba* but with white throat. *M. a. ocularis* has grey mantle, and is much as *alba* but has black eyestripe in all plumages. *M. a. baicalensis* has grey mantle and is much as *alba*, but has white chin and upper throat contrasting with black breast. There is much variation in non-breeding and first-winter plumages, although non-breeding birds of some races retain characteristics of breeding plumage. Juvenile *alboides* has grey head, mantle and breast with whitish supercilium. **Voice** Call is a loud *tslee-vit*; song is a lively twittering and chattering with call-like notes. **HH** Usually near water in open country, including cultivated areas.

♂ br

♀ / ♂ non-br

1st-winter

Grey Wagtail

Forest Wagtail

ad

ad

juv

White-browed Wagtail

♀
alboides

♂ br
alboides

♂ br
personata

juv
alboides

♂
baicalensis

1st-winter
personata

♂
alba

♂
ocularis

♂
leucopsis

White Wagtail

Common Rosefinch *Carpodacus erythrinus* 14.5–15cm

Scarce but apparently increasing winter visitor mainly to E, but also recorded in SW and C. **ID** Compact, with short, stout bill. Female and first-year birds rather plain and best told from female House Sparrow by streaked underparts, rather plain face with beady black eye, and double wingbars. Male has red on head, breast and rump. **Voice** Distinctive, clear, rising *ooeet* call; alarm is *charpchay-eeee*; song is monotonous, clear whistling *weeeja-wu-weeeja*. **HH** Gregarious outside breeding season. Forages by hopping on ground and clambering among foliage of bushes and trees. Sometimes perches inactively for long periods. Has undulating flight. Cultivation with bushes and open wooded country, most often in hilly areas.

Chestnut-eared Bunting *Emberiza fucata* 16cm

Rare winter visitor to coast and NE. **ID** Adult has chestnut ear-coverts, black breast streaking and usually some chestnut on breast-sides. Some first-winter birds rather nondescript, but plain head with warm brown ear-coverts and prominent eye-ring distinctive. **Voice** Rapid, twittering *zwee zwizwezwizizi trup-trup* song; explosive *pzick*, higher-pitched *zii* and lower-pitched *chutt* calls. **HH** Difficult to see; forages on ground and when disturbed perches in bushes or tall grasses. Tall grasses with bushes, usually in wetlands.

Yellow-breasted Bunting *Emberiza aureola* 15cm

In the past a locally common winter visitor and passage migrant to E and main rivers, now declining and rare. **ID** A stocky, comparatively short-tailed bunting. Has more direct, less undulating flight compared with other buntings so appears more weaver- or sparrow-like. Male has black face and chestnut breast-band (obscured when fresh), and white inner wing-coverts. Female has striking head pattern (with broad yellowish supercilium and pale crown-stripe), boldly streaked mantle (with pale 'braces' often apparent), and prominent white median-covert bar. Juvenile as female, but underparts are paler yellowish-buff with fine, dense streaking on breast and flanks. **Voice** Calls include soft *chup* and metallic *tick* like Little. **HH** Typical bunting habits, see Chestnut-eared. Cultivation, stubbles and tall grassland in or near wetlands. **Threatened** Globally (CR) and nationally (VU).

Little Bunting *Emberiza pusilla* 13cm

Rare winter visitor to NE, SE and SW. **ID** Small, with comparatively fine, pointed bill. Has chestnut ear-coverts (and often supercilium and crown-stripe), uniform grey-brown mantle, lightly streaked dark brown (lacking pale 'braces'), finely streaked breast and flanks, and prominent pale median- and greater-covert wingbars. **Voice** Sharp *tzic* call. **HH** Typical bunting habits, see Chestnut-eared. Bare earth, stubble, ploughed or grassy fields.

Black-faced Bunting *Emberiza spodocephala* 15cm

Local and uncommon winter visitor to NE. **ID** Male has greenish-grey head with blackish lores and chin, and yellow underparts. Non-breeding male has yellow submoustachial stripe and throat. Female has yellowish supercilium, yellow throat, olive rump and white on tail. **Voice** Soft *tsip* or sharper *tzit* call. **HH** Typical bunting habits, see Chestnut-eared. Swamp-thickets, reeds, scrub and wet grassland in large wetlands.

Common Rosefinch

Chestnut-eared Bunting

♂

♀

♀

♂

1st-winter

♂ non-br

♂ br

♀

juv

Yellow-breasted Bunting

♂

1st-winter

♂ non-br

♂ br

♀

Little Bunting

Black-faced Bunting

APPENDIX 1: VAGRANTS

The following species have been recorded in Bangladesh and are regarded as vagrants with fewer than ten records (most fewer than five) to the end of December 2019. This distinction is challenged by new discoveries and a growth in observer effort. Some species with few records but which have bred or are strictly sedentary are described in the main text. Almost all vagrants have been recorded in winter. With further fieldwork and visits to the more remote parts of the country, a number of those included in this appendix may prove to be regular rare visitors, and some species recorded in the Chittagong Hill Tracts may prove to be resident. Species only recorded in one century are noted; where this is not indicated, the species has been recorded with certainty only since the 1970s and in both the 20th and 21st centuries.

Greater White-fronted Goose *Anser albifrons* 66–86cm

One late 20th-century record from Jamuna River (C). **ID** Adult best told from Greylag Goose (Plate 4) by broad white band at front of head, browner coloration, black barring on belly, and orange legs and feet. Has more uniform upperwing in flight, darker back and rump and darker base to tail than Greylag. Juvenile lacks white frontal band and barring on belly; is more similar to Greylag, and best told by smaller size and less stocky build, browner coloration, darker feathering at base of bill, dark tip (nail) to bill, and orange legs and feet. **Voice** Has cackling and honking flight call which contains distinctive musical *lyo-lyok* phrase. **HH** Habits similar to Greylag. Large rivers and lakes.

Lesser White-fronted Goose *Anser erythropus* 53–66cm

One 21st-century record in NE. **ID** Adult distinguished from Greater White-fronted Goose by half-moon white frontal band on head (white typically extending as point at front of crown), and yellow eye-ring. Has slightly darker head and neck, and less extensive black barring on belly. Also has 'squarer' head, with more steeply rising forehead, and is smaller and more compact, with stout triangular bill. Juvenile lacks white frontal band and black barring on belly, and is best distinguished from juvenile Greater by yellow eye-ring, slightly darker head and neck, and structural differences described above. **Voice** Call is similar to Greater although higher pitched and includes repeated *kyu-yu-yu* phrase. **HH** Habits similar to Greylag (Plate 4). Wet grassland and lakes. **Threatened** Globally (VU) and nationally (VU).

Common Goldeneye *Bucephala clangula* 42–50cm

One early 20th-century record. **ID** Stocky, with bulbous head. Male has dark green head, with large white patch on lores, and black-and-white patterned upperparts. Female has brown head, indistinct whitish collar and grey body, with white wing patch usually visible at rest. Has 'golden' eye (as does male) and pink band at tip of bill. Immature male resembles female but shows pale loral spot and has some white in scapulars. Eclipse male resembles female, but wing pattern is as breeding male. In flight, both sexes show distinctive white pattern on wing. **HH** Swims with body flattened, and partially spreads wings when diving. Fast flight, the wings producing a distinctive whistling sound. Open waterbodies.

Smew *Mergellus albellus* 38–44cm

One late 20th-century record in NE. **ID** A small, stocky 'sawbill' with squarish head. In flight, both sexes show dark upperwing with white wing-covert patch. Male is mainly white, with black face, crest-stripe, breast-stripes and back. Flanks are grey. Female, first-winter and eclipse male have chestnut cap and white cheeks, and mainly dark grey body. **HH** Feeds diurnally, mainly by diving. Freshwater lakes and rivers.

Goosander *Mergus merganser* 58–72cm

Records from NW, C and NE. **ID** Male has dark green head and whitish breast and flanks (with variable pink wash). Shows extensive white patch on wing-coverts and secondaries in flight. Female and eclipse/immature male have chestnut head and upper neck with shaggy crest, which contrasts with white throat and greyish neck, and show white secondaries in flight. Eclipse male has upperwing pattern like breeding male. **HH** Feeds mainly by diving, usually after scanning with head submerged. Lakes and fast- and slow-moving rivers.

Red-breasted Merganser *Mergellus serrator* 52–58cm

One 21st-century record from SE coast. **ID** Male has spiky crest, white collar, ginger breast and grey flanks. Female and eclipse/immature male more closely resemble respective plumages of Goosander and are best told by slimmer appearance, with slimmer bill, and narrower head with weaker and more ragged crest. Chestnut of head and upper neck is duller and contrasts less with grey lower neck and breast, throat is only slightly paler and body is browner. In flight, white wing patch is broken by black bar, unlike on Goosander. **HH** Habits similar to Goosander. Coastal waters.

Mandarin Duck *Aix galericulata* 41–49cm

One late 20th-century record from NE. **ID** Male is spectacular. Most striking features are reddish bill, orange 'mane' and 'sails', white stripe behind eye, and black-and-white stripes on side of breast. Female and eclipse male are mainly greyish with white 'spectacles' and white spotting on breast and flanks. In flight, shows dark upperwing and underwing, with white trailing edge, and white belly. **HH** Large wetlands.

Greater Scaup *Aythya marila* 40–51cm

Four late-20th century records from NE haors. **ID** Larger and stockier than Tufted Duck (Plate 6), with more rounded head lacking any sign of crest. Bill is larger and wider, and has smaller black nail at tip than Tufted. Male has grey upperparts contrasting with black rear end and green gloss to blackish head. Female has broad white face patch, which is less extensive on juvenile/immature. Female usually has greyish-white vermiculations ('frosting') on upperparts and flanks. Eclipse/immature male has brownish-black head, neck and breast, and variable patch of grey on upperparts. **HH** Feeds mainly by diving, loafs in open water when not feeding. Large lakes and rivers.

Baikal Teal *Sibirionetta formosa* 39–43cm

Six records from NE, NW and C. **ID** Male has striking head pattern, black-spotted pinkish breast, white vertical stripe down sides of breast, black undertail-coverts and chestnut-edged scapulars. Female has complex (albeit variable) head pattern: typical birds show dark-bordered white loral spot, buff supercilium broken above eye by dark crown, and white throat which curves up to form half-moon-shaped cheek-stripe. Both sexes have grey forewing and broad white trailing edge to wing in flight (recalling Northern Pintail, Plate 8). Eclipse male is similar to female but has darker and more rufous fringes to mantle, rufous breast and flanks, and less well-defined loral spot. **HH** Large wetlands. **TN** Formerly in genus *Anas* but now in a monospecific genus.

Red-necked Grebe *Podiceps grisegena* 40–50cm

Four 21st-century records from SW and NE. **ID** Slightly smaller than Great Crested Grebe (Plate 8) with stouter neck, squarer head, and stockier body which is often puffed up at rear end. Black-tipped yellow bill. Unlike Great Crested, Red-necked often leaps clear of water when diving. Black crown extends to eye (including lores); dusky cheeks and foreneck in non-breeding plumage. Whitish cheeks and reddish foreneck in breeding plumage. Juvenile is similar to non-breeding but has brown striping on cheeks and rufous foreneck. **HH** Similar to Great Crested Grebe. Wetlands.

Black-necked Grebe *Podiceps nigricollis* 28–34cm

Two 21st-century records from NE and C. **ID** Steep forehead, with crown typically peaking at front or centre. In non-breeding plumage, head pattern more contrasting than Little (Plate 8); black of crown extends below eye, ear-coverts are dusky grey, and striking white throat curves up behind ear-coverts. Has yellow ear-tufts, black neck and breast, and rufous flanks in breeding plumage. Juvenile as non-breeding but may show buff wash to cheeks and foreneck, and more closely resembles Little. **HH** Similar to Little Grebe. Large wetlands.

Greater Flamingo *Phoenicopterus roseus* 125–145cm

Two 21st-century records in NW and SE. **ID** Larger than Lesser Flamingo *Phoeniconaias minor* (not recorded in Bangladesh), with longer, thinner neck and longer legs. Bill larger and less prominently kinked. Adult from adult Lesser by paler pink bill with prominent dark tip, and pink facial skin. Has pinkish-white head, neck and body, although Lesser can be similar. Comparatively uniform crimson-pink upperwing-coverts contrast in flight with paler, whitish body. Immature has greyish-white head, neck and body; brown-streaked coverts, bill grey tipped black, and legs grey (pinker with age). Juvenile has brownish head, neck and body, with heavy brown streaking to upperparts. **HH** Immerses head in shallow water with bill inverted and filters food. Shallow brackish lakes, mudflats and saltpans; coastal and riverine islands in Bangladesh.

Red-billed Tropicbird *Phaethon aethereus* 48cm

Two 21st-century records. **ID** Adult has red bill, white tail-streamers, black barring on mantle and scapulars, and much black on primaries. Juvenile has yellow bill with black tip, and black band across nape; shows more black on primaries, with black primary coverts, compared with juveniles of extralimital Red-tailed and White-tailed Tropicbirds *P. rubricauda* and *P. lepturus*. **HH** Aerial seabird. Has graceful pigeon-like flight with flapping and circling alternating with long glides. Feeds by first hovering to locate fish or squid, then plunge-diving on half-closed wings. Pelagic.

Pale-capped Pigeon *Columba punicea* 36cm

Three records in NE in 1980s. **ID** Pale cap, vinous-chestnut underparts, and maroon-brown mantle and wing-coverts with green-and-purple gloss. Sexes are similar, but female is darker with darker grey crown. Juvenile initially lacks cap; upperparts browner, with chestnut fringes, and underparts mixed grey and rufous-buff. **Voice** A soft mew. **HH** Evergreen forest. **Threatened** Globally (VU) and nationally (CR).

Laughing Dove *Spilopelia senegalensis* 27cm

One late 20th-century record in NE. **ID** Slim, small, with fairly long tail. Brownish-pink head and underparts, uniform upperparts and black stippling on upper breast. Juvenile duller, lacks black stippling, and has whitish fringes to scapulars and coverts. **Voice** A soft *coo-rooroo-rooroo* or *cru-do-do-do-do*. **HH** Similar to those of Oriental Turtle-dove (Plate 9). Dry cultivation and scrub-covered hills. **TN** Formerly placed in genera *Streptopelia* and *Stigmatopelia*.

Barred Cuckoo-dove *Macropygia unchall* 41cm

One 21st-century record in NE. **ID** Long, graduated tail, slim body and small head. Face, belly and vent pale. Upperparts and tail rufous, barred with dark brown. Male has unbarred head and neck with extensive purple-and-green gloss. Female is heavily barred on head, neck and underparts, with gloss restricted to nape and sides of neck. Juvenile is more uniformly dark and heavily barred. **Voice** A very deep *croo-umm*, the second syllable a booming note, audible a long way off; heard in the distance as low, muffled single *umm*, repeated at short intervals. **HH** Usually found in pairs and small flocks. Feeds on berries, acorns and shoots in forest trees, clambering about and sometimes swinging upside-down to reach food items. Evergreen forest.

Crested Treeswift *Hemiprocne coronata* 23cm

Five records from C and SW. **ID** Large size with sickle-shaped wings and long, deeply forked tail which is usually held closed and pointed in flight. In flight, appears mainly blue-grey with darker upperwing and tail, and whitish abdomen and undertail-coverts. At rest, both sexes show prominent dark green-blue crest, and wing-coverts are glossed with blue. Male has dull orange ear-coverts. Female has dark grey ear-coverts, forming dark mask, bordered below by whitish moustachial stripe. Juvenile has extensive white fringes to upperparts (especially noticeable on lower back and rump); feathers of underparts are fringed with white and have grey-brown subterminal bands. **Voice** Harsh *whit-tuck... whit-tuck* in flight. **HH** Perches readily in trees. Typically flies above canopy of open forest with mixture of rapid, rather heavy fluttering and periods of banking and gliding.

White-rumped Spinetail *Zoonavena sylvatica* 11cm

One late 20th-century record from SW. **ID** Small and stocky, with broad wings, pinched in at base and pointed at tip. Flight is fast with rapid wingbeats, banking from side to side, interspersed with short glides on slightly bowed wings. Upperparts mainly blue-black with contrasting white rump; throat and breast are grey-brown, merging into whitish lower belly. Long white undertail-coverts contrast with black of sides and tip of undertail. 'Spines' at tip of tail visible at close range. Wing shape and flight action different from House Swift (Plate 13), and lacks white throat. **Voice** Twittering *chick-chick* in flight. **HH** Usually in flocks. Hawks over forest with great manoeuvrability.

Silver-backed Needletail *Hirundapus cochinchinensis* 20cm

Two late 20th-century records in SW. **ID** From larger Brown-backed Needletail (Plate 13) by pale brown or grey to greyish-white throat. Lacks white patch on lores. Juvenile has dark fringes to white undertail-coverts. **Voice** Soft rippling *trp-trp-trp-trp-trp*. **HH** Extremely fast and powerful flight. Mainly hawks over forest and forested hills.

Sirkeer Malkoha *Taccocua leschenaultii* 42cm

Recorded in early 20th-century in SE and possibly in NW. **ID** Adult mainly sandy grey-brown, with yellow-tipped red bill and white-edged dark facial skin giving masked appearance. Black shaft streaking on crown, mantle and breast, throat buff and belly rufous-buff. Has long, graduated, white-tipped tail. Immature very similar, but has indistinct buff fringes to wing-coverts, scapulars and tertials. Juvenile has broad dark brown streaking on head, mantle, throat and breast, and buff fringes to mantle, wing-coverts and tertials. **Voice** Normally silent. **HH** Largely terrestrial. Sometimes clambers among shrubs and small trees or hops from branch to branch. Thorn scrub and acacia bushes. **TN** Formerly placed in genus *Phaenicophaeus*.

Whistling Hawk-cuckoo *Hierococcyx nisicolor* 29cm

One record from C in 1980s. **ID** Smaller than Common Hawk-cuckoo (Plate 16), with stouter bill. Upperparts are darker slate-grey, with slate-grey chin, more extensive rufous on underparts, and unbarred white belly and flanks. Throat and breast may show dark grey streaking. Has more pronounced rufous tip to tail, and frequently shows a single pale inner tertial (on both wings on some) not present on Common. Juvenile has darker brown and more uniform upperparts than juvenile Common, and broader (squarer) spots on underparts. Immature has dark grey chin, ear-coverts and crown, rufous barring to upperparts and strongly streaked underparts. **Voice** Shrill repeated *gee-whiz*, becoming more frantic and high-pitched. **HH** Usually in low trees or bushes but moves higher when calling. Broadleaved forest. **TN** Hodgson's Hawk-cuckoo *H. fugax* recently considered to comprise several species, with northern migratory populations to which this record belongs now treated as Whistling Hawk-cuckoo *H. nisicolor*.

Oriental Cuckoo *Cuculus saturatus* 30–32cm

One record in NE in 1980s. **ID** Compared with Eurasian Cuckoo (Plate 16), has broader black barring on buffish-white (rather than pure white) underparts, and darker grey upperparts, sometimes showing contrast with paler head, and can be distinctly smaller. Female and juvenile, like Eurasian, occur as grey and rufous morphs, but have broader black barring, especially on breast, back, rump and tail than in Eurasian. **Voice** Resonant *ho... ho... ho... ho* very similar to Common Hoopoe (Plate 52). **HH** Forest and well-wooded country. **TN** 'Himalayan Cuckoo' *C. saturatus*, to which this record refers, is treated by some authorities as a separate species from Palearctic breeding form 'Oriental Cuckoo' *C. optatus*.

Lesser Cuckoo *Cuculus poliocephalus* 25cm

Four records in NE and SE. **ID** Generally smaller than Oriental (although Oriental can be similar-sized) with finer bill. Plumage almost identical, but has darker rump and uppertail-coverts, contrasting less with tail. Hepatic morph of female prevails and is typically more rufous than hepatic Oriental (some with almost unmarked rufous crown, nape, rump and uppertail-coverts). Juvenile like juvenile Oriental but with dark grey-brown upperparts and whiter (broadly barred) underparts. **Voice** Strong, cheerful *pretty-peel-lay-ka-beet*. **HH** Forest and well-wooded country.

Spotted Crake *Porzana porzana* 22–24cm

One 21st-century record from NE. **ID** Profuse white spotting on head, neck and breast. Stout bill, irregularly barred flanks and unmarked buff undertail-coverts. Adult has yellowish bill with red at base, and grey head and breast. Sexes similar, but female has less grey on head, neck and breast, and is more profusely spotted with white. Juvenile similar to female but with buffish-brown head and breast, and bill is brown. **Voice** Song may be heard on migration: a swishing *h-wet... hwet,* resembling whiplash. **HH** Reed-swamp bordering large wetlands.

Demoiselle Crane *Anthropoides virgo* 90–100cm

One record in NE in 1980s. **ID** Small crane, with short, fine bill. Adult has black head and neck with white tuft behind eye, and grey crown; black neck feathers extend as point beyond breast, and elongated tertials project as shallow arc beyond body, giving rise to distinctive shape. Immature is initially almost entirely grey, with slate-grey on foreneck and shorter all-grey tertials. By first winter (on arrival in Subcontinent) is similar to adult, but head and neck are dark grey and less contrasting, tuft behind eye is grey and less prominent, upperparts have brown cast, and elongated feathers of foreneck and tertials are shorter. In flight, black breast helps separate from Common Crane at distance; also, legs and neck appear relatively shorter, and wings are shorter and broader-based. **Voice** A *garrooo* flight call, higher pitched than Common. **HH** Cultivation and large rivers. **TN** Often included in genus *Grus*.

Common Crane *Grus grus* 110–120cm

Three 21st-century records on coast and in NW. **ID** Adult has mainly black head and foreneck, with white stripe behind eye extending down side of neck; red patch on crown is visible at close range. Immature has brown markings on upperparts with buff or grey head and neck; adult head pattern apparent on some by first winter and is as adult by second winter. In flight, both adult and immature show black primaries and secondaries, which contrast with grey wing-coverts. **Voice** Loud, trumpeting *krrooah* flight call; similar bugling calls including typical *kroo-krii-kroo-krii* duet on ground. **HH** Coast and large rivers.

Short-tailed Shearwater *Ardenna tenuirostris* 41–43cm

Two 21st-century records from SE. **ID** Sooty-brown, with pale grey underwing-coverts. Distinguished from very similar but extralimital Sooty Shearwater *A. grisea* by more slender and shorter bill; steeper rounded forehead; shorter extension of rear body and tail behind wings, with feet extending beyond tail in flight; pale throat and lower breast; and less striking pale panel on underwing. Flight action is strong and deliberate, with comparatively fast, stiff-winged flapping followed by long glides. In strong winds, banks high above ocean in steep arcs. **HH** Coastal waters. **TN** Formerly in genus *Puffinus*.

Greater Adjutant *Leptoptilos dubius* 120–150cm

Historic records from C and NW. **ID** Larger than Lesser Adjutant (Plate 20); stouter, conical bill has convex ridge to culmen. Adult breeding from adult Lesser Adjutant by larger size, bluish-grey (rather than glossy black) mantle, prominent silvery-grey panel across greater coverts and tertials, mainly white neck-ruff (lacking or with less pronounced black patch on sides of breast in flight) and grey undertail-coverts. Further, has blackish face and forehead (with appearance of dried blood), more sparsely feathered head and neck (lacking small crest) and larger neck pouch (visible only when inflated). Non-breeding adult plumage has darker grey mantle and wing-coverts (which barely contrast with rest of wing). Immature is similar to non-breeding adult, but upperparts including wings are browner, iris is brownish (rather than whitish), and head and neck are more densely feathered. **HH** Usually seen singly. Habits are similar to those of Lesser Adjutant but feeds partly on carrion. Marshes and open fields. **Threatened** Globally (EN).

White Stork *Ciconia ciconia* 100–125cm

Historic record from C, 20th-century record in NE. **ID** Mainly white, with black flight feathers and striking red bill and legs. Generally has cleaner black-and-white appearance than Asian Openbill (Plate 20); note tail is white (black in Asian Openbill). Juvenile is similar to adult but with brown greater coverts and duller brownish-red bill and legs. **HH** Habits are similar to those of other storks. Flight is a few flaps followed by glide; appears leisurely but is fast and strong. Stalks deliberately on dry or moist ground in search of prey. Grassland and fields.

Eurasian Bittern *Botaurus stellaris* 70–80cm

Four records in NE. **ID** Stocky, broad-winged, with stout-looking neck and head. Golden-brown and cryptically patterned, with boldly streaked neck and breast, and black crown and moustachial stripe. In flight, wing-coverts appear paler than brown (barred) flight feathers. **HH** Habits are those of typical bittern. Normally crepuscular. Usually remains hidden in aquatic vegetation and is most often seen briefly flying low over wetlands. Freezes when in danger, pointing bill and neck upwards. Large well-vegetated wetlands. **AN** Great Bittern.

White-bellied Heron *Ardea insignis* 127cm

One record in 1980s in NE. **ID** Large size, very long neck, huge dark bill, and dark legs and feet. Grey head with white throat, and white-striped grey foreneck and breast contrast with white belly. In flight, has uniform dark grey upperwing, and white underwing-coverts contrast with dark grey flight feathers. In breeding plumage, has greyish-white nape plumes, grey back plumes and white-striped breast plumes; lores and orbital skin are yellowish-green. Juvenile has browner upperparts, with streaked appearance. **Voice** Has loud, very donkey-like croaking bray, *ock, ock, ock, ock, urrrrr.* **HH** Shy. Found singly or in pairs. Foraging methods are similar to those of other diurnal herons. Rivers, marshes and lakes in forest. **Threatened** Globally (CR).

Goliath Heron *Ardea goliath* 135–150cm

Three 20th-century records from coast and C. **ID** Recalls giant Purple Heron (Plate 23); in all plumages, distinguished from that species by much larger size, thicker head and neck, huge and thick bill, and dark legs and feet. Adult has rufous head and neck (lacking black head-stripes of Purple), dark bill and lores (yellow in Purple), grey upperparts, broken blackish stripes down foreneck, and deep purplish-chestnut underparts and underwing-coverts. Immature is similar to adult, but has dark grey forehead, yellow on bill and lores, chestnut underparts streaked with buff, dark grey flanks, and dark grey mottling on underwing-coverts. Juvenile has black forehead and crown, paler rufous hindneck and indistinct neck-stripes (compared with adult), rufous fringes to feathers of mantle and upperwing-coverts, grey underparts streaked with pale chestnut, and dark grey underwing-coverts. **HH** Diurnal, solitary and shy. Mangroves, wetlands.

Pacific Reef-egret *Egretta sacra* 58cm

At least two records from SE coast. **ID** Legs shorter and stouter than Little Egret (Plate 24, with short leg/feet extension in flight), and is stockier, with shorter and thicker neck. Occurs in dark grey, intermediate and white colour morphs. **Voice** A grunted *ork* when feeding and harsh *squak* when disturbed. **HH** Rocky coasts, coral beds and sandy shores. Feeds by stalking prey in shallow water and rock pools. **AN** Pacific Reef Heron.

Great White Pelican *Pelecanus onocrotalus* 40–175cm

Two 20th-century records in SE and NE. **ID** Adult and immature have black underside to primaries and secondaries which contrast strongly with white (or largely white) underwing-coverts. Orbital skin is more extensive and contiguous with bill compared with Spot-billed Pelican (Plate 25). Pouch yellow or orange-yellow (except when young). Adult is generally cleaner and whiter than Spot-billed. Adult breeding has white body and wing-coverts tinged with pink, bright orange-yellow pouch and pinkish skin around eye, and short drooping crest. Adult non-breeding has duller bare parts and lacks pink tinge and white crest. Immature has variable amounts of brown on wing-coverts and scapulars. Juvenile has largely brown head, neck and upperparts; upperwing appears more uniform brown, and underwing shows pale central panel contrasting with dark inner coverts and flight feathers; greyish pouch becomes yellower with age. **HH** Large lakes.

Lesser Frigatebird *Fregata ariel* 70–80cm

One 21st-century record from SW. **ID** Smaller and more finely built than the other two frigatebirds occurring in South Asia, which are unrecorded in Bangladesh. Adult male entirely black except for white spur extending from breast-sides onto inner underwing. Adult female has black head including throat, white neck-sides, white spur extending from white breast onto inner underwing, and black belly and vent. Juvenile and immature have rufous or white head, blackish breast-band and much white on underparts, which are gradually replaced by adult plumage; always show white spur on underwing (lacking on unrecorded Great Frigatebird *F. minor*). **HH** A real seabird, rarely landing on water; powerful and agile flyer, and can soar for long periods. Pelagic.

Masked Booby *Sula dactylatra* 81–92cm

One 21st-century record from coastal waters. **ID** Large and robust booby. Adult largely white, with black mask and black flight feathers and tail. Juvenile has brown head, neck and upperparts, with whitish collar and whitish scaling on upperparts; underparts white and underwing-coverts show much white. Head, upperparts and upperwing-coverts of immature become increasingly white with age. **HH** Forages on the wing and plunge-dives for food. Flight is direct, with alternating periods of flapping and gliding. Pelagic.

Long-billed Plover *Charadrius placidus* 19–21cm

Four late 20th-century records from central coast, SE and NE. **ID** Like large Little Ringed Plover (Plate 27) but has longer tail with clearer dark subterminal bar, and more prominent white wingbar; ear-coverts never black and eye-ring less distinct than Little Ringed. White forehead and supercilium more prominent in non-breeding plumage compared with Little Ringed. **Voice** A clear, penetrating *piwee* in flight. **HH** Habits are similar to those of Little Ringed, but usually solitary. Large wetlands.

Oriental Plover *Charadrius veredus* 22–25cm

One 21st-century record from central coast. **ID** Medium-sized, long-legged and long-winged plover. In non-breeding plumage has prominent buffish supercilium and brownish-buff breast. Adult male breeding plumage has mainly whitish head and neck and chestnut breast with black lower border. From similar Caspian Plover *C. asiaticus* (not recorded in Bangladesh) by larger size, longer neck and legs, brown underwing-coverts and axillaries (giving uniformly dark underwing) and uniformly dark upperwing typically lacking any sign of white wingbar. Yellowish or pinkish legs. **Voice** Sharp, whistled *chip-chip-chip*. **HH** Short grassland.

White-tailed Lapwing *Vanellus leucurus* 26–29cm

One 21st-century record from NW. **ID** Blackish bill, large dark eyes and very long yellow legs. Plain sandy-brown head and upperparts, and white underparts. In flight, white secondaries contrast with black primaries. Tail all white, lacking black band of other *Vanellus* plovers. Juvenile has dark subterminal marks and pale fringes to feathers of upperparts, and paler neck and breast than adult. **Voice** Calls include *pet-oo-wit* and *pee-wick*. **HH** Usually feeds in shallow water by pecking at surface and by foot-dabbling. Freshwater marshes and marshy lake edges.

Far Eastern Curlew *Numenius madagascariensis*   60–66cm

One 20th-century record from SE coast. **ID** Large size and very long, downcurved bill. Dark back and rump and heavily barred underwing-coverts and axillaries are best features from Eurasian Curlew (Plate 30). Underparts washed with buff in adult breeding and juvenile (ground colour of Eurasian's breast and belly is whiter), although underparts in non-breeding adult are paler and more like Eurasian. **Voice** Typically flatter *coor-ee* compared with Eurasian. **HH** Coastal. **AN** Eastern Curlew. Threatened Globally (EN).

Long-billed Dowitcher *Limnodromus scolopaceus*   27–30cm

One 21st-century record in NE. **ID** Rather snipe-like in shape and feeding action. Superficially resembles Bar-tailed Godwit or Asian Dowitcher (both Plate 30) but is smaller and has shorter legs (greyish, yellowish or greenish, rather than black). In flight, shows clear white trailing edge to wing, barred rump and tail, and striking white back. In all plumages has pronounced white supercilium. In breeding plumage has rufous underparts, with some barring and spotting, and dark upperparts have narrow rufous fringes. In non-breeding plumage, has grey upperparts and breast, and white belly. Juvenile recalls non-breeding adult, but has rufous fringes to mantle and scapulars, and buff wash to underparts. From non-breeding Short-billed Dowitcher *L. griseus* (unrecorded in Subcontinent) by lack of primary projection beyond tertials, long bill and little flank barring. **Voice** Call is high, thin *keek*. **HH** Large freshwater wetlands.

Eurasian Woodcock *Scolopax rusticola*   33–35cm

One 19th-century record and one in 20th century in NE. **ID** Bulky and rufous-brown in coloration with broad, rounded wings. Crown and nape banded black and buff; lacks sharply defined mantle and scapular stripes. **Voice** Usually silent when flushed; sometimes gives harsh *schaap*. **HH** Solitary, crepuscular and nocturnal, passing the day in thick cover. When flushed, flies off without calling with wings making swishing sound; zigzags away with wavering wing movements and quickly drops into cover again. Moist forest with thick undergrowth and marshy glades.

Wood Snipe *Gallinago nemoricola*   28–32cm

One record from 1990s in NE. **ID** Large, with heavy and direct flight and broad wings. Bill relatively short and broad-based. Boldly marked, with buff and blackish head-stripes, broad buff stripes on blackish mantle and scapulars (white in juvenile), and warm buff neck and breast with brown streaking. Legs greenish. **Voice** When flushed, *che-dep, che-dep*. **HH** If flushed, has slow, heavy, wavering flight and soon settles. Forest marshes. **Threatened** Globally (VU).

Swinhoe's Snipe *Gallinago megala*   25–27cm

One record from 1990s in NE. **ID** Almost identical to Pintail Snipe (Plate 33). In flight, is heavier, with longer bill and more pointed wings, and feet only just project beyond tail. At rest, tail extends noticeably beyond folded wings. Some birds in spring can appear quite dusky on sides of head, neck, breast and flanks (and are distinct from Pintail), but plumage often as Pintail. **Voice** Can be silent when flushed. Call similar in pitch to Pintail, but thinner. **HH** Habits and habitat very similar to Pintail.

Jack Snipe *Lymnocryptes minimus*   17–19cm

19th-century record and at least six records since 1990 in NE and NW. **ID** Small, with short bill. Flight weaker and slower than that of Common Snipe (Plate 33), with rounded wingtips. Has divided supercilium but lacks pale crown-stripe. Mantle and scapular stripes very prominent. **Voice** Invariably silent when flushed. **HH** When feeding has characteristic habit of bobbing its body. Usually crepuscular and nocturnal like other snipes. Marshes and wet fields.

Red-necked Phalarope *Phalaropus lobatus*   18–19cm

One 21st-century record (of a small flock) in NW. **ID** Typically seen swimming. More delicately built than Red Phalarope, with finer bill. Adult breeding has white throat and red stripe down side of grey neck. Adult non-breeding has prominent black mask and cap, with dark line running down hindneck; has darker grey upperparts than Red, with white edges to mantle and scapular feathers forming fairly distinct lines. Juvenile has dark grey upperparts with orange-buff mantle and scapular lines. **Voice** Utters single *twick*. **HH** Swims buoyantly, spins around, darts erratically here and there. Winters at sea; on migration also on saltpans, rivers, shallow pools and lakes.

Red Phalarope *Phalaropus fulicarius* 20–22cm

One 21st-century record in C. **ID** Typically seen swimming. Stockier than Red-necked with stouter bill that is often pale or yellowish at base. Breeding adult has red neck and underparts, and white face patch. Non-breeding adult has more uniform and paler grey mantle, scapulars and rump than Red-necked. Juvenile has dark upperparts evenly fringed with buff (lacking mantle and scapular stripes of Red-necked). **HH** Similar to Red-necked. On migration, wetlands and large rivers. **AN** Grey Phalarope.

Crab-plover *Dromas ardeola* 38–41cm

One 20th-century record from central coast. **ID** Black-and-white plumage, with stout black bill and very long blue-grey legs. Juvenile like washed-out version of adult with black streaking on nape and grey rather (than black) mantle. **Voice** Nasal, yappy *kirruc* flight call; also, *kwerk-kwerk-kwerk-kwerk*. **HH** Mainly crepuscular and usually very wary. Feeds chiefly on crabs. Flies with neck and legs extended. Intertidal mudflats.

Black-legged Kittiwake *Rissa tridactyla* 38–40cm

One 21st-century record from central coast. **ID** In all plumages outer primaries on upper and underwing are black-tipped, underwing is white. Adult has white head in summer with grey hindneck, dark grey upperparts shading to whitish before black tip; yellowish bill and blackish legs. First-year plumage has dark 'W' pattern across wings and black tail-band; juvenile and some first-winter birds have white head with black ear-spot and black half-collar; 'W' is much faded in first-summer. **HH** Pelagic. **Threatened** Globally (VU).

Black-naped Tern *Sterna sumatrana* 35cm

One 20th-century record from SE coast. **ID** Adult very pale greyish-white, with black bill and legs, and black mask and nape-band. Has whiter mantle and wings than Common (Plate 39), with distinct black outer edge to outermost primary, and lacks obvious white trailing edge to upperwing. Nape-band is paler and not so well defined in non-breeding plumage. Juvenile has black subterminal marks to upperpart feathers, and black streaking on crown. **Voice** Call is a sharp *tsii-chee-ch-chip* and a hurried *chit-chit-chit-er.* **HH** Inshore waters around islands and lagoons.

Sandwich Tern *Thalasseus sandvicensis* 36–41cm

One 21st-century record from central coast. **ID** Slim black bill with yellow tip, and more rakish appearance than Common Gull-billed (Plate 38) with narrower, more pointed wings appearing set forward and sharply angled, and longer, tapering body. White rump and tail contrast with greyer back. Adult breeding plumage has black cap with crest. Adult non-breeding has white forehead and crown, and black crest forming U-shaped patch. First-winter and first-summer plumages as adult non-breeding, but with darker lesser-covert and secondary bars, and dark corners to tail. Juvenile more heavily marked than juvenile Common Gull-billed, with dark subterminal bars to wing-coverts and mantle, and dark pattern to tertials; has black rear crown and nape lacking in juvenile Common Gull-billed. **Voice** Gives an upward-inflected hoarse *kree-it.* **HH** A marine tern; frequents coasts, tidal creeks and open sea. **TN** Formerly placed in genus *Sterna*.

Arctic Jaeger *Stercorarius parasiticus* 45cm

One 1980s record from coastal waters. **ID** Smaller and more lightly built than Pomarine Jaeger (Plate 37) with slimmer bill and narrower-based wings. Adult breeding has pointed tip to elongated central tail feathers. Occurs in both pale and dark morphs. Adult non-breeding plumage is as Pomarine but has more pointed tail-tip. Juvenile more variable than juvenile Pomarine, ranging from grey and buff with heavy barring to completely blackish-brown, and many have rusty-orange to cinnamon-brown cast to head and nape (not found on Pomarine); except for all-dark juveniles, further distinctions from Pomarine are dark streaking on head and neck, and pale tips to primaries. **HH** Often associates with terns and gulls. Normal flight is buoyant, with jorky wingbeats alternating with glides; swift, dashing and falcon-like when in pursuit. Coastal waters, often near mouths of major creeks, rivers and lagoons with roosting and feeding gulls and terns. **AN** Parasitic Jaeger, Arctic Skua.

Egyptian Vulture *Neophron percnopterus* 60–70cm

Two records from NW and SE. **ID** Small vulture with long, pointed wings, small and pointed head, and wedge-shaped tail. Adult mainly dirty white, with bare yellowish face and black flight feathers. Juvenile blackish-brown with bare grey face. With maturity, tail, body and wing-coverts become whiter and face yellower. **HH** An opportunistic scavenger; less dependent on large carcasses than other vultures. Spends the day soaring and gliding in search of food and perched around habitation and rubbish dumps. Gathers at carcasses and rubbish dumps with other vultures. **Threatened** Globally (EN).

Short-toed Snake-eagle *Circaetus gallicus* 62–67cm

Six records since 1970s from C, SW, SE and NE. **ID** Long and broad wings, pinched in at base, and rather long tail. Head broad and rounded. Soars with wings flat or slightly raised; frequently hovers. When perched, appears big-headed, with wingtips reaching tail-end, and shows long, unfeathered tarsus. Plumage variable, often with dark head and breast, barred underparts, dark trailing edge to underwing, and broad subterminal tail-band; can be very pale on head, underparts and underwing. On upperside, in all plumages, pale brown inner wing-coverts contrast with dark greater coverts and flight feathers. Juvenile similar in plumage to adult. **HH** Spends most of day soaring, searching for prey. Typically hunts 15–30m above ground, occasionally higher, frequently hovering with gently beating wings; on spotting prey, plummets down almost vertically. Feeds mainly on snakes. Open country and wetlands.

Griffon Vulture *Gyps fulvus* 95–105cm

Three records from NE. **ID** Larger than Slender-billed Vulture (Plate 44), with stockier head and neck and stouter bill. Key features of adult are yellowish bill with blackish cere, whitish head and neck, fluffy white ruff, rufescent-buff upperparts, rufous-brown underparts and thighs with prominent pale streaking, and dark grey legs and feet. Rufous-brown underwing-coverts usually show prominent whitish banding, especially across median coverts. Immature is richer rufous-brown on upperparts and upperwing-coverts (with prominent pale streaking) than adult; has rufous-brown feathered neck-ruff, more whitish down covering grey head and neck, blackish bill, and dark iris (pale yellowish-brown in adult). **HH** Seen at carcasses, often with other vultures. Regularly makes long flights in search of carcasses. Open country.

Bonelli's Eagle *Aquila fasciata* 65–72cm

One late 20th-century record in NE. **ID** Medium-sized eagle with long and broad wings, distinctly protruding head, and long square-ended tail. Soars with wings flat. Adult has pale underparts and forewing, blackish carpals and band along underwing-coverts, greyish underside to flight feathers, whitish patch on mantle, and pale greyish tail with broad dark terminal band. Juvenile has pronounced curve to trailing edge of wing. Has ginger-buff to reddish-brown underparts and underwing-coverts (with variable dark band along greater coverts), and narrow greyish barring on underside of wings and tail, which both lack dark trailing edge. Also shows pale inner primaries on underwing, uniform upperwing, pale crescent across uppertail-coverts and patch on back. **Voice** Usually rather silent; mellow fluting calls, *klu-klu-klu-kluee* or *kluu-klu-klu-klu-klu* in display and near nest. **HH** Forested hills. **TN** Formerly placed in genus *Hieraaetus*.

Pallid Harrier *Circus macrourus* 40–48cm

19th-century records and four since 1990 in NW, NE and SE. **ID** Slim-winged and fine-bodied, with buoyant flight. Folded wings fall short of tail-tip and legs longer than on Montagu's. Male has pale grey upperparts, dark wedge on primaries, very pale grey head and underparts, and lacks black secondary bars. Immature male may show rusty breast-band and juvenile facial markings. Female has distinctive underwing pattern: pale primaries, irregularly barred and lacking dark trailing edge, contrast with darker secondaries which have pale bands narrower than on female Montagu's and tapering towards body (although first-summer Montagu's is more similar in this respect), and lacks prominent barring on axillaries. Typically, female has stronger head pattern than Montagu's, with more pronounced pale collar, dark ear-coverts and dark eyestripe, and upperside of flight feathers is barred and lacks banding; from female Hen (Plate 49) by narrower wings with more pointed hand, stronger head pattern and pattern on underside of primaries. Juvenile has unstreaked orange-buff underparts and underwing-coverts; on underwing, primaries evenly barred (lacking pronounced dark fingers), without dark trailing edge, and usually has pale crescent at base; head pattern more pronounced than Montagu's, with narrower white supercilium, more extensive dark ear-covert patch and broader pale collar contrasting strongly with dark neck-sides. **HH** Quarters ground with buoyant flapping, interspersed with gliding on raised wings, occasionally dropping into vegetation to catch prey. Open grassy areas near wetlands.

Montagu's Harrier *Circus pygargus* 43–47cm

19th-century record and four since 1990 in NW, C and SE. **ID** Folded wings reach tail-tip and legs are shorter than on Pallid. Male has black band across secondaries, extensive black on underside of primaries, and rufous streaking on belly and underwing-coverts. Female differs from female Pallid in distinctly and evenly barred underside to primaries with dark trailing edge, broader and more pronounced pale bands across secondaries, barring on axillaries, less pronounced head pattern, and distinct dark banding on upperside of remiges. Juvenile has unstreaked rufous underparts and underwing-coverts, and darker secondaries than female; differs from juvenile Pallid in having broad

dark fingers and dark trailing edge to hand on underwing, and paler face with smaller dark ear-covert patch and less distinct collar. **HH** Similar to Pallid Harrier.

Japanese Sparrowhawk *Accipiter gularis* 25–31.5cm

Two 21st-century records in NE. **ID** Very small. Long primary projection (extending nearly halfway down tail). In flight, wings a shade narrower and more pointed than on Eurasian Sparrowhawk and Besra (both Plate 49). Underside of flight feathers and underwing-coverts are distinctly barred. In all plumages, pale bars on tail generally broader than dark bars (reverse on Besra). Underpart pattern of adults rather different from Besra; juveniles more similar. Male has dark bluish-grey upperparts, pale rufous to pale grey underparts (some with fine grey barring), and dark crimson iris. Has very indistinct gular stripe (often not visible in field). Female has browner upperparts, whitish underparts with distinct greyish-brown barring, indistinct gular stripe, and yellow iris. Juvenile has dark greyish-brown upperparts with narrow rufous fringes, white supercilium, brown to rufous streaking on breast, heart-shaped spots on lower breast and belly, and barring on flanks and thighs. **HH** Habits are like Besra. Secondary forest.

Northern Goshawk *Accipiter gentilis* 50–61cm

Four records in C and SE. **ID** Very large, with heavy, deep-chested appearance. Wings comparatively long, with bulging secondaries. Male has grey upperparts (greyer than female Eurasian Sparrowhawk, Plate 49), white supercilium and finely barred underparts. Female considerably larger with browner upperparts. Juvenile has heavily streaked buff underparts. **Voice** Most common call is shrill chatter, and female has disyllabic *hee-aa*. **HH** Hunting habits similar to Eurasian Sparrowhawk, but more powerful. Open wooded areas.

White-tailed Eagle *Haliaeetus albicilla* 70–90cm

One late 20th-century record in NW. **ID** Huge, with broad parallel-edged wings, short wedge-shaped tail, protruding head and neck, and heavy bill. Soars and glides with wings level. Adult has yellow bill, pale head and white tail. Juvenile is mainly blackish-brown with whitish centres to tail feathers, pale patch on axillaries and variable pale band across underwing-coverts; bill becomes yellow with age. **Voice** Rather vocal; usual call is quick series of metallic yapping notes. **HH** Spends most of time perched on stump or ground near water. Captures fish, its main prey, by flying low over water and seizing them near surface; also, occasionally hunts ducks and small mammals by flying low along shores. Large rivers.

Indian Grey Hornbill *Ocyceros birostris* 50cm

Two 21st-century records in NW. **ID** Small hornbill with sandy-grey upperparts. Broad greyish-white supercilium with dark grey ear-coverts. Has prominent blackish casque, and extensive black at base of bill. Tail has white tips, dark grey subterminal band and elongated central feathers. In flight, shows white tips to primaries and secondaries. Female is similar to male but has smaller casque with less pronounced tip. Immature has bill as female's, but smaller and with smaller casque; lacks white wing-tips. **Voice** Calls include loud cackling *k-k-k-ka-e*, rapid piping *pi-pi-pi-pi-pipipieu-pipipieu-pipipieu* and kite-like *chee-oowww*. **HH** Open wooded areas with fruiting trees.

Wreathed Hornbill *Rhyticeros undulatus* 75–85cm

Two 21st-century records from Chittagong Hill Tracts in SE. **ID** Male has rufous-brown crown and nape, and whitish sides of head, foreneck and upper breast. Tail is white. Has yellow gular pouch (with dark bar) and reddish circumorbital skin. Female is black with white tail. Has blue pouch and reddish circumorbital skin. Both sexes have prominent corrugated casque. In flight, both sexes have all-black wings. Immature is similar to adult male, but initially lacks casque and corrugations on bill; eyes are pale blue rather than red (adult male) or brown (adult female). **Voice** Gives very loud, gasping *uk-hweerk*, with emphasised second note; less harsh and grating than similar calls of Great Hornbill (Plate 52). **HH** Evergreen forest. **TN** Formerly placed in genus *Aceros*. **Threatened** Globally (VU).

Blyth's Kingfisher *Alcedo hercules* 22cm

Three records in NE and SE. **ID** From Common and Blue-eared (both Plate 54) by considerably larger size, larger and longer bill, darker greenish-blue (almost brownish-black) coloration to scapulars and wings (with prominent turquoise spotting on lesser and median coverts), almost blackish crown with sharply contrasting turquoise spotting, and less distinct orange loral spot. Female has red on lower mandible. In flight, dark mantle and wings contrast with brilliant blue back and rump. **Voice** Flight call is shrill, single note, more powerful and with deeper tone than other *Alcedo* kingfishers. **HH** Habits similar to Common, but shy and poorly known. Shaded streams in forest.

Crested Kingfisher *Megaceryle lugubris* 41–43cm

Three 21st-century records from Chittagong Hill Tracts in SE. **ID** Very large with prominent crest, often held open. From much smaller Pied Kingfisher (Plate 54) by lack of white supercilium, complete white neck-collar, finely spotted breast-band (sometimes mixed with rufous), dark grey back and rump finely spotted with paler grey, and dark grey wings and tail finely barred with white (lacking Pied's prominent white wing patches). Female and juvenile are similar to male but have pale rufous underwing-coverts. **Voice** Usually silent, although has squeaky *aik* flight call and loud, hoarse, grating trill. **HH** Keeps to favourite stretches of river; spends most of day perched on branches close to and overhanging water or on rocks at river's edge (or in middle where current is swift). Unlike Pied, does not hover but dives from perch to catch fish. Rocky, fast-flowing hill rivers.

Heart-spotted Woodpecker *Hemicircus canente* 16cm

Two late 20th-century records in NE and SE. **ID** Looks top-heavy both at rest and in flight, with large crested head and very short tail (the latter obscured by wings at rest). Easily identified by prominent black crest, black heart-shaped spotting on white tertials and some coverts and scapulars, and white throat, becoming dusky olive or grey on rest of underparts. Shows white rump in flight. Female has white forehead and forecrown. Juvenile is similar to female, but forehead and forecrown are spotted with black, and feathers of upperparts have narrow whitish fringes. **Voice** Calls include weak, nasal *ki-yew,* high-pitched *kee...kee* and quarrelsome *kirrrick.* **HH** Evergreen forest.

Indian Pygmy Woodpecker *Picoides nanus* 13cm

Two 21st-century records in NW. **ID** From Grey-capped Woodpecker (Plate 59) by brown crown (warmer-coloured than mantle), browner upperparts, and greyish-white to brownish underparts which are streaked (sometimes faintly) with brown. Also, throat is evenly mottled and streaked with brown, and stripe behind eye is brown and concolorous with crown. Male has small crimson patch on each side of hindcrown. **Voice** A rapid, metallic rattle. **HH** Open woodland. **AN** Brown-capped Pygmy Woodpecker.

Yellow-crowned Woodpecker *Leiopicus mahrattensis* 17–18cm

Two records (mid-20th and early 21st centuries) in SE. **ID** Has yellowish forehead and forecrown, white-spotted mantle and wing-coverts, bold white barring on central tail feathers, whitish rump, and diffuse brown moustachial stripe and patch on sides of neck. Underparts are rather dirty grey, with fairly heavy brown streaking, and has small red patch on lower belly. Male has red hindcrown and nape. Female has brownish hindcrown and nape. **Voice** Shrill *peek-peek* calls and rapidly repeated *kik-kik-kik-r-r-r-h.* **HH** Similar to Fulvous-breasted (Plate 59). Open wooded areas. **TN** Previously in genus *Dendrocopos*.

Lesser Kestrel *Falco naumanni* 29–32cm

Two 21st-century records of migrating flocks in NE and SE. **ID** Slightly smaller and slimmer than Common Kestrel (Plate 60). Claws whitish (black on Common). Male has uniform blue-grey head (without dark moustachial stripe), unmarked rufous upperparts, blue-grey greater coverts, and almost plain orange-buff underparts. In flight, underwing whiter with more clearly pronounced wingtips; tail often looks more wedge-shaped. First-year male more like Common but has unmarked rufous mantle and scapulars. Female and juvenile have less distinct moustachial stripe than Common and lack any suggestion of dark eyestripe; underwing less heavily marked with more pronounced dark wingtips. **Voice** A *kee-chee-chee* or *chet-che-che,* less piercing and more slurred than Common. **HH** Hunting manner rather like Common, but usually in small parties or larger flocks, and is more agile and graceful in flight. Roosts communally. Open country.

Merlin *Falco columbarius* 25–30cm

♂ ♀

One 21st-century record on central coast. **ID** Small and compact, with short, pointed wings. Fine supercilium and weak moustachial apparent in all plumages. Male has blue-grey upperparts, broad black subterminal tail-band, diffuse patch of rufous-orange on nape and rufous-orange streaking on underparts. Female and juvenile have brown upperparts with variable buffish markings, heavily streaked underparts and strongly barred uppertail. **Voice** Usually silent away from breeding grounds. **HH** Normally found singly. Chiefly hunts in low flight with fast wingbeats and short glides. Open country.

Saker Falcon *Falco cherrug* 50–58cm

One late 20th-century record in C. **ID** Large falcon with long wings and long tail. Wingbeats slow in level flight, with lazier flight action than Peregrine (Plate 61). At rest, tail extends noticeably beyond closed wings (wings fall just short of tail-tip on Laggar, Plate 61, and are equal to tail on Peregrine). Additional differences include paler crown, less distinct moustachial and less heavily marked underparts (with flanks and thighs usually clearly streaked and not appearing wholly brown – although some overlap exists). **HH** Spends long periods perched on ground. When hunting, flies fast and low and strikes prey on ground; also stoops on aerial prey like Peregrine. Open areas. **Threatened** Globally (EN).

Slender-billed Oriole *Oriolus tenuirostris* 27cm

Three 21st-century records in NE. **ID** Bill is more slender than on Black-naped (Plate 64) and black nape-band is typically not as broad. Male has olive-yellow coloration to mantle and wing-coverts (brighter yellow in male Black-naped). Female and immature not separable from Black-naped by plumage. Male from male Indian Golden (Plate 64) by black nape-band, and olive-yellow coloration to mantle and wing-coverts. Female similar to male but has dull black nape-band, duller yellow head and underparts, and duller olive-yellow upperparts; nape-band is best feature from female Indian Golden. Immature initially has dark bill and eye, more uniformly olive wings, and whitish underparts with black streaking; nape-band and eyestripe can be very diffuse. Diffuse nape-band if apparent, and longer bill, are best features from immature Indian Golden. **Voice** Mellow, fluty song *wheeow* or *chuck, tarry-you*; diagnostic, high-pitched woodpecker-like *kick* call. **HH** Habits like Indian Golden. Well-wooded areas.

Long-tailed Minivet *Pericrocotus ethologus* 20cm

Four late 20th-century records in C, NE and SE. **ID** Smaller and slimmer in build than Scarlet Minivet (Plate 65) with different shape to wing patch (with red in male or yellow in female extending as narrow panel along tertials and secondaries). Female has narrow area of yellow on forehead and supercilium; ear-coverts are more uniformly grey than in female Scarlet and throat is distinctly paler yellow (than breast). **Voice** A distinctive, sweet double whistle *pi-ru*, the second note lower than the first. **HH** Evergreen and deciduous forest.

White-bellied Drongo *Dicrurus caerulescens* 24cm

Two 21st-century records in NW. **ID** Whitish from belly downwards. Smaller than Black and Ashy Drongos (both Plate 68), and tail is shorter with typically shallower fork. Upperparts are glossy slate-grey, much as Ashy (and therefore less black than Black). Throat and breast are browner in first-winter plumage compared with adult, and border between breast and white belly is less clearly defined. **Voice** Similar to Black, but more continuous, richer and with fewer harsh notes. **HH** Open forest and well-wooded areas.

Crow-billed Drongo *Dicrurus annectans* 28cm

Five records in NE and SE. **ID** From Black (Plate 68) by much stouter bill, more extensive area of rictal bristles (resulting in tufted forehead not unlike that of Lesser Raquet-tailed, Plate 68), and shorter, broader tail which is widely splayed at tip but not deeply forked (with outer feathers more noticeably curving outwards). Note also habitat differences from Black. First-winter has white spotting from breast to undertail-coverts (recalling first-winter Greater Raquet-tailed). Juvenile has uniform brownish-black upperparts and underparts lack gloss. **Voice** Loud, musical whistles and churrs; has characteristic descending series of harp-like notes. **HH** Usually hunts from middle storey of forest or lower canopy. Evergreen forest.

Chinese Paradise-flycatcher *Terpsiphone incei* 20cm

One 21st-century record from SW coast. **ID** Has smaller bill and smaller crest than Indian Paradise-flycatcher (Plate 67); female has clearly defined dark hood. Bill shorter and shallower than in Oriental Paradise-flycatcher (Plate 67). Rufous male and female differ further in having darker chestnut upperparts and white vent; female has clearly defined dark hood. **Voice** As Indian Paradise-flycatcher. **HH** As Indian Paradise-flycatcher. **TN** Previously treated as conspecific with Indian and Oriental Paradise-flycatchers as Asian Paradise-flycatcher.

Burmese Shrike *Lanius collurioides* 20cm

Six 21st-century records in SW, C, NE and SE. **ID** Male has dark grey crown and nape, dark chestnut mantle, white tail-sides and chestnut rump. Female has whitish forehead, and paler chestnut mantle than male. Juvenile has two-toned upperparts, with buff barring on greyish or brown crown and nape, and rest of upperparts rufous with brown barring. **Voice** Alarm call is loud, rapid, harsh chattering. **HH** Second growth and bushes in cultivation.

Great Grey Shrike *Lanius excubitor* 24cm

One late 20th-century record in C. **ID** Adult told from Grey-backed and Long-tailed (both Plate 69) by pale grey mantle and white scapulars, bold white markings on black wings and tail, greyish rump, and white breast and flanks. Black of mask extends over forehead and down sides of neck. Has extensive white patch at base of primaries and, with inner webs of secondaries and tips of outer webs also largely white, shows much white in wing at rest and in flight. Juvenile has sandy cast to grey crown and mantle, with very indistinct barring on crown, buff tips to tertials and median and greater coverts, and faint buffish wash to underparts; mask is grey and does not extend over forehead as on adult. **Voice** Chattering song with harsh notes; mimics other birds; drawn-out whistle *kwiet* call. **HH** Habits like Brown (Plate 69). Dry country, open scrub, cultivation edges. **TN** Form recorded in Bangladesh is treated as separate species, Southern Grey Shrike *L. meridionalis*, by some authorities.

Red-billed Blue Magpie *Urocissa erythrorhyncha* 65–68cm

Two 20th-century records in NE and SE. **ID** Has red or orange-red bill, mainly black head with extensive white nape, and turquoise-blue mantle and wings. Tail is very long, showing white and black at tip. Juvenile has duller blue-grey upperparts, dull brownish-red bill, and more extensive white crown. **Voice** A piercing *quiv-pig-pig*, softer *beeee-trik*, subdued *kluk* and sharp *chwenk-chwenk*. **HH** Evergreen forests.

Sultan Tit *Melanochlora sultanea* 20.5cm

Three 21st-century records, all from Chittagong Hill Tracts in SE. **ID** A huge, bulbul-like tit. Male is largely glossy blue-black, with bright yellow crest and yellow underparts below black breast. Female is similar, but black of plumage is duller blackish-olive (especially pronounced on throat and breast). Juvenile has shorter crest and fine yellowish-white tips to greater coverts. **Voice** Song a series of five loud *chew* notes; call a loud, squeaky whistle, *tcheery-tcheery-tcheery*. **HH** Evergreen forest.

Green-backed Tit *Parus monticolus* 12.5cm

One late 20th-century record in NE. **ID** From Great Tit (Plate 69) by bright green mantle and back, yellow on breast-sides and flanks, and double white wingbars. Wings look bluish owing to blue edges to remiges. Female has duller black throat and narrower stripe down centre of belly. Juvenile is duller than adult, and white cheeks and wingbars are washed with yellow. **Voice** Song includes loud, pleasant, ringing *whitee... whitee*; calls resemble Great. **HH** Forages chiefly in lower, also in middle storey; occasionally on ground. Highly acrobatic, often hanging upside-down from twigs. Wooded country.

Hume's Lark *Calandrella acutirostris* 14cm

One 21st-century record. **ID** Greyer and less heavily streaked upperparts than Eastern Short-toed Lark (Plate 71), with pinkish uppertail-coverts. Head pattern usually less pronounced than Eastern, with rather uniform ear-coverts, dark lores (pale in Eastern), and less pronounced supercilium and eyestripe. Bill yellowish with pronounced dark culmen and tip. As Eastern, dark breast-side patch usually apparent, but with greyish-buff breast-band. **Voice** Full, rolling *tiurr* flight call. **HH** Habits like Eastern Short-toed. Fallow cultivation and open wasteland. **AN** Hume's Short-toed Lark.

Booted Warbler *Iduna caligata* 12cm

At least eight records in C, NE and SE. **ID** Small, with square-ended tail, and short undertail-coverts. Upperparts are brownish, with buff on sides of breast and flanks in fresh plumage; becomes greyish-brown on upperparts, and whiter on underparts in worn plumage. Often shows faint whitish edges and tip to tail and fringes to tertials. Supercilium usually reasonably distinct and square-ended behind eye and may be bordered above by diffuse dark line. Possible confusion with Siberian Chiffchaff (Plate 79), but lacks any greenish or olive tones, and has pale base to lower mandible and pale legs. See Sykes's Warbler for differences from that species. **Voice** Hard *chur chur* call. **HH** Forages at all levels from canopy down to bushes, undergrowth and sometimes on ground. Scrub and bushes at cultivation edges. **TN** Formerly in genus *Hippolais*.

Sykes's Warbler *Iduna rama* 12.5cm

One 21st-century record in NE. **ID** Typically greyer and distinctly longer-billed than Booted, and more arboreal in habits. Confusable with larger Blyth's Reed-warbler (Plate 74), but has more distinct supercilium behind eye, paler greyish-brown upperparts, pale sides to tail and edges to remiges, longer-looking, square-ended tail, and shorter undertail- and uppertail-coverts. **Voice** Scratchy, rattling song; rapidly repeated *tut-tut-tut* call. **HH** Bushy areas. **TN** Formerly in genus *Hippolais*.

Large-billed Reed-warbler *Acrocephalus orinus* 13.5cm

One 21st-century record in NE. **ID** Very similar in appearance to Blyth's Reed-warbler (Plate 74). In fresh plumage, is less olive, more rufous-tinged. Structurally has longer and slightly broader bill, more rounded wings, and longer and more graduated tail with more pointed tail feathers. Smaller, with weaker bill and feet than Clamorous Reed-warbler (Plate 74). **HH** Swamp-thickets.

Pygmy Cupwing *Pnoepyga pusilla* 9cm

Two or more 21st-century records: multiple sightings in winter at two sites in NE. **ID** Boldly scaled underparts. Occurs in both white and fulvous colour morphs (background coloration to scaled underparts). From similar Scaly-breasted Cupwing *P. albiventer* (not recorded in Bangladesh) by smaller size and distinctive song. Spotting on upperparts is confined to lower back and wing-coverts (lacking well-defined buff spotting on crown and neck, which is usually present on Scaly-breasted). **Voice** Song is loud, slowly drawn-out *see-saw*, repeated monotonously; *tzit* and *tzook* calls. **HH** On or near ground in dense forest understorey. **AN** Pygmy Wren-babbler. **TN** This and others in genus *Pnoepyga* recently found to be unrelated to other wren-babblers.

Lanceolated Warbler *Locustella lanceolata* 12cm

Nine records in NE. **ID** From Common Grasshopper-warbler by bold, fine streaking (almost spotting) on throat, breast and flanks, and stronger and better-defined streaking on upperparts. In addition, is smaller, has shorter tail and stouter bill, and in fresh plumage has warmer brown upperparts and warmer buff flanks, typically lacking yellow on underparts. Some can be very similar to Common Grasshopper-warbler: streaking on undertail-coverts is less extensive but blacker and more clear-cut, and tertials are darker with clear-cut pale edges. **Voice** Calls include sharp, metallic *pit,* faint *tack* and shrill *cheek-cheek-cheek-cheek.* **HH** Very skulking, keeping close to ground in tall grassland and swamp-thickets.

Common Grasshopper-warbler *Locustella naevia* 13cm

Five records from NW and NE. **ID** From Pallas's Grasshopper-warbler (Plate 75) by olive-brown coloration to upperparts, including rump and tail, and less prominent supercilium. From Lanceolated by (usually) unmarked or only lightly streaked throat and breast, and less heavily streaked upperparts (especially rump and uppertail-coverts). Some Common Grasshopper-warblers are as heavily streaked as lightly streaked Lanceolated (see latter for subtle differences). First-winter can have yellowish wash on underparts. **Voice** Calls include hard *sit.* **HH** Tall grassland, reedbeds and paddy-fields. **AN** Grasshopper Warbler.

Wire-tailed Swallow *Hirundo smithii* 14cm

Two records in C and NE; also 20th-century published claims from SW and SE. **ID** From Barn Swallow (Plate 76) by chestnut crown, brighter blue upperparts, glistening white underparts, and fine filamentous projections to outer tail feathers. White underwing-coverts contrast more strongly with dark underside of flight feathers. Can have pinkish or buffish wash to breast. Wire-like tail-streamers are frequently broken, entirely lost or can be difficult to see, tail then appearing square-ended. Juvenile has brownish cast to blue upperparts and dull brownish crown. Possibly confusable with Streak-throated (Plate 76) but is larger and proportionately longer-winged/shorter-tailed, and has whiter, unstreaked underparts and underwing-coverts. White throat and breast and squarer tail help separate it from juvenile Barn Swallow. **Voice** Twittering song; calls include double *chirrik-weet, chit-chit* contact call and *chichip chichip* alarm call. **HH** Habits like Barn, but more closely associated with water. Open country near water.

Pale Sand Martin *Riparia diluta* 13cm

One 21st-century record in NE but may be under-recorded due to confusion with Collared Sand Martin (Plate 77). **ID** Very similar to Collared, but upperparts are paler and greyer, the breast-band is pale and weakly defined, the throat is greyish-white and grades into the pale greyish-brown ear-coverts, and the tail-fork is shallower. Juvenile has brownish throat, recalling Asian Plain Martin (Plate 77), but with stronger contrast between breast and belly, and lacks pinkish-buff wash across breast of juvenile Asian Plain. There is confusing subspecies variation; compared with nominate, *indica* is smaller and has very shallow tail-fork, and *tibetana* is comparatively large and as dark as Collared. **HH** Habits like Asian Plain. Around large waterbodies. **AN** Pale Martin. **TN** Until recently treated as subspecies of Collared Sand Martin.

Flavescent Bulbul *Pycnonotus flavescens* 22cm

One late 20th-century record in NE. **ID** Has short white supercilium contrasting with black lores, black bill, slight crest, olive-brown upperparts with greenish cast, brownish-grey underparts (mixed with yellow), yellow undertail-coverts, and olive-yellow wings and tail. Juvenile has less prominent supercilium, browner upperparts, rufous edges to flight feathers and paler bill. **Voice** Song is jolly phrase of usually 3–6 notes; harsh alarm call. **HH** Arboreal. Keeps in flocks of 6–30 birds. Rather quiet. Forages inside bushes and trees, rather than perching conspicuously on top. Evergreen forest with plenty of undergrowth.

Hume's Leaf-warbler *Phylloscopus humei* 10–11cm

Two 21st-century records in NW and E, but may be under-recorded due to similarity with Yellow-browed (Plate 79). **ID** Compared with Yellow-browed has greyish-olive upperparts, with variable yellowish-green on mantle and back, and browner crown, while supercilium, ear-coverts and greater-covert wingbar are buffish-white. Median-covert bar poorly defined, but double wingbar can be apparent. Bill all dark, legs blackish-brown. Supercilium, wingbars and underparts white when worn, upperparts much greyer. *P. h. mandellii* (recorded in E) has darker olive upperparts and slightly more prominent pale median crown-stripe, and darker lateral crown-stripes. **Voice** Song a repeated *wesoo,* often followed by descending high-pitched *zweeeeeeeeooo;* disyllabic *whit-hoo* and sparrow-like *chwee* calls. *P. h. mandelli* has strikingly disyllabic *tjis-ip* call. **HH** Very active and frequent member of mixed foraging flocks. Forest and woodland.

Lemon-rumped Leaf-warbler *Phylloscopus chloronotus* 9cm

Two late 20th-century records from C. **ID** From similar species by combination of broad yellowish-white supercilium and crown-stripe (contrasting with dark olive sides of crown), double yellowish-white wingbars, well-defined yellowish (sometimes almost whitish) rump and whitish underparts. Lacks white on tail. **Voice** High-pitched *uist* call. **HH** Forest and second growth. **AN** Lemon-rumped Warbler.

Smoky Warbler *Phylloscopus fuliginventer* 10cm

Three late 20th-century records in C and NE. **ID** From Dusky (Plate 79) by smaller size, short-looking tail, darker sooty-olive upperparts (with greenish tinge in fresh plumage), short, indistinct supercilium (with bold white crescent below eye), and mainly dusky-olive underparts (with oily yellow centre). **Voice** Monotonous *tsli-tsli-tsli-tsli-tsli* song; throaty *thrup thrup* call. **HH** Dense undergrowth near water.

White-spectacled Warbler *Phylloscopus intermedius* 11–12cm

Three late 20th-century records in C and NE. **ID** Has white eye-ring, yellow lores, grey supercilium and crown-stripe contrasting with well-defined dark grey lateral crown-stripes, and greenish (rather than grey) lower ear-coverts. **Voice** Sharp *che-wheet* call. Song is combination of sweet variable phrases, each consisting of 5–8 rapidly delivered notes. **HH** Deciduous and evergreen forest. **TN** Formerly in genus *Seicercus*.

Chestnut-crowned Warbler *Phylloscopus castaniceps* 9.5cm

One late 20th-century record in C. **ID** Best told by combination of small size, chestnut crown with diffuse dark brown lateral crown-stripes, bright lemon-yellow rump, grey sides of head with white eye-ring, grey throat and upper breast contrasting with white lower breast and belly, and bright yellow flanks. **Voice** Song has 5–7 notes, extremely high-pitched, sibilant and slightly undulating. **HH** Restless in middle and upper storeys of forest. Deciduous forest. **TN** Formerly in genus *Seicercus*.

Pale-legged Leaf-warbler *Phylloscopus tenellipes* 10cm

One 21st-century record in NE. **ID** Superficially resembles Greenish Warbler (Plate 80). Always distinguished by very pale grey-pink legs and feet, whitish tip to dark bill with pale (but not orange-tinted) base, and distinctive, high-pitched metallic *pink* call. Often has buff wash to ear-coverts and olive-brown cast to upperparts, especially rump. Crown is distinctly greyer and usually contrasts with olive mantle. Underparts are whitish (never showing any yellow). **HH** Usually keeps close to ground in forest understorey. **TN** Sakhalin Leaf-warbler *P. borealoides* has been split from this species; although not recorded in Bangladesh it is almost indistinguishable apart from longer primary projection and different song.

Green Warbler *Phylloscopus nitidus* 10–11cm

Two late 20th-century records in C and NE. **ID** In fresh plumage, upperparts are brighter and purer green than Greenish (Plate 80), and has one, or sometimes two, slightly broader and yellower wingbars; supercilium and cheeks are noticeably yellow and underparts have stronger yellow suffusion. When worn, upperparts are duller but still brighter than in Greenish, and supercilium and underparts retain yellowish wash. **Voice** Trisyllabic *chis-ru-weet* call. **HH** Similar to Greenish, wooded areas.

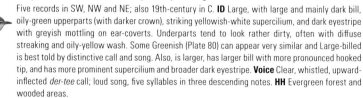

Large-billed Leaf-warbler *Phylloscopus magnirostris* 13cm

Five records in SW, NW and NE; also 19th-century in C. **ID** Large, with large and mainly dark bill, oily-green upperparts (with darker crown), striking yellowish-white supercilium, and dark eyestripe with greyish mottling on ear-coverts. Underparts tend to look rather dirty, often with diffuse streaking and oily-yellow wash. Some Greenish (Plate 80) can appear very similar and Large-billed is best told by distinctive call and song. Also, is larger, has larger bill with more pronounced hooked tip, and has more prominent supercilium and broader dark eyestripe. **Voice** Clear, whistled, upward-inflected *der-tee* call; loud song, five syllables in three descending notes. **HH** Evergreen forest and wooded areas.

Grey-hooded Warbler *Phylloscopus xanthoschistos* 10cm

Three late 20th-century records in NE. **ID** Best told by combination of greyish-white supercilium, and grey crown and mantle. Has diffuse pale grey central crown-stripe, and darker lateral crown-stripes. **Voice** High-pitched *psit-psit*, plaintive *tyee-tyee* call. Song is a brief, incessantly repeated high-pitched warble, *ti-tsi-ti-wee-tee*. **HH** Active warbler, hunts restlessly and feeds by gleaning, making short aerial sallies. Forest edge and semi-evergreen forest. **TN** Formerly in genus *Seicercus*.

Slaty-bellied Tesia *Tesia olivea* 9cm

Two late 20th-century records in NE. **ID** Best told from Grey-bellied Tesia (Plate 80) by uniform dark slate-grey underparts, and yellowish-green crown which is distinctly brighter than mantle (although some birds have duller and less contrasting crown and may show suggestion of brighter supercilium). Additional features are less prominent black stripe behind eye, and finer bill with brighter orange or orange-red lower mandible which lacks dark tip. Juvenile is said to be similar to juvenile Grey-bellied, but with darker olive-green underparts. **Voice** Song comprises 4–6 measured whistles followed by an explosive tuneless jumble of notes; calls include sharp *tchirik*. **HH** Habits like Grey-bellied. Thick low undergrowth in dense evergreen forest.

Chestnut-crowned Bush-warbler *Cettia major* 13cm

Five 21st-century records from NE. **ID** From Grey-sided Bush-warbler by larger size and more robust appearance, larger bill, longer supercilium (indistinct and rufous-buff in front of eye) and whiter underparts (particularly throat and centre of breast). Juvenile lacks chestnut on crown, and is more olive on upperparts and underparts, with greyish-buff supercilium behind eye; whiter throat and belly separate it from juvenile Grey-sided. **Voice** Song comprises an introductory note followed by explosive 3–4-note warble. Call very similar to Grey-sided. **HH** Typical bush-warbler behaviour. Winters in swamp-thickets.

Grey-sided Bush-warbler *Cettia brunnifrons* 10cm

One 21st-century record from NE. **ID** From Chestnut-crowned by smaller size, smaller bill, shorter supercilium (whitish-buff and well defined in front of eye) and greyer underparts. Juvenile lacks chestnut crown, with rufous-brown upperparts and brownish-olive underparts. **Voice** Call is bunting-like *pseek*. Song is loud wheezing *sip ti ti sip*, repeated continually. **HH** Winters in swamp-thickets.

Chestnut-headed Tesia *Cettia castaneocoronata* 8cm

Two late 20th-century records in NE and SE. **ID** Adult has bright chestnut 'hood', prominent white crescent behind eye, dark olive-green mantle and wings, and bright yellow underparts with olive-green sides of breast and flanks. Juvenile has dark olive upperparts with brownish cast (lacking chestnut head of adult), and dark rufous underparts. **Voice** Song an explosive *cheep-cheeu-chewit*; sharp, explosive *whit* call. **HH** Typically skulks among thick undergrowth in evergreen forest, usually close to ground. Active and inquisitive. Weak flight. **TN** Formerly in genera *Tesia* and *Oligura*.

Asian Stubtail *Urosphena squameiceps* 11cm

Six records in NE. **ID** Very short tail. Has rufescent upperparts, white underparts, very prominent buffish supercilium contrasting with brownish-black eyestripe which almost reaches hindcrown, and long pale pinkish legs and large feet. **Voice** Sharp *zit* and *tshk-tik* calls. **HH** Skulks in undergrowth of evergreen forest.

Mountain Tailorbird *Phyllergates cucullatus* 13cm

Two records in SE. **ID** From other tailorbirds by brighter orange-rufous forecrown, yellowish supercilium contrasting with dark grey eyestripe, grey ear-coverts, grey nape and sides of breast, and bright yellow belly and undertail-coverts. Juvenile has olive-green crown and nape (concolorous with mantle) and diffuse eyestripe. **Voice** Song is thin, high-pitched and melodious whistle of 4–6 notes. **HH** Evergreen forest and second growth in hills. **TN** Formerly placed in genus *Orthotomus*.

Brownish-flanked Bush-warbler *Horornis fortipes* 12cm

One late 20th-century record in C. **ID** Has rufous-brown upperparts, brownish-buff coloration to throat and breast, and buffish supercilium. Juvenile has yellow underparts; confusable with Aberrant Bush-warbler, but upperparts browner. **Voice** Song is loud whistle, *weeee*, followed by explosive *chiwiyou*; calls are *chuk* and *tyit-tyu-tyu*. **HH** Typically keeps well hidden in undergrowth. Second growth in deciduous forest. **TN** Formerly placed in genus *Cettia*.

Aberrant Bush-warbler *Horornis flavolivaceus* 12cm

Six 21st-century records in NE. **ID** From other bush-warblers by yellowish-green cast to olive upperparts, yellowish supercilium, and buffish-yellow to olive-yellow underparts. Confusable with Tickell's Leaf-warbler (Plate 79), but has longer, rounded tail, which appears to be loosely attached, usually held slightly cocked; more rounded wings; and grating call accompanied by much wing-flicking. **Voice** Song is short warble, followed by long inflected whistle, *dir dir-tee teee-weee*. Call is *brrrt-brrrt*, different from any *Phylloscopus*. **HH** Swamp-thickets. **TN** Formerly placed in genus *Cettia*.

Eastern Orphean Warbler *Sylvia crassirostris* 15cm

One late 20th-century record in SE. **ID** Larger and bigger-billed than Lesser Whitethroat; more ponderous movements and heavier appearance in flight. Adult has blackish crown, pale grey mantle, blackish tail, and pale iris, white in male (always dark in Lesser). First-year has crown concolorous with mantle, darker grey ear-coverts and dark iris, and can appear similar to Lesser Whitethroat. **Voice** Strong, varied thrush-like warbling song. **HH** Scrub and groves. **TN** Formerly considered race of Orphean Warbler *S. hortensis*.

Lesser Whitethroat *Sylvia curruca* 13cm

Four records in NE. **ID** Has brownish-grey upperparts, dull whitish underparts (can have pinkish flush to breast in fresh plumage), slate-grey crown (greyer and slightly darker than mantle), and darker lores and ear-coverts (forming diffuse mask). Bill is blackish and legs and feet grey. Can show suggestion of paler supercilium and pale buffish fringes to tertials and secondaries in fresh plumage. **Voice** A soft, low-pitched, rather scratchy warbling song and dry rattle, the two often running together. **HH** Chiefly keeps low in vegetation. Mainly scrub.

Rufous-headed Parrotbill *Psittiparus bakeri* 18cm

Two late 20th-century records in NE. **ID** From similar Pale-billed Parrotbill *Chleuasicus atrosuperciliaris* (not recorded in Bangladesh) by larger size, longer and less stubby bill, lack of black eyebrow, and deep rufous-orange ear-coverts and lores (which are well demarcated from throat). From juvenile White-hooded Babbler (Plate 83) by much stouter bill. **Voice** A strong, whistled *swee-swee-swee-swo* and a harsh, buzzing, metallic *dzaw-dzaw*. **HH** Bamboo stands, moist broadleaved forest undergrowth. **TN** Previously grouped with extralimital White-breasted Parrotbill as Greater Rufous-headed Parrotbill *P. ruficeps*.

Red-billed Scimitar-babbler *Pomatorhinus ochraceiceps* 23cm

One late 20th-century record in Chittagong Hill Tracts, in SE. **ID** Long, downcurved reddish bill and striking white supercilium. Upperparts are olive-brown, and throat is white merging into buffish underparts. Red bill and warm rufous crown are best features from White-browed Scimitar-babbler (Plate 82). **Voice** Utters hurried, hollow piping; also, very rapid human-like whistle. **HH** Dense undergrowth in broadleaved evergreen forest and bamboo thickets.

Spot-breasted Scimitar-babbler *Erythrogenys mcclellandi* 25cm

One late 20th-century record in SE. **ID** Likely to be confused only with Large Scimitar-babbler (Plate 82) and best told by smaller size, finer bill, yellow eye, rufous forehead and ear-coverts, bold brown spotting on white breast, uniform olive-brown breast-sides and flanks, paler olive-brown upperparts, and brownish rather than lead-grey legs and feet. **Voice** Low-pitched, persistent, fluty *tiuu-tuu*, first note stressed, followed by higher-pitched *tiuuk*. **HH** Semi-evergreen forest undergrowth and second growth. **TN** Formerly in genus *Pomatorhinus* as *P. erythrocnemis*.

Streaked Wren-babbler *Turdinus brevicaudatus* 12cm

Two 21st-century records in Chittagong Hill Tracts, in SE. **ID** A large wren-babbler with prominent tail. Has grey lores, supercilium and ear-coverts (resulting in grey-faced appearance), whitish throat diffusely streaked with grey, olive-brown breast, becoming brighter rufous-brown on flanks and vent, olive-brown upperparts with dark brown fringes (resulting in untidy streaked or scaled appearance), and prominent buff tips to wing-coverts and tertials. Greyish edges to primaries form panel on wing. **Voice** Song of variable, clear, ringing whistles; sometimes a single *pweeee*. **HH** Moist forest on rocky ground and ravines. **TN** Previously in genus *Napothera*.

Yellow-throated Laughingthrush *Garrulax galbanus* 23cm

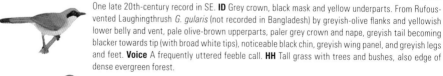

One late 20th-century record in SE. **ID** Grey crown, black mask and yellow underparts. From Rufous-vented Laughingthrush *G. gularis* (not recorded in Bangladesh) by greyish-olive flanks and yellowish lower belly and vent, pale olive-brown upperparts, paler grey crown and nape, greyish tail becoming blacker towards tip (with broad white tips), noticeable black chin, greyish wing panel, and greyish legs and feet. **Voice** A frequently uttered feeble call. **HH** Tall grass with trees and bushes, also edge of dense evergreen forest.

Long-tailed Sibia *Heterophasia picaoides* 30cm

One late 20th-century record in NE. **ID** Long tail with greyish-white tips, grey head and upperparts (with slightly darker lores), paler grey underparts, and dark grey wings with white patch on secondaries. From Grey Treepie (Plate 70) by thinner bill, grey face and pale vent. **Voice** Calls include thin, metallic, high-pitched *tsittsit* and *tsic* notes, interspersed with dry, even-pitched rattling. **HH** Arboreal; mainly frequents middle levels and canopy. Often feeds on nectar of flowering trees. Evergreen forest.

Bar-tailed Treecreeper *Certhia himalayana*   12cm

One early 20th-century record in SE. **ID** The only treecreeper recorded in Bangladesh; separated from other treecreepers by (generally) longer, more downcurved bill, dark cross-barring on tail, less distinct supercilium and pale banding across wings, white throat and dull whitish or dirty greyish-buff underparts. **Voice** Song is high-pitched trill *chi-chi-chi-chiu-chiu-chiu-chu*; weak, high-pitched, thin *tsi-tsi* call. **HH** Open forests and forest edge.

Indian Nuthatch *Sitta castanea*   12cm

Records in 20th century in Sundarbans in SE. **ID** Male always shows striking white cheek patch contrasting with dark chestnut underparts. Female is similar, but underparts are paler and cheek patch less striking. In both sexes, scalloping on undertail-coverts is grey (same colour as mantle), and crown and nape are distinctly paler than mantle (scalloping white, and crown concolourous with mantle in Chestnut-bellied Nuthatch, Plate 86). **Voice** Songs include rapid trill, descending in pitch. **HH** Forages on tree trunks and branches, often head down. Mangrove forest. **TN** Merged with Chestnut-bellied in some taxonomies.

Rosy Starling *Pastor roseus*   21cm

Nine records in SE, SW and NE including a wintering flock. **ID** Adult has blackish head with shaggy crest, pinkish mantle and underparts, and blue-green gloss to wings. In non-breeding and first-winter plumages, much duller; pink of plumage partly obscured by buff fringes; black by greyish fringes. Juvenile mainly sandy-brown, with stout yellowish bill and broad pale fringes to wing feathers. **Voice** Flight call is loud clear *ki-ki-ki*; also *shrr*, and rattling *chik-ik-ik-ik* when feeding. **HH** Cultivation and damp grassland.

Purple-backed Starling *Agropsar sturninus*   19cm

At least four 21st-century records in C, SW and NE, including one migrating flock. **ID** A small stocky starling with short tail and stout bill. Adult male has pale grey head, nape and underparts, purplish-black hindcrown patch and mantle, white tips to median coverts and rear scapulars forming prominent white V from behind, and glossy greenish-black wings with greyish-white tips to inner greater coverts and tertials. Female and juvenile are duller; wingbars and tips to scapulars are less prominent in juvenile. **HH** Open wooded areas, flowering trees. **AN** Daurian Starling. **TN** Formerly placed in genus *Sturnus*.

Chestnut-cheeked Starling *Agropsar philippensis*   19cm

One 21st-century record in NE. **ID** A small stocky starling with short tail and stout bill. Adult male has creamy-white head with chestnut ear-coverts, blackish back and scapulars with violet gloss, dark green wings with white shoulder patch, buff rump and black tail. Female grey-buff with plain face, dark brown back and wings, with white shoulder patch. **HH** Open wooded areas, flowering trees. **TN** Formerly placed in genus S*turnus*.

Spot-winged Starling *Saroglossa spiloptera*   19cm

Five 21st-century records in NE. **ID** White wing patch and whitish iris. Male has blackish mask, reddish-chestnut throat, pale rusty-orange breast, dark-scalloped greyish upperparts and rufous tail. Female has browner upperparts, and whitish underparts with greyish-brown markings on throat and breast. Juvenile is similar to female, but has buff wingbar, more uniform upperparts and dark eye. **Voice** Continuous harsh, unmusical jumble of discordant notes; calls include explosive scolding *kwerrh* and grating nasal *schaik*. **HH** Prefers to feed on nectar. Found in noisy flocks, often with mynas and drongos in flowering or fruiting trees. Open evergreen forest, well-wooded areas.

'Plain-backed Thrush' *Zoothera mollissima*   27cm

One 21st-century record in C. **ID** From very similar Long-tailed Thrush *Z. dixoni* (not recorded in Bangladesh) by indistinct (or absent) wingbars (can show narrow buff tips to median and greater coverts). In addition, belly and flanks are generally more clearly scaled with black, has more rufescent coloration to upperparts, especially to uppertail-coverts/rump and tail, less pronounced pale wing panel, and shorter tail. In flight, shows broad whitish banding across underwing (as does Long-tailed). Lower-altitude eastern populations recently treated as separate species, Himalayan Forest Thrush *Z. salimali*. Bangladesh record is of this form or *Z. mollissima* (renamed Alpine Thrush). Himalayan Forest Thrush is best told from Alpine Thrush by warmer, rufous-brown upperparts, longer all-dark bill, pinkish legs and feet, and subtly different face pattern. Area of pale on lores is more restricted and has dark bar between bill and eye which can extend below eye; less paleness to ear-coverts also and usually does not show dark patch at rear. **Voice** Song of Alpine Thrush consists of short, hurried strophes of highly variable, complex notes. Mainly rasping, grating, scratchy, cracked voice and a few squeaky, clearer notes. Song of Himalayan Forest Thrush is much more musical and 'thrush-like' than Alpine: a mix of rich, drawn-out, clear notes and shorter, thinner ones, with hardly any harsh scratchy notes. **HH** Secretive and shy. Forest and open country with bushes.

Dark-sided Thrush *Zoothera marginata* 25cm

Five records in SE and NE. **ID** From Long-billed by smaller size, smaller bill, rufous-brown upperparts and wing panel, and paler underparts with prominent scaling on breast and flanks; also more strongly patterned sides of head (variable, but usually with paler lores, more distinct dark and pale patches on ear-coverts, pale crescent behind). **Voice** Song a thin whistle; soft, deep, guttural *tchuck* call. **HH** Dense forest near streams.

Long-billed Thrush *Zoothera monticola* 28cm

Two 21st-century records in NE and SE. **ID** From Dark-sided by larger size and bill, more uniform head-sides (dark lores, diffuse dark malar stripe and narrow white throat patch), dark slaty-olive upperparts, darker and more uniform breast and flanks (both with diffuse dark spotting), and dark spotting on whitish belly. **Voice** Song comprises loud, slow plaintive whistle of 2–3 notes; alarm is loud *zaaaaaaaa*. **HH** Crepuscular, skulks on forest floor. Dense moist forests, usually near streams.

Purple Cochoa *Cochoa purpurea* 30cm

Two records in C and SE. **ID** Adult male is dull purplish-grey with lilac-blue crown, black mask, lilac panels on wing, and lilac tail with black tip. Adult female recalls male (with similar pattern to wings and tail) but has rusty-brown upperparts and brownish-orange underparts. Juvenile has black scaling to crown, indistinct buff streaking and spotting on upperparts, orange-buff underparts with bold black barring, and buff tips to wing-coverts; wings and tail as adult. **Voice** Song is flute-like *peeeee*; also *peeee-you-peeee*, like music of shepherd's bamboo flute; low chuckling call. **HH** Quiet, unobtrusive and rather lethargic. Evergreen forest.

Grey-winged Blackbird *Turdus boulboul* 28cm

Three records in NE. **ID** Adult male is black, with pale grey panel on wing. In fresh plumage, has prominent whitish fringes to belly and vent. Bill is orange and legs are yellowish. Female is olive-brown; has paler rufous-brown panel on wing (with greater coverts becoming paler buffish or greyish towards tips, contrasting with dark brown primary coverts). **Voice** Rich melodious song, with repeated two-note whistles; *chook-chook-chook* call. **HH** Evergreen forest and wooded areas.

Eyebrowed Thrush *Turdus obscurus* 23cm

Seven records in C, SE and NE. **ID** Striking features are white supercilium and white crescent below eye, contrasting with dark lores. Has peachy-orange flanks contrasting with white belly. Adult male has blue-grey head, including throat, with just small area of white on chin. Female has olive-brown crown and nape, browner ear-coverts, white throat and submoustachial stripe, dark malar stripe, narrow grey gorget across upper breast, and duller orange breast and flanks. First-winter similar to female but has fine greater-covert wingbar; first-winter males are brighter, with more grey on ear-coverts and upper breast. **Voice** Thin drawn-out *tseep* call. **HH** Open wooded areas; also evergreen forest.

Tibetan Blackbird *Turdus maximus* 26–29cm

One late 20th-century record in C. **ID** Male is entirely black, with yellow bill. Female is uniform dark brown, lacking paler throat. **Voice** Rattling *chak-chak-chak* call. **HH** Forest and thickets. Shy, flying off rapidly if alarmed. Deciduous forest. **TN** Formerly treated as subspecies of Eurasian Blackbird *T. merula*.

White-collared Blackbird *Turdus albocinctus* 27cm

One late 20th-century record in C. **ID** Adult male is mainly black, with white throat and broad white collar; bill and legs are yellow. Female has variable pale greyish-white to buffish collar, and rest of plumage is rufous-brown with pale feather fringes on underparts. **Voice** Coarse chuckling chatter call. **HH** Deciduous forest, especially clearings and edges.

Dusky Thrush *Turdus eunomus* 24cm

One late 20th-century record in SE. **ID** Adult male has broad white supercilium and throat contrasting with dark crown and ear-coverts, chestnut wing panel, rufous-brown mantle with dark feather centres, double gorget of blackish spotting across breast, and bold spotting on flanks contrasting with the white of underparts. Female is similar to male but is usually duller and less strikingly patterned, and typically shows more distinct black-streaked malar stripe. First-winter plumage is

variable, and duller than adult: crown and ear-coverts are greyer and supercilium less pronounced; double gorget of spotting less distinct; upperparts greyer; and wing panel is browner (and less distinct). **Voice** Calls include shrill *shrree* and rather harsh *chack-chack*. **HH** Areas with scattered trees and forest edge. **TN** Formerly treated as conspecific with Naumann's Thrush *T. naumanni*.

Rufous-throated Thrush   *Turdus ruficollis*   26.5cm

One 21st-century record in NW. **ID** Adult male has red supercilium, throat, breast and outer tail feathers, grey upperparts and whitish underparts. Female is similar to male, but has buff supercillium and throat, black-streaked malar stripe and black spotting across breast. First-winter similar to Black-throated (Plate 88) but has reddish-buff supercilium. **Voice** Calls include shrill rattle in alarm and high-pitched squeaky contact notes. **HH** Forages on ground in open wooded areas. **AN** Red-throated Thrush. **TN** Formerly treated as conspecific with Black-throated Thrush, under the name Dark-throated Thrush.

Ferruginous Flycatcher   *Muscicapa ferruginea*   13cm

One 21st-century record in SW. **ID** Compact, with large head, large eye and prominent white eye-ring. Adult has blue-grey cast to head (with darker malar stripe), rufous-brown mantle, rufous-orange rump and tail-sides, rufous-orange underparts, and prominent rufous fringes to greater coverts and tertials. **Voice** Call is quiet trill. **HH** Partly crepuscular, unobtrusive and very quiet, usually keeping to middle or lower levels in trees.

Spotted Flycatcher   *Muscicapa striata*   15cm

One 21st-century record in SW. **ID** From Dark-sided Flycatcher (Plate 93) by larger size, longer bill, paler grey-brown upperparts, faint dark streaking on forehead and crown, indistinct eye-ring, and diffuse grey-brown streaking on throat and breast. **Voice** Thin, scratchy calls. **HH** Open forest.

Small Niltava   *Niltava macgrigoriae*   13cm

Seven records in NE. **ID** Small size. Male dark blue, with brilliant blue forehead and neck patch. Female is dusky brown with indistinct blue neck patch and rufescent wings and tail; lacks oval throat patch of female Rufous-bellied Niltava (Plate 93). **Voice** Calls include high-pitched *see-see* (second note lower) and metallic scolding and churring notes. **HH** Evergreen forest.

Large Niltava   *Niltava grandis*   21cm

One late 20th-century record in NE. **ID** A very large, stocky niltava. Male is dark blue (often appearing entirely black in poor light), with blackish face and tufted forehead. Has brilliant blue crown, neck patch, shoulder patches and rump. Female has blue patch on side of neck (which can be obscured), dark olive-brown upperparts with rufescent wings and tail, clearly defined (narrow) buff throat, and rufous-buff forecrown and lores. Lacks white patch on lower throat of female Rufous-bellied Niltava (Plate 93). **Voice** Harsh rattle and unobtrusive nasal *dju-ee*. **HH** Evergreen forest, especially near streams.

Large Blue-flycatcher   *Cyornis magnirostris*   15cm

One 21st-century record in SE. **ID** From Tickell's and Hill Blue-flycatchers (Plate 94) by larger size, longer bill (with prominent hooked tip) and longer primary projection. Upperparts of male are deeper blue than Tickell's. Throat shows less contrast with breast, and undertail-coverts creamy (throat concolorous with breast, and undertail-coverts white, in Hill). Female from female Blue-throated (Plate 94) and Hill by combination of larger size and bill, sharper demarcation between ear-coverts and creamy throat, with throat paler than breast, and creamy undertail-coverts. **Voice** Unknown in region. **HH** Evergreen forest. **TN** Formerly treated as conspecific with Hill Blue-flycatcher.

Tickell's Blue-flycatcher   *Cyornis tickelliae*   14cm

Five records in SE, NW and NE. **ID** Male has clear horizontal division between orange breast and white flanks and belly; from Blue-throated (Plate 94) by orange throat. Female has blue-grey cast to upperparts (especially rump and tail), orange breast, and white belly and flanks. **Voice** Gives *tick tick* call. **HH** Open forests.

Siberian Blue Robin   *Larvivora cyane*   15cm

Seven 21st-century records in NE and SE. **ID** Male has dark blue upperparts, black sides to throat and breast and white underparts. Female has olive-brown upperparts, pale buff throat and breast, the latter faintly scaled with dark brown, and usually has blue on uppertail-coverts and tail. Female is similar to female Indian Blue Robin (Plate 90) but lacks orange-buff across breast and on flanks of that species. First-winter male similar to female, but with blue on mantle. **Voice** Subdued *tuk* or *tak* call. **HH** Undergrowth in evergreen forest. **TN** Formerly in genus *Luscinia*.

Rufous-breasted Bush-robin *Tarsiger hyperythrus* 15cm

One late 20th-century record in NE. **ID** Carriage and profile as Himalayan Bush-robin. Long legs help separate from blue-flycatchers. Male has dark blue upperparts, blackish ear-coverts, glistening blue supercilium and shoulders, and rufous-orange underparts. Female has blue tail; compared with female Himalayan Bush-robin has orange-buff throat, and browner breast and flanks. **Voice** Alarm call is *duk-duk-duk-squeak*; lisping warbling song *zeew..zee..zee..zee*. **HH** Evergreen forest undergrowth.

Himalayan Bush-robin *Tarsiger rufilatus* 14cm

Four 21st-century records in SE and NE. **ID** White throat, orange flanks, blue tail and redstart-like stance. Male has blue upperparts and breast-sides. Female has olive-brown upperparts and breast-sides. **Voice** Call is deep croaking *tock-tock*; song is rather soft and weak *churrh-cheee* or *dirrh-tutu-dirrh*. **HH** Forest understorey. **AN** Himalayan Bluetail. **TN** Treated by some authorities as conspecific with Orange-flanked Bush-robin (=Red-flanked Bluetail) *T. cyanurus*.

White-browed Bush-robin *Tarsiger indicus* 15cm

Two late 20th-century records in NE. **ID** Upright stance, long tail (frequently cocked) and dark legs are good features to separate from Indian Blue Robin (Plate 90). Male also has longer and finer supercilium, greyer upperparts and entirely rufous-orange underparts. Female has long (sometimes partly concealed) buffish-white supercilium, which curves down behind eye, and orange-buff throat concolorous with underparts. **Voice** Call is repeated *trrr*. **HH** Feeds mainly on ground in dense evergreen forest understorey.

Slaty-backed Forktail *Enicurus schistaceus* 25cm

Three late 20th-century records in NE. **ID** From Black-backed Forktail (Plate 91) by slate-grey (rather than black) crown and mantle, contrasting with black throat and wing-coverts. Also, bill is generally larger, and forehead shows less white. **Voice** Mellow *cheet* or metallic *teenk* calls. **HH** Fast-flowing streams in forest.

White-crowned Forktail *Enicurus leschenaulti* 28cm

Three records in SE and NE. **ID** Resembles Black-backed (Plate 91) but is larger, with longer tail, has prominent white forehead and forecrown, and black of throat extends to breast. **Voice** Call is harsh *scree* or *scree chit chit*; also has elaborate, high-pitched whistling song. **HH** Fast-flowing rivers and streams in evergreen forest.

Slaty-backed Flycatcher *Ficedula erithacus* 13cm

One 21st-century record in SE. **ID** Small, long-tailed flycatcher with very short bill. Male has deep blue upperparts (blacker on face) and bright orange underparts (becoming whiter on belly), and has black tail with white patches at base. Lacks any glistening blue in plumage. Female is rather nondescript, with olive-brown upperparts and greyish-olive underparts, and poorly defined whitish throat, lores and eye-ring. First-year male resembles female. **Voice** Particularly sharp rattle call, *terrht*. **HH** Evergreen forests. **TN** Previously *Ficedula hodgsonii*.

Rufous-gorgeted Flycatcher *Ficedula strophiata* 14cm

Seven records in C, SE and NE. **ID** Male has dark olive-brown upperparts, blackish face and throat, prominent white forehead and eyebrow, small rufous patch in centre of grey breast (can be difficult to see), and large white patches at sides of tail. Female is similar, but has less distinct eyebrow, duller and less distinct rufous 'gorget', and paler grey face and throat. **Voice** Calls include *pee-tweet*, metallic *pink* and harsh *trrt*. **HH** Evergreen and deciduous forests.

Sapphire Flycatcher *Ficedula sapphira* 11cm

Three records in NE. **ID** Breeding male has bright blue upperparts and sides of breast (with glistening blue crown and rump), orange throat and centre to breast, and white belly and undertail-coverts. Non-breeding and immature male have brown head and mantle, and brownish sides of breast. Female has olive-brown upperparts, orange throat and breast, and rufous rump and tail. Small size, slim appearance and tiny bill help separate female from female *Cyornis* flycatchers. **Voice** Calls comprise low *tit-it-it* rattle and dry rattled *trrrt*. **HH** Canopy and mid-levels of evergreen forest.

Ultramarine Flycatcher *Ficedula superciliaris* 12cm

Two records in SE and NE. **ID** Small, compact, arboreal flycatcher with small bill. Male has deep blue upperparts and sides of neck/breast, and white underparts. Female has greyish-brown upperparts and whitish underparts, with greyish patches on sides of breast (mirroring pattern of male); some have blue cast to uppertail-coverts and tail. First-year male resembles female, but with blue cast to mantle, wings and tail. **Voice** Calls include rising squeak. **HH** Evergreen forest.

Rusty-tailed Flycatcher *Ficedula ruficauda* 14cm

Two 21st-century records in NW and C. **ID** Has rufous uppertail-coverts and tail, resulting in (female) redstart-like appearance. Larger than Asian Brown Flycatcher (Plate 93), with flatter forehead, and crown feathers are often slightly raised, giving crested appearance to nape. Further, has rather plain face, with only faint supercilium (in front of eye) and indistinct eye-ring, and has entirely orange lower mandible and cutting edges to upper mandible. **Voice** Song is a drawn-out, rising and falling mournful whistle followed by rapid warbling. Calls include mournful *peu-peu* and short, deep churring. **HH** Wooded areas.

Red-breasted Flycatcher *Ficedula parva* 11.5–12.5cm

One 21st-century record in C, but may be under-recorded. **ID** Very similar to Red-throated Flycatcher (Plate 94) with white sides to long blackish tail. Has brownish (rather than black) uppertail-coverts. Bill has distinctly paler base to lower mandible (all black in Red-throated). Male has red of throat extending onto upper breast (red restricted to throat and with grey breast-band in Red-throated), and rest of underparts creamy-white. Female and first-winter plumages have warmer brown upperparts than Red-throated, and underparts are creamy-white, suffused with buff on breast. First-winter has orange-buff greater-covert wingbar. **Voice** Calls include a *tic,* unlike usual call of Red-throated, also quiet, dry *trrt, trrt.* **HH** Frequently cocks tail. Open wooded areas. **TN** Formerly treated as conspecific with Red-throated Flycatcher (= Taiga Flycatcher).

Blue-fronted Redstart *Phoenicurus frontalis* 15cm

Four records in SE and C. **ID** Orange rump and tail-sides, with black centre and tip to tail in all plumages. Male has blue head and upperparts and chestnut-orange underparts, which are duller in non-breeding and first-winter plumages due to fresh rufous-brown feather fringes. Female has dark brown upperparts and underparts, with orange wash to belly; tail pattern best feature from other female redstarts. **Voice** Calls include *ee-tit.* **HH** Unlike most redstarts, flicks tail up and down but does not vibrate it. Bushes and open forest.

Daurian Redstart *Phoenicurus auroreus* 15cm

Two 21st-century records in SE. **ID** Adult male (worn) has prominent white wing patch, blackish mantle, and black of throat does not extend onto breast. Adult male (fresh) and first-winter male have black of mantle and coverts partly obscured by brown fringes (mantle appears brown, diffusely streaked with black), and grey of crown and black of throat are duller owing to dark grey fringes. Adult female is similar to female Black Redstart (Plate 92) but has prominent white wing patch and darker centre to rufous tail. **HH** Forest edge.

Blue-capped Rock Thrush *Monticola cinclorhyncha* 17cm

Five 21st-century records in C and NE. **ID** Male has white wing patch and blue-black tail; also, blue crown and throat and orange rump and underparts; pattern and coloration obscured by pale fringes in non-breeding and first-winter plumages. Female has olive-brown upperparts, with barred rump, and whitish underparts which are boldly scaled and barred with brown; lacks buff neck patch of Chestnut-bellied, and blue cast to upperparts of Blue (Plate 92). **Voice** Calls include a *trigoink.* **HH** Chiefly arboreal, feeding on trunks and branches, occasionally descending to ground. Wooded areas.

Chestnut-bellied Rock Thrush *Monticola rufiventris* 23cm

Three 21st-century records in SE. **ID** Male has chestnut-red underparts and blue upperparts, including rump, uppertail-coverts and tail; lacks white on wing. Female has orange-buff lores and neck patch, dark malar stripe, dark barring on slaty olive-brown upperparts, and heavy scaling on underparts. Non-breeding and first-winter male plumages are very similar to breeding male but have fine buff fringes to mantle, scapulars and throat. **Voice** Calls include rasping jay-like notes. **HH** Perches high in tall forest trees, slowly jerking tail up and down from time to time. Forages mainly on ground; sometimes makes aerial sallies after insects. Open evergreen forest.

Jerdon's Bushchat *Saxicola jerdoni* 15cm

Eight records in C and NE. **ID** Male has blue-black upperparts, including rump and tail, and white underparts. Female and first-winter male similar to female Grey Bushchat (Plate 92), but lack prominent supercilium and have longer, more graduated tail lacking rufous at sides. **Voice** Call is short, plaintive whistle, higher-pitched than other chats. **HH** Habits similar to Common Stonechat (Plate 92) but less active on the wing. Sometimes also forages on ground at base of reeds and grass like a babbler. Tall moist/marshy grassland and tall grass in wetlands and tea estates.

White-throated Bushchat *Saxicola insignis* 17cm

Two 21st-century records in NE and NW. **ID** Larger than Common Stonechat (Plate 92), with bigger-looking head and bill. Male has white throat extending to form almost complete white collar and more white on wing than Common Stonechat. Female has broad buffish-white wingbars. **Voice** Metallic *teck-teck* call. **HH** Habits similar to Common Stonechat, but often solitary and rather shy. Tall grass and reeds. **AN** Hodgson's Bushchat. **Threatened** Globally (VU).

Desert Wheatear *Oenanthe deserti* 14–15cm

Two late 20th-century records in SE. **ID** Comparatively small and well-proportioned wheatear, with largely black tail and contrasting white rump. Male has black throat (partly obscured when fresh) and buff mantle. Female has blackish centres to wing-coverts and tertials in fresh plumage, and largely black wings when worn. **Voice** Hard *check-check* call. **HH** Forages chiefly by running or hopping on ground a short distance, then stopping to pick up prey. Open areas on coast.

Jerdon's Leafbird *Chloropsis jerdoni* 20cm

Three 21st-century records from same area of NW, where potentially resident. **ID** Has greenish wings and tail (lacking blue panels) compared with Blue-winged (Plate 95). Male lacks golden forehead of Golden-fronted (Plate 95) and has smaller black throat patch. Female has broad, diffuse yellow border to turquoise throat (female Golden-fronted has black throat). Juvenile has turquoise throat, brighter in moustachial region. **Voice** As Blue-winged. **HH** As Blue-winged. **TN** Formerly treated as conspecific with Blue-winged Leafbird.

Yellow-bellied Flowerpecker *Dicaeum melanoxanthum* 13cm

Two late 20th-century records in NE. **ID** A large, stout-billed flowerpecker. Has white spots at tip of undertail. Male has bluish-black upperparts and breast-sides, white centre of throat and breast, and yellow rest of underparts. Has bright red eye. Female is a dull version of male, with upperparts olive-brown, sides of breast olive-grey, and belly and vent dull olive-yellow. Juvenile male is similar to female but has brighter yellow underparts and blue-black cast to mantle and back. **Voice** Agitated *zit-zit-zit-zit* call. **HH** Evergreen forest.

Fire-breasted Flowerpecker *Dicaeum ignipectus* 9cm

One record in SE. **ID** Male has dark metallic blue or green upperparts, buff-coloured underparts, scarlet breast patch, and black centre to belly. Female has olive-green upperparts and orange-buff underparts with olive breast-sides and flanks. Juvenile has whiter throat merging into pale greyish-olive of underparts, and duller olive upperparts than female (more similar to Plain, Plate 96, but lacks supercilium and pale ear-coverts, and usually found at higher elevation). **Voice** Shrill *titty-titty-titty* song; clicking *chip* call. **HH** Evergreen forest and second growth in hills.

Olive-backed Sunbird *Cinnyris jugularis* 11cm

Three late 20th-century records in SE. **ID** Has olive-green upperparts, metallic purple-and-green throat and yellow belly and vent. Female similar to female Purple (Plate 97) but is greener above and brighter yellow below. Eclipse male as female but has broad blackish stripe down centre of throat and breast. **HH** Forest and scrub. **TN** Formerly placed in genus *Nectarinia*.

Fire-tailed Sunbird *Aethopyga ignicauda* 12cm

One 21st-century record in NW. **ID** Male has scarlet nape and mantle, and very long scarlet tail. Female similar to female Green-tailed, but has straighter bill, squarer tail (lacking white tips) with trace of brownish-orange at sides, and more noticeable olive-yellow on rump (not forming prominent band). Eclipse male similar to female, but has brighter yellow belly, and scarlet uppertail-coverts and tail-sides. **Voice** High-pitched monotonous *dzidzi-dzidzidzidzi* song. **HH** Wooded areas.

Green-tailed Sunbird *Aethopyga nipalensis* 11cm

One late 20th-century record in NE. **ID** Male from Gould's by maroon mantle and olive-green back, dark metallic blue-green crown and throat, and blackish sides of head. Has blue-green uppertail-coverts and tail (can appear blue, but not purplish-blue as on Gould's). Female has greyish-olive throat and breast (very grey on some birds), becoming yellowish-olive on belly and flanks. Lacks well-defined yellow rump-band (although rump and uppertail-coverts are yellowish-green). **Voice** Loud *chit chit* call. **HH** Evergreen forest.

Gould's Sunbird *Aethopyga gouldiae* 10cm

Five records in NE, C and SE. **ID** Male has metallic purplish-blue crown, ear-coverts and throat; crimson sides of neck, mantle and back (reaching yellow rump); yellow belly, and blue tail. Female has pale yellow rump-band, yellow belly, short bill and prominent white on tail. Juvenile male is similar to female but has bright yellow breast and belly. **Voice** Fast-repeated *tzip* call; *tshi-stshi-ti-ti-ti* and a lisping *squeeeeee*, which rises in middle in alarm. **HH** Evergreen forest. **AN** Mrs Gould's Sunbird.

Tree Pipit *Anthus trivialis* 15cm

Recorded in 19th century in C, one 21st-century record in NW. **ID** Buffish-brown colour to upperparts (lacking greenish-olive cast of Olive-backed Pipit, Plate 100), and buffish edges to wing feathers (edges greenish-olive in Olive-backed). Head pattern typically less marked, although can appear similar. **Voice** Call comprises harsher *teez* than Olive-backed, but much overlap. **HH** Groves, grasslands, fallow fields and stubbles.

Tawny Pipit *Anthus campestris* 16cm

Three records in NW, NE and SE. **ID** Adult and first-winter plumages have plain or only very faintly streaked upperparts and unstreaked or only very lightly streaked breast. Juvenile plumage can be retained until midwinter, and upperparts and breast are noticeably streaked. In juvenile, subtle difference from Paddyfield (Plate 100) are dark lores and eyestripe contrasting with supercilium, which tends to be broader and square-ended, and more uniform buffish-white underparts. **Voice** Distinctive loud *tchilip* call; softer *chep* similar to Paddyfield and Blyth's (Plate 100) calls. **HH** Cultivation.

Long-billed Pipit *Anthus similis* 20cm

Old specimen record from NW. **ID** Considerably larger than Tawny, with larger and darker bill and shorter-looking legs. Like Tawny has dark lores. Lacks distinct dark malar and moustachial, and has darker and greyer upperparts, deeper orange-buff underparts, rufous fringes to tertials and coverts, and rufous-buff outer edge to tail. **Voice** Deep *chup* and loud ringing *che-vlee* calls. **HH** Grassland, scrub and cultivation.

Crested Bunting *Emberiza lathami* 17cm

At least three records in 21st century in NW (including a flock of 20) and NE. **ID** Always has crest and chestnut on wings and tail; tail lacks white. Male has bluish-black head and body (with paler fringes when fresh). Female and first-winter male streaked on upperparts and breast; first-winter male darker and more heavily streaked than female, with olive-grey ground colour to underparts. **Voice** Call is *tip* or *pink*. **HH** Dry grassland and fallow cultivation. **TN** Formerly in separate genus, *Melophus*.

Black-headed Bunting *Emberiza melanocephala* 16–18cm

Eight 21st-century records in NW and SW. **ID** Larger than Red-headed Bunting *E. bruniceps* (not recorded in Bangladesh) with longer bill. Male has black on head and chestnut on mantle. Female, when worn, may show ghost pattern of male; fresh female almost identical to Red-headed, but indicative features (although not always apparent) include rufous fringes to mantle and/or back, slight contrast between throat and greyish ear-coverts and more uniform yellowish underparts. Immature has buff underparts and yellow undertail-coverts. **Voice** A *pyiup* or *tyilp* call; *plut* flight call. **HH** Crop fields, grasslands.

Grey-necked Bunting *Emberiza buchanani* 15cm

Three 21st-century records in SW and NE. **ID** In all plumages, has pinkish-orange bill and rather plain head with whitish eye-ring. Male has blue-grey head with buffish submoustachial stripe and throat, deep rusty-pink breast and belly, and diffusely streaked sandy-brown mantle with pronounced rufous scapulars. Rump is sandy-grey. Female is very similar to male, but generally paler, with buffish cast to grey head and nape (often with some streaking). First-winter/juvenile plumages often have only slight greyish cast to head, underparts are warm buff (with variable rufous cast), and crown and underparts are faintly streaked. **Voice** Call is soft click. **HH** As vagrant, bushes and mangrove edge.

Tristram's Bunting *Emberiza tristrami* 15cm

One 21st-century record in NE. **ID** In all plumages, has striking head pattern with whitish crown-stripe and supercilium contrasting with dark sides to crown and ear-coverts. Mantle is grey-brown, streaked with black; chestnut rump. Male in breeding plumage has black throat. **Voice** Call is explosive *tzick*, usually repeated irregularly. **HH** Evergreen forest.

APPENDIX 2: SPECIES LIST INCLUDING BANGLA NAMES

Th = BirdLife/IUCN Red List (from HBW and BirdLife International 2019), where NT = Near Threatened; VU = Vulnerable; EN = Endangered; and CR = Critically Endangered.

BDTh = National Red List for Bangladesh (IUCN Bangladesh 2015), where in a national context RE = Regionally Extinct; DD = Data Deficient; NT = Near Threatened; VU = Vulnerable; EN = Endangered; and CR = Critically Endangered. Brackets have been used for alternative and previously used English names, including some names used prior to taxonomic changes.

Bangla names are provided by Enam Ul Haque.

	English Name	Genus	Species	Bangla Name	Th	BDTh
1	Rufous-throated Partridge	Arborophila	rufogularis	Laalgola Batai		RE
2	White-cheeked Partridge	Arborophila	atrogularis	Dholagaal Batai	NT	NT
3	Indian Peafowl	Pavo	cristatus	Deshi Moyur		RE
4	Green Peafowl	Pavo	muticus	Shobooj Moyur	EN	RE
5	Grey Peacock-pheasant	Polyplectron	bicalcaratum	Metey Kathmoyur		VU
6	Common Quail	Coturnix	coturnix	Pati Botera		DD
7	Rain Quail	Coturnix	coromandelica	Brishti Botera		DD
8	Asian Blue (King) Quail	Synoicus	chinensis	Raaj Botera		DD
9	Black Francolin	Francolinus	francolinus	Kala Titir		EN
10	Grey Francolin	Francolinus	pondicerianus	Metey Titir		RE
11	Swamp Francolin	Francolinus	gularis	Bada Titir	VU	RE
12	Red Junglefowl	Gallus	gallus	Laal Bonmurgi		
13	Kalij Pheasant	Lophura	leucomelanos	Kala Mothura		VU
14	Fulvous Whistling-duck	Dendrocygna	bicolor	Raaj Shorali		
15	Lesser Whistling-duck	Dendrocygna	javanica	Pati Shorali		
16	Bar-headed Goose	Anser	indicus	Dagi Rajhash		
17	Greylag Goose	Anser	anser	Metey Rajhash		
18	Greater White-fronted Goose	Anser	albifrons	Boro Dholakopal Rajhash		
19	Lesser White-fronted Goose	Anser	erythropus	Choto Dholakopal Rajhash	VU	VU
20	Common Goldeneye	Bucephala	clangula	Pati Shonachokh		
21	Smew	Mergellus	albellus	Smew Hash		
22	Goosander	Mergus	merganser	Pati Merganser		
23	Red-breasted Merganser	Mergellus	serrator	Laalbook Merganser		
24	Common Shelduck	Tadorna	tadorna	Pati Chokachoki		
25	Ruddy Shelduck	Tadorna	ferruginea	Khoyra Chokachoki		
26	African Comb (Knob-billed) Duck	Sarkidiornis	melanotos	Nakta Hash		NT
27	Cotton Pygmy-goose	Nettapus	coromandelianus	Dhola Balihash		
28	Mandarin Duck	Aix	galericulata	Mandarin Hash		
29	White-winged Duck	Asarcornis	scutulata	Badi Hash	EN	RE
30	Red-crested Pochard	Netta	rufina	Laaljhuti Bhutihash		
31	Common Pochard	Aythya	ferina	Pati Bhutihash	VU	
32	Baer's Pochard	Aythya	baeri	Beyarer Bhutihash	CR	CR
33	Ferruginous Duck	Aythya	nyroca	Morcheyrong Bhutihash	NT	NT
34	Tufted Duck	Aythya	fuligula	Tiki Hash		
35	Greater Scaup	Aythya	marila	Boro Scaup		DD
36	Pink-headed Duck	Rhodonessa	caryophyllacea	Golapi Hash	CR	RE
37	Garganey	Spatula	querquedula	Giria Hash		
38	Northern Shoveler	Spatula	clypeata	Utturey Khuntehash		

	English Name	Genus	Species	Bangla Name	Th	BDTh
39	Baikal Teal	*Sibirionetta*	*formosa*	Baikal Tilihash		DD
40	Falcated Duck	*Mareca*	*falcata*	Fuluri Hash	NT	NT
41	Gadwall	*Mareca*	*strepera*	Piang Hash		
42	Eurasian Wigeon	*Mareca*	*penelope*	Euresio Shithihash		
43	Indian Spot-billed Duck	*Anas*	*poecilorhyncha*	Deshi Metehash		
44	Mallard	*Anas*	*platyrhynchos*	Neelmatha Hash		
45	Northern Pintail	*Anas*	*acuta*	Utturey Lanjahash		
46	Common Teal	*Anas*	*crecca*	Pati Tilihash		
47	Little Grebe	*Tachybaptus*	*ruficollis*	Choto Duburi		
48	Red-necked Grebe	*Podiceps*	*grisegena*	Laalgola Duburi		
49	Great Crested Grebe	*Podiceps*	*cristatus*	Boro Khopaduburi		
50	Black-necked Grebe	*Podiceps*	*nigricollis*	Kalaghar Duburi		
51	Greater Flamingo	*Phoenicopterus*	*roseus*	Boro Flamingo		
52	Red-billed Tropicbird	*Phaethon*	*aethereus*	Laalthot Bishubia		
53	Rock Dove (Common Pigeon, Feral Pigeon)	*Columba*	*livia*	Gola Paira, Kobutor		
54	Pale-capped Pigeon	*Columba*	*punicea*	Dholatupi Paira	VU	CR
55	Oriental Turtle-dove	*Streptopelia*	*orientalis*	Udoyee Rajghughu		
56	Eurasian Collared-dove	*Streptopelia*	*decaocto*	Euresio Konthighughu		
57	Red Turtle-dove (Collared-dove)	*Streptopelia*	*tranquebarica*	Lal Rajghughu		
58	Eastern Spotted Dove	*Spilopelia*	*chinensis*	Poober Tilaghughu		
59	Western Spotted Dove	*Spilopelia*	*suratensis*	Teela Ghughu		
60	Laughing Dove	*Spilopelia*	*senegalensis*	Hashir Ghughu		DD
61	Barred Cuckoo-dove	*Macropygia*	*unchall*	Dagi Kokilghughu		DD
62	Grey-capped Emerald Dove	*Chalcophaps*	*indica*	Pati Shamaghughu		
63	Orange-breasted Green-pigeon	*Treron*	*bicinctus*	Komlabook Horial		
64	Ashy-headed Green-pigeon	*Treron*	*phayrei*	Choto Horial		
65	Thick-billed Green-pigeon	*Treron*	*curvirostra*	Thotmota Horial		
66	Yellow-footed Green-pigeon	*Treron*	*phoenicopterus*	Holdeypa Horial		
67	Pin-tailed Green-pigeon	*Treron*	*apicauda*	Lanja Horial		
68	Wedge-tailed Green-pigeon	*Treron*	*sphenurus*	Gejlej Horial		
69	Green Imperial-pigeon	*Ducula*	*aenea*	Shobooj Dhumkol		
70	Mountain Imperial-pigeon	*Ducula*	*badia*	Pahari Dhumkol		
71	Hodgson's Frogmouth	*Batrachostomus*	*hodgsoni*	Hojsoni Byangmukho		DD
72	Great Eared-nightjar	*Lyncornis*	*macrotis*	Boro Kaanchora		NT
73	Grey Nightjar	*Caprimulgus*	*jotaka*	Metey Raatchora		
74	Sykes's Nightjar	*Caprimulgus*	*mahrattensis*	Sykeser Raatchora		
75	Large-tailed Nightjar	*Caprimulgus*	*macrurus*	Lanja Raatchora		
76	Indian Nightjar	*Caprimulgus*	*asiaticus*	Deshi Raatchora		
77	Savanna Nightjar	*Caprimulgus*	*affinis*	Metho Raatchora		DD
78	Crested Treeswift	*Hemiprocne*	*coronata*	Jhutial Gachbatashi		
79	White-rumped Spinetail	*Zoonavena*	*sylvatica*	Dholakomor Suibatashi		
80	Silver-backed Needletail	*Hirundaps*	*cochinchinensis*	Chadipith Suibatashi		
81	Brown-backed Needletail	*Hirundapus*	*giganteus*	Khoirapith Shuibatashi		
82	Himalayan Swiftlet	*Aerodramus*	*brevirostris*	Himaloyee Kootibatashi		
83	Asian Palm-swift	*Cypsiurus*	*balasiensis*	Aesio Talbatashi		
84	Pacific (Fork-tailed) Swift	*Apus*	*pacificus*	Cheralej Batashi		

	English Name	Genus	Species	Bangla Name	Th	BDTh
85	House Swift	Apus	nipalensis	Ghar Batashi		
86	Greater Coucal	Centropus	sinensis	Boro Koobo, Kanakua		
87	Lesser Coucal	Centropus	bengalensis	Choto Koobo		
88	Sirkeer Malkoha	Taccocua	leschenaultii	Metey Malkoa		
89	Green-billed Malkoha	Phaenicophaeus	tristis	Shoboojthot Malkoa		
90	Jacobin (Pied) Cuckoo	Clamator	jacobinus	Pakra Papia		
91	Chestnut-winged Cuckoo	Clamator	coromandus	Khoirapakh Papia		
92	Western (Asian) Koel	Eudynamys	scolopaceus	Aeshio Kokil		
93	Asian Emerald Cuckoo	Chrysococcyx	maculatus	Aesio Shamapapia		
94	Violet Cuckoo	Chrysococcyx	xanthorhynchus	Beguni Papia		
95	Banded Bay Cuckoo	Cacomantis	sonneratii	Dagi Tamapapia		
96	Plaintive Cuckoo	Cacomantis	merulinus	Koroon Papia		
97	Grey-bellied Cuckoo	Cacomantis	passerinus	Meteypet Papia		
98	Square-tailed Drongo-cuckoo	Surniculus	lugubris	Aeshio Fingepapia		
99	Large Hawk-cuckoo	Hierococcyx	sparverioides	Boro Chokhgelo		
100	Common Hawk-cuckoo	Hierococcyx	varius	Pati Chokhgelo		
101	Whistling (Hodgson's) Hawk-cuckoo	Hierococcyx	nisicolor	Hojsoni Chokhgelo		
102	Indian Cuckoo	Cuculus	micropterus	Boukothakou Papia		
103	Eurasian (Common) Cuckoo	Cuculus	canorus	Pati Papia		DD
104	Oriental Cuckoo	Cuculus	saturatus	Udoyee Papia		
105	Lesser Cuckoo	Cuculus	poliocephalus	Choto Papia		
106	Masked Finfoot	Heliopais	personatus	Kalamukh Parapakhi	EN	EN
107	Slaty-legged Crake	Rallina	eurizonoides	Meteypa Jhilli		
108	Eastern Water (Brown-cheeked) Rail	Rallus	indicus	Panta Jhilli		
109	Slaty-breasted Rail	Lewinia	striata	Meteybook Jhilli		
110	Spotted Crake	Porzana	porzana	Tila Jhilli		
111	Ruddy-breasted Crake	Zapornia	fusca	Lalbook Gurguri		
112	Brown Crake	Zapornia	akool	Khoyra Gurguri		
113	Baillon's Crake	Zapornia	pusilla	Beillon Gurguri		
114	White-breasted Waterhen	Amaurornis	phoenicurus	Dholabook Dahook		
115	Watercock	Gallicrex	cinerea	Kora		
116	Purple Swamphen	Porphyrio	porphyrio	Beguni Kalem		
117	Common Moorhen	Gallinula	chloropus	Pati Paanmurgi		
118	Eurasian (Common) Coot	Fulica	atra	Pati Coot		
119	Sarus Crane	Antigone	antigone	Deshi Sarus	VU	RE
120	Demoiselle Crane	Anthropoides	virgo	Demoiselle Sarus		
121	Common Crane	Grus	grus	Pati Sarus		
122	Bengal Florican	Houbaropsis	bengalensis	Bangla Dahor	CR	RE
123	Lesser Florican	Sypheotides	indicus	Pati Dahor	EN	RE
124	Short-tailed Shearwater	Ardenna	tenuirostris	Chotolej Panikata		
125	Greater Adjutant	Leptoptilos	dubius	Boro Modontak, Hargila	EN	RE
126	Lesser Adjutant	Leptoptilos	javanicus	Choto Modontak	VU	VU
127	Painted Stork	Mycteria	leucocephala	Ranga Manikjor	NT	CR
128	Asian Openbill	Anastomus	oscitans	Aesio Shamkhol		
129	Black Stork	Ciconia	nigra	Kala Manikjor		VU

	English Name	Genus	Species	Bangla Name	Th	BDTh
130	Asian Woollyneck (Woolly-necked Stork)	Ciconia	episcopus	Dholagola Manikjor	VU	CR
131	White Stork	Ciconia	ciconia	Dhola Manikjor		
132	Black-necked Stork	Ephippiorhynchus	asiaticus	Kalagola Manikjor	NT	EN
133	Eurasian Spoonbill	Platalea	leucorodia	Euresio Chamochthuti		CR
134	Black-headed Ibis	Threskiornis	melanocephalus	Kalamatha Kastechora	NT	VU
135	Red-naped Ibis	Pseudibis	papillosa	Kala Kastechora		
136	Glossy Ibis	Plegadis	falcinellus	Khoira Kastechora		
137	Eurasian (Great) Bittern	Botaurus	stellaris	Bagha Bogla		
138	Yellow Bittern	Ixobrychus	sinensis	Holdey Bogla		
139	Cinnamon Bittern	Ixobrychus	cinnamomeus	Khoyra Bogla		
140	Black Bittern	Ixobrychus	flavicollis	Kala Bogla		NT
141	Malay (Malayan) Night-heron	Gorsachius	melanolophus	Malayee Nishibok		
142	Black-crowned Night-heron	Nycticorax	nycticorax	Kalamatha Nishibok		
143	Green-backed (Striated) Heron	Butorides	striata	Khudey Bok		
144	Indian Pond-heron	Ardeola	grayii	Deshi Kanibok		
145	Chinese Pond-heron	Ardeola	bacchus	China Kanibok		
146	Cattle Egret	Bubulcus	ibis	Go Boga		
147	Grey Heron	Ardea	cinerea	Dhoopni Bok		
148	White-bellied Heron	Ardea	insignis	Dholapet Bok	CR	RE
149	Goliath Heron	Ardea	goliath	Doytto Bok		DD
150	Purple Heron	Ardea	purpurea	Laalchey Bok		
151	Great White Egret	Ardea	alba	Boro Boga		
152	Intermediate Egret	Ardea	intermedia	Majhla Boga		
153	Little Egret	Egretta	garzetta	Choto Boga		
154	Pacific Reef-egret (Reef Heron)	Egretta	sacra	Proshanto Shoiloboga		
155	Spot-billed Pelican	Pelecanus	philippensis	Chitithuti Gaganber	NT	RE
156	Great White Pelican	Pelecanus	onocrotalus	Boro Dhola Gaganber		
157	Lesser Frigatebird	Fregeta	ariel	Pati Frigatepakhi		
158	Masked Booby	Sula	dactylatra	Kalamukh Booby		
159	Little Cormorant	Microcarbo	niger	Choto Pankouri		
160	Great Cormorant	Phalacrocorax	carbo	Boro Pankouri		
161	Indian Cormorant	Phalacrocorax	fuscicollis	Deshi Pankouri		
162	Oriental Darter	Anhinga	melanogaster	Goyar	NT	NT
163	Indian Thick-knee	Burhinus	indicus	Deshi Motahatu		
164	Great Thick-knee	Esacus	recurvirostris	Boro Motahatu	NT	NT
165	Eurasian Oystercatcher	Haematopus	ostralegus	Euresio Jhinukmar	NT	VU
166	Pied Avocet	Recurvirostra	avosetta	Pakra Ultothuti		
167	Black-winged Stilt	Himantopus	himantopus	Kalapakh Thengi		
168	Grey Plover	Pluvialis	squatarola	Metey Jiria		
169	Pacific Golden Plover	Pluvialis	fulva	Proshanta Sonajiria		
170	Long-billed Plover	Charadrius	placidus	Lombathuto Jiria		DD
171	Little Ringed Plover	Charadrius	dubius	Choto Nothjiria		
172	Kentish Plover	Charadrius	alexandrinus	Kentish Jiria		
173	Lesser Sandplover	Charadrius	mongolus	Choto Dhuljiria		
174	Greater Sandplover	Charadrius	leschenaultii	Boro Dhuljiria		
175	Oriental Plover	Charadrius	veredus	Udoyee Jiria		

	English Name	Genus	Species	Bangla Name	Th	BDTh
176	Northern Lapwing	*Vanellus*	*vanellus*	Utturey Titi	NT	
177	River Lapwing	*Vanellus*	*duvaucelii*	Nodi Titi	NT	NT
178	Yellow-wattled Lapwing	*Vanellus*	*malarbaricus*	Holdeygal Titi		NT
179	Grey-headed Lapwing	*Vanellus*	*cinereus*	Meteymatha Titi		
180	Red-wattled Lapwing	*Vanellus*	*indicus*	Hot Titi		
181	White-tailed Lapwing	*Vanellus*	*leucurus*	Dholalej Titi		
182	Greater Painted-snipe	*Rostratula*	*benghalensis*	Bangla Rangachega		
183	Pheasant-tailed Jacana	*Hydrophasianus*	*chirurgus*	Neo Pipi		
184	Bronze-winged Jacana	*Metopidius*	*indicus*	Dol Pipi		
185	Whimbrel	*Numenius*	*phaeopus*	Nata Gulinda		
186	Eurasian Curlew	*Numenius*	*arquata*	Euresio Gulinda	NT	NT
187	Far Eastern Curlew	*Numenius*	*madagascariensis*	Poober Gulinda	EN	DD
188	Bar-tailed Godwit	*Limosa*	*lapponica*	Dagilej Jourali	NT	NT
189	(Western) Black-tailed Godwit	*Limosa*	*limosa*	Kalalej Jourali	NT	NT
190	Ruddy Turnstone	*Arenaria*	*interpres*	Laal Nooribatan		
191	Great Knot	*Calidris*	*tenuirostris*	Boro Knot	EN	EN
192	Red Knot	*Calidris*	*canutus*	Laal Knot	NT	NT
193	Ruff	*Calidris*	*pugnax*	Geoala Batan		
194	Broad-billed Sandpiper	*Calidris*	*falcinellus*	Motathuto Batan		
195	Curlew Sandpiper	*Calidris*	*ferruginea*	Gulinda Batan	NT	
196	Temminck's Stint	*Calidris*	*temminckii*	Temmincker Chapapkhi		
197	Long-toed Stint	*Calidris*	*subminuta*	Lombangul Chapakhi		NT
198	Spoon-billed Sandpiper	*Calidris*	*pygmaea*	Chamochthuto Batan	CR	CR
199	Red-necked Stint	*Calidris*	*ruficollis*	Laalghar Chapakhi	NT	
200	Sanderling	*Calidris*	*alba*	Sanderling		
201	Dunlin	*Calidris*	*alpina*	Dunlin		
202	Little Stint	*Calidris*	*minuta*	Choto Chapakhi		
203	Asian Dowitcher	*Limnodromus*	*semipalmatus*	Aesio Dowitcher	NT	EN
204	Long-billed Dowitcher	*Limnodromus*	*scolopaceus*	Lombathuto Dowitcher		
205	Eurasian Woodcock	*Scolopax*	*rusticola*	Euresio Bonchega		
206	Wood Snipe	*Gallinago*	*nemoricola*	Bon Chega	VU	DD
207	Pintail (Pin-tailed) Snipe	*Gallinago*	*stenura*	Lanja Chega		
208	Swinhoe's Snipe	*Gallinago*	*megala*	Swinhoer Chega		
209	Common Snipe	*Gallinago*	*gallinago*	Pati Chega		
210	Jack Snipe	*Lymnocryptes*	*minimus*	Jack Chega		DD
211	Red-necked Phalarope	*Phalaropus*	*lobatus*	Laalghar Phalarope		
212	Red (Grey) Phalarope	*Phalaropus*	*fulicarius*	Laal Phalarope		
213	Terek Sandpiper	*Xenus*	*cinereus*	Terek Batan		
214	Common Sandpiper	*Actitis*	*hypoleucos*	Pati Batan		
215	Green Sandpiper	*Tringa*	*ochropus*	Shobuj Batan		
216	Grey-tailed Tattler	*Tringa*	*brevipes*	Meteylej Tattler	NT	NT
217	Spotted Redshank	*Tringa*	*erythropus*	Teela Laalpa		
218	Common Greenshank	*Tringa*	*nebularia*	Pati Shobujpa		
219	Common Redshank	*Tringa*	*totanus*	Pati Laalpa		
220	Wood Sandpiper	*Tringa*	*glareola*	Bon Batan		
221	Marsh Sandpiper	*Tringa*	*stagnatilis*	Beel Batan		

	English Name	Genus	Species	Bangla Name	Th	BDTh
222	Spotted (Nordmann's) Greenshank	Tringa	guttifer	Nordmann Shoboojpa	EN	CR
223	Common (Small) Buttonquail	Turnix	sylvaticus	Pati Nataboter		DD
224	Yellow-legged Buttonquail	Turnix	tanki	Holdepa Nataboter		
225	Barred Buttonquail	Turnix	suscitator	Dagi Nataboter		
226	Crab-plover	Dromas	ardeola	Kakrajiria		
227	Oriental Pratincole	Glareola	maldivarum	Udoyee Babubatan		
228	Little (Small) Pratincole	Glareola	lactea	Choto Baboobatan		
229	Indian Skimmer	Rynchops	albicollis	Deshi Gangchosha	EN	CR
230	Black-legged Kittiwake	Rissa	tridactyla	Kalapa Kittiwake	VU	
231	Slender-billed Gull	Larus	genei	Shoruthuto Gangchil		
232	Brown-headed Gull	Larus	brunnicephalus	Khoiramatha Gangchil		
233	Black-headed Gull	Larus	ridibundus	Kalamatha Gangchil		
234	Pallas's (Great Black-headed) Gull	Larus	ichthyaetus	Pallasi Gangchil		
235	Lesser Black-backed (Heuglin's) Gull	Larus	fuscus	Heugliner Gangchil		
236	Little Tern	Sternula	albifrons	Choto Panchil		
237	Common Gull-billed Tern	Gelochelidon	nilotica	Kalathot Panchil		
238	Caspian Tern	Hydroprogne	caspia	Caspian Panchil		
239	Whiskered Tern	Chlidonias	hybrida	Julfi Panchil		
240	White-winged Tern	Chlidonias	leucopterus	Dholapakh Panchil		DD
241	River Tern	Sterna	aurantia	Nodia Panchil	VU	NT
242	Black-naped Tern	Sterna	sumatrana	Kalaghar Panchil		
243	Common Tern	Sterna	hirundo	Pati Panchil		
244	Black-bellied Tern	Sterna	acuticauda	Kalapet Panchil	EN	CR
245	Lesser Crested Tern	Thalasseus	bengalensis	Bangla Tikipanchil		
246	Sandwich Tern	Thalasseus	sandvicensis	Sandwitch Panchil		
247	Greater Crested (Swift) Tern	Thalasseus	bergii	Boro Tikipanchil		
248	Arctic (Parasitic) Jaeger (Skua)	Stercorarius	parasiticus	Porojibi Jaeger		
249	Pomarine Jaeger (Skua)	Stercorarius	pomarinus	Pomarine Jaeger		
250	Common Barn Owl	Tyto	alba	Lokkhi Pecha		
251	Brown Boobook (Hawk-Owl)	Ninox	scutulata	Khoyra Shikrepecha		
252	Collared Owlet	Glaucidium	brodiei	Dagighar Kutipecha		
253	Asian Barred Owlet	Glaucidium	cuculoides	Aesio Dagipecha		
254	Spotted Owlet	Athene	brama	Khurule Pecha		
255	Collared Scops-owl	Otus	lettia	Konthi Nimpecha		
256	Indian Scops-owl	Otus	bakkamoena	Deshi Nimpecha		DD
257	Mountain Scops-owl	Otus	spilocephalus	Pahari Nimpecha		DD
258	Oriental Scops-owl	Otus	sunia	Udoyee Nimpecha		
259	Short-eared Owl	Asio	flammeus	Sotokaan Pecha		
260	Brown Wood-owl	Strix	leptogrammica	Khoyra Gachpecha		
261	Spot-bellied Eagle-owl	Bubo	nipalensis	Chitpet Hootompecha		
262	Dusky Eagle-owl	Bubo	coromandus	Metey Hootompecha		
263	Brown Fish-owl	Ketupa	zeylonensis	Khoyra Mesopecha		
264	Tawny Fish-owl	Ketupa	flavipes	Tamatey Mesopecha		DD
265	Buffy Fish-owl	Ketupa	ketupu	Metey Mesopecha		DD
266	Osprey	Pandion	haliaetus	Masmural		

	English Name	Genus	Species	Bangla Name	Th	BDTh
267	Black-winged Kite	*Elanus*	*caeruleus*	Katua Chil		
268	Oriental Honey-buzzard	*Pernis*	*ptilorhynchus*	Udoyee Modhubaaj		
269	Jerdon's Baza	*Aviceda*	*jerdoni*	Jerdoner Baaj		
270	Black Baza	*Aviceda*	*leuphotes*	Kala Baaj		
271	Egyptian Vulture	*Neophron*	*percnopterus*	Dhola Shokoon	EN	DD
272	Crested Serpent-eagle	*Spilornis*	*cheela*	Teela Naageagle		
273	Short-toed Snake-eagle	*Circaetus*	*gallicus*	Dholapet Shaapeagle		
274	Red-headed Vulture	*Sarcogyps*	*calvus*	Raaj Shokoon	CR	RE
275	Himalayan Griffon (Vulture)	*Gyps*	*himalayensis*	Himaloyee Gridhini	NT	
276	White-rumped Vulture	*Gyps*	*bengalensis*	Bangla Shokoon	CR	CR
277	Slender-billed Vulture	*Gyps*	*tenuirostris*	Shoroothuti Shokoon	CR	DD
278	(Eurasian) Griffon Vulture	*Gyps*	*fulvus*	Euresio Gridhini		
279	Cinereous Vulture	*Aegypius*	*monachus*	Kala Shokoon	NT	NT
280	Mountain Hawk-eagle	*Nisaetus*	*nipalensis*	Pahari Shikreyeagle		VU
281	Changeable Hawk-eagle	*Nisaetus*	*cirrhatus*	Bohurupi Shikreyeagle		
282	Rufous-bellied Eagle	*Lophotriorchis*	*kienerii*	Lalpet Eagle	NT	VU
283	Black Eagle	*Ictinaetus*	*malaiensis*	Kala Eagle		DD
284	Indian Spotted Eagle	*Clanga*	*hastata*	Deshi Gutieagle	VU	EN
285	Greater Spotted Eagle	*Clanga*	*clanga*	Boro Gutieagle	VU	VU
286	Tawny Eagle	*Aquila*	*rapax*	Tamatey Eagle	VU	
287	Steppe Eagle	*Aquila*	*nipalensis*	Nepali Eagle	EN	
288	Eastern Imperial Eagle	*Aquila*	*heliaca*	Aesio Shahieagle	VU	VU
289	Bonelli's Eagle	*Aquila*	*fasciata*	Bonelli Eagle		
290	Booted Eagle	*Hieraaetus*	*pennatus*	Bootpa Eagle		
291	Western (Eurasian) Marsh-harrier	*Circus*	*aeruginosus*	Poshchima Paankapashi		
292	Eastern Marsh-harrier	*Circus*	*spilonotus*	Puber Paankapashi		
293	Hen Harrier	*Circus*	*cyaneus*	Murgi Kapashi		DD
294	Pallid Harrier	*Circus*	*macrourus*	Dhola Kapashi	NT	DD
295	Pied Harrier	*Circus*	*melanoleucos*	Pakra Kapashi		
296	Montagu's Harrier	*Circus*	*pygargus*	Montagur Kapashi		
297	Crested Goshawk	*Accipiter*	*trivirgatus*	Jhootial Godashikrey		
298	Shikra	*Accipiter*	*badius*	Pati Shikrey		
299	Japanese Sparrowhawk	*Accipiter*	*gularis*	Japani Choruishikrey		
300	Besra	*Accipiter*	*virgatus*	Besra Shikrey		
301	Eurasian Sparrowhawk	*Accipiter*	*nisus*	Euresio Choruishikrey		
302	Northern Goshawk	*Accipiter*	*gentilis*	Utturey Godashikrey		DD
303	White-bellied Sea-eagle	*Haliaeetus*	*leucogaster*	Dholapet Sindhueagle		
304	Pallas's Fish-eagle	*Haliaeetus*	*leucoryphus*	Pallasi Kuraeagle	EN	EN
305	White-tailed Sea-eagle	*Haliaeetus*	*albicilla*	Dholalej Sindhueagle		
306	Grey-headed Fish-eagle	*Ichthyophaga*	*ichthyaetus*	Meteymatha Kuraeagle	NT	NT
307	Brahminy Kite	*Haliastur*	*indus*	Shongkho Cheel		
308	Black Kite	*Milvus*	*migrans*	Bhubon Cheel		
309	White-eyed Buzzard	*Butastur*	*teesa*	Dholachokh Tisabaaj		
310	Japanese (Eurasian?) Buzzard	*Buteo*	*japonicus*	Pati Tisabaaj		
311	Long-legged Buzzard	*Buteo*	*rufinus*	Lombapa Tisabaaj		
312	Red-headed Trogon	*Harpactes*	*erythrocephalus*	Laalmatha Kuchkuchi		

	English Name	Genus	Species	Bangla Name	Th	BDTh
313	Great Hornbill	Buceros	bicornis	Raaj Dhonesh	VU	VU
314	Indian Grey Hornbill	Ocyceros	birostris	Deshi Meteydhonesh		
315	Oriental Pied Hornbill	Anthracoceros	albirostris	Udoyee Pakradhonesh		
316	Wreathed Hornbill	Rhyticeros	undulatus	Patathuti Dhonesh	VU	DD
317	Common Hoopoe	Upupa	epops	Pati Hood-hood		
318	Blue-bearded Bee-eater	Nyctyornis	athertoni	Neeldari Shuichora		
319	Asian Green Bee-eater	Merops	orientalis	Shobooj Shuichora		
320	Chestnut-headed Bee-eater	Merops	leschenaulti	Khoyramatha Shuichora		
321	Blue-tailed Bee-eater	Merops	philippinus	Neel-lej Shuichora		
322	Indian Roller	Coracias	benghalensis	Bangla Neelkantho		
323	Indochinese Roller	Coracias	affinis	Chinbharati Neelkantho		DD
324	Oriental Dollarbird	Eurystomus	orientalis	Pahari Neelkantho		
325	Oriental Dwarf-kingfisher	Ceyx	erithaca	Udoyee Bamonranga		EN
326	Blue-eared Kingfisher	Alcedo	meninting	Neelkan Machranga		
327	Blyth's Kingfisher	Alcedo	hercules	Blyther Machranga	NT	DD
328	Common Kingfisher	Alcedo	atthis	Pati Machranga		
329	Crested Kingfisher	Megaceryle	lugubris	Jhutial Machranga		DD
330	Pied Kingfisher	Ceryle	rudis	Pakra Machranga		
331	Stork-billed Kingfisher	Pelargopsis	capensis	Megh-hou Machranga		
332	Brown-winged Kingfisher	Pelargopsis	amauroptera	Khoirapakh Machranga	NT	VU
333	Ruddy Kingfisher	Halcyon	coromanda	Laal Machranga		
334	White-breasted (White-throated) Kingfisher	Halcyon	smyrnensis	Dholagola Machranga		
335	Black-capped Kingfisher	Halcyon	pileata	Kalatupi Machranga		
336	Collared Kingfisher	Todiramphus	chloris	Dholaghar Machranga		
337	Coppersmith Barbet	Psilopogon	haemacephalus	Shekra Boshonto		
338	Blue-eared Barbet	Psilopogon	cyanotis	Neelkan Boshonto		
339	Great Barbet	Psilopogon	virens	Boro Boshonto		NT
340	Lineated Barbet	Psilopogon	lineatus	Dagi Boshonto		
341	Blue-throated Barbet	Psilopogon	asiaticus	Neelgola Boshonto		
342	Eurasian Wryneck	Jynx	torquilla	Euresio Gharbetha		
343	White-browed Piculet	Sasia	ochracea	Dholavroo Kootikurali		
344	Speckled Piculet	Picumnus	innominatus	Teela Kootikurali		
345	Heart-spotted Woodpecker	Hemicircus	canente	Koljebooti Kathkurali		DD
346	Bay Woodpecker	Blythipicus	pyrrhotis	Tamatey Kathkurali		
347	Greater Flameback (Goldenback)	Chrysocolaptes	guttacristatus	Boro Kath-thokra		
348	Himalayan Flameback (Goldenback)	Dinopium	shorii	Himalayee Kath-thokra		DD
349	Common Flameback (Goldenback)	Dinopium	javanense	Pati Kath-thokra		
350	Black-rumped Flameback (Lesser Goldenback)	Dinopium	benghalense	Bangla Kath-thokra		
351	Pale-headed Woodpecker	Gecinulus	grantia	Dholamatha Kathkurali		
352	Rufous Woodpecker	Micropternus	brachyurus	Khoyra Kathkurali		
353	Greater Yellownape	Chrysophlegma	flavinucha	Boro Holdeykurali		
354	Lesser Yellownape	Picus	chlorolophus	Choto Holdeykurali		
355	Streak-throated Woodpecker	Picus	xanthopygaeus	Dagigola Kathkurali		
356	Streak-breasted Woodpecker	Picus	viridanus	Dagibook Kathkurali		

	English Name	Genus	Species	Bangla Name	Th	BDTh
357	Black-naped (Grey-headed) Woodpecker	Picus	guerini	Meteymatha Kathkurali		
358	Great Slaty Woodpecker	Mulleripicus	pulverulentus	Boro Meteykurali	VU	NT
359	Grey-capped (Pygmy) Woodpecker	Picoides	canicapillus	Meteytupi Batkurali		
360	Indian (Brown-capped) Pygmy Woodpecker	Picoides	nanus	Deshi Batkurali		
361	Yellow-crowned Woodpecker	Leiopicus	mahrattensis	Holdeytupi Kathkurali		DD
362	Fulvous-breasted Woodpecker	Dendrocopos	macei	Batabi Kathkurali		
363	Lesser Kestrel	Falco	naumanni	Choto Kestrel		
364	Common Kestrel	Falco	tinnunculus	Pati Kestrel		
365	Red-headed (Red-necked) Falcon	Falco	chicquera	Laalghar Shaheen	NT	
366	Amur Falcon	Falco	amurensis	Amur Shaheen		
367	Merlin	Falco	columbarius	Merlin		
368	Eurasian Hobby	Falco	subbuteo	Euresio Tikashaheen		
369	Oriental Hobby	Falco	severus	Udoyee Tikashaheen		
370	Laggar Falcon	Falco	jugger	Laggar Shaheen	NT	VU
371	Saker Falcon	Falco	cherrug	Saker Shaheen	EN	DD
372	Peregrine Falcon	Falco	peregrinus	Peregrine Shaheen		
373	Vernal Hanging-parrot	Loriculus	vernalis	Bashanti Lotkontia		
374	Grey-headed Parakeet	Psittacula	finschii	Meteymatha Tia	NT	VU
375	Blossom-headed Parakeet	Psittacula	roseata	Phoolmatha Tia	NT	NT
376	Plum-headed Parakeet	Psittacula	cyanocephala	Laalmatha Tia		
377	Red-breasted Parakeet	Psittacula	alexandri	Modna Tia	NT	
378	Alexandrine Parakeet	Psittacula	eupatria	Chondona Tia	NT	
379	Rose-ringed Parakeet	Psittacula	krameri	Shobooj Tia		
380	Blue-naped Pitta	Hydrornis	nipalensis	Neelghar Shoomcha		
381	Blue Pitta	Hydrornis	cyaneus	Neel Shoomcha		
382	Indian Pitta	Pitta	brachyura	Deshi Shoomcha		
383	Mangrove Pitta	Pitta	megarhyncha	Para Shoomcha	NT	
384	Western Hooded Pitta	Pitta	sordida	Khoyramatha Shoomcha		
385	Long-tailed Broadbill	Psarisomus	dalhousiae	Lanja Motathuti		DD
386	Grey-browed (Silver-breasted) Broadbill	Serilophus	rubropygius	Chadibook Motathuti		
387	Maroon Oriole	Oriolus	traillii	Tamarong Benebou		
388	Black-hooded Oriole	Oriolus	xanthornus	Kalamatha Benebou		
389	Indian Golden Oriole	Oriolus	kundoo	Deshi Sonabou		
390	Black-naped Oriole	Oriolus	chinensis	Kalaghar Benebou		
391	Slender-billed Oriole	Oriolus	tenuirostris	Shoruthot Benebou		
392	Mangrove Whistler	Pachycephala	cinerea	Nonabon Sheeshmar		
393	White-bellied Erpornis (Yuhina)	Erpornis	zantholeuca	Dholapet Yuhina		
394	Small Minivet	Pericrocotus	cinnamomeus	Choto Saheli		
395	Long-tailed Minivet	Pericrocotus	ethologus	Lanja Saheli		
396	Scarlet Minivet	Pericrocotus	flammeus	Shidurey Saheli		
397	Ashy Minivet	Pericrocotus	divaricatus	Metey Saheli		
398	Brown-rumped (Swinhoe's) Minivet	Pericrocotus	cantonensis	Swinhoer Saheli		
399	Rosy Minivet	Pericrocotus	roseus	Golapi Saheli		

	English Name	Genus	Species	Bangla Name	Th	BDTh
400	Large Cuckooshrike	*Coracina*	*javensis*	Boro Kabashi		
401	Black-winged Cuckooshrike	*Lalage*	*melaschistos*	Kalapakh Kabashi		
402	Black-headed Cuckooshrike	*Lalage*	*melanoptera*	Kalamatha Kabashi		
403	Ashy Woodswallow	*Artamus*	*fuscus*	Metey Bonababil		
404	Bar-winged Flycatcher-shrike	*Hemipus*	*picatus*	Dagipakh Chutkilatora		
405	Large Woodshrike	*Tephrodornis*	*virgatus*	Boro Bonlatora		
406	Common Woodshrike	*Tephrodornis*	*pondicerianus*	Pati Bonlatora		
407	Common Iora	*Aegithina*	*tiphia*	Kalaghar Rajon		
408	White-throated Fantail	*Rhipidura*	*albicollis*	Dholagola Satighuruni		
409	Black Drongo	*Dicrurus*	*macrocercus*	Kala Fingey		
410	Ashy Drongo	*Dicrurus*	*leucophaeus*	Metey Fingey		
411	White-bellied Drongo	*Dicrurus*	*caerulescens*	Dholapet Fingey		
412	Crow-billed Drongo	*Dicrurus*	*annectans*	Kakthuto Fingey		DD
413	Bronzed Drongo	*Dicrurus*	*aeneus*	Bronze Fingey		
414	Lesser Racquet-tailed Drongo	*Dicrurus*	*remifer*	Choto Racketfingey		
415	Hair-crested Drongo	*Dicrurus*	*hottentotus*	Keshori Fingey		
416	Greater Racquet-tailed Drongo	*Dicrurus*	*paradiseus*	Boro Racketfingey		
417	Black-naped Monarch	*Hypothymis*	*azurea*	Kalaghar Rajon		
418	Indian Paradise-flycatcher	*Terpsiphone*	*paradisi*	Deshi Shabulbuli		
419	Chinese Paradise-flycatcher	*Terpsiphone*	*incei*	China Shabulbuli		
420	Oriental Paradise-flycatcher	*Terpsiphone*	*affinis*	Udoyee Shabulbuli		
421	Brown Shrike	*Lanius*	*cristatus*	Khoyra Latora		
422	Burmese Shrike	*Lanius*	*collurioides*	Bormi Latora		
423	Bay-backed Shrike	*Lanius*	*vittatus*	Tamapith Latora		
424	Long-tailed Shrike	*Lanius*	*schach*	Lanja Latora		
425	Grey-backed Shrike	*Lanius*	*tephronotus*	Meteypith Latora		
426	Great Grey Shrike	*Lanius*	*excubitor*	Boro Meteylatora		
427	Rufous Treepie	*Dendrocitta*	*vagabunda*	Khoyra Harichacha		
428	Grey Treepie	*Dendrocitta*	*formosae*	Metey Harichacha		
429	Red-billed Blue Magpie	*Urocissa*	*erythrorhyncha*	Laalthot Neeltaura		
430	Common Green Magpie	*Cissa*	*chinensis*	Pati Shoboojtaura		
431	House Crow	*Corvus*	*splendens*	Pati Kaak		
432	Large-billed Crow	*Corvus*	*macrorhynchos*	Daar Kaak		
433	Grey-headed Canary-flycatcher	*Culicicapa*	*ceylonensis*	Meteymatha Canarychutki		
434	Sultan Tit	*Melanochlora*	*sultanea*	Sultan Tit		
435	Green-backed Tit	*Parus*	*monticolus*	Shoboojpith Tit		
436	Great Tit	*Parus*	*major*	Boro Tit		
437	Ashy-crowned Sparrow-lark	*Eremopterix*	*griseus*	Meteychadi Choruibhorot		
438	Horsfield's (Singing) Bushlark	*Mirafra*	*javanica*	Shurela Jharbhorot		
439	Bengal Bushlark	*Mirafra*	*assamica*	Bangla Jharbhorot		
440	Sand Lark	*Alaudala*	*raytal*	Bali Bhorot		
441	Hume's (Short-toed) Lark	*Calandrella*	*acutirostris*	Humer Bhorot		
442	Eastern (Greater) Short-toed Lark	*Calandrella*	*dukhunensis*	Boro Votabhorot		DD
443	Oriental Skylark	*Alauda*	*gulgula*	Udoyee Ovrobhorot		
444	Zitting Cisticola	*Cisticola*	*juncidis*	Vomra Choton		
445	Golden-headed Cisticola	*Cisticola*	*exilis*	Dholamatha Choton		

	English Name	Genus	Species	Bangla Name	Th	BDTh
446	Rufescent Prinia	Prinia	rufescens	Laalchey Prina		
447	Grey-breasted Prinia	Prinia	hodgsonii	Meteybook Prina		
448	Graceful Prinia	Prinia	gracilis	Shundori Prina		
449	Yellow-bellied Prinia	Prinia	flaviventris	Holdeypet Prina		
450	Ashy Prinia	Prinia	socialis	Kalchey Prina		DD
451	Plain Prinia	Prinia	inornata	Nirol Prina		
452	Common Tailorbird	Orthotomus	sutorius	Pati Tuntuni		
453	Dark-necked Tailorbird	Orthotomus	atrogularis	Kalagola Tuntuni		
454	Thick-billed Warbler	Arundinax	aedon	Motathot Futki		
455	Booted Warbler	Iduna	caligata	Bootpa Futki		
456	Sykes's Warbler	Iduna	rama	Sykeser Futki		
457	Black-browed Reed-warbler	Acrocephalus	bistrigiceps	Kalavru Nolfutki		
458	Large-billed Reed-warbler	Acrocephalus	orinus	Borothot Nolfutki	DD	DD
459	Blyth's Reed-warbler	Acrocephalus	dumetorum	Blyther Nolfutki		
460	Paddyfield Warbler	Acrocephalus	agricola	Dhani Futki		
461	Oriental Reed-warbler	Acrocephalus	orientalis	Udoyee Nolfutki		
462	Clamorous Reed-warbler	Acrocephalus	stentoreus	Bachal Nolfutki		
463	Pygmy Cupwing (Wren-babbler)	Pnoepyga	pusilla	Bamon Tunisatarey		DD
464	Pallas's Grasshopper-warbler (Rusty-rumped Warbler)	Locustella	certhiola	Pallasi Foringfutki		
465	Lanceolated Warbler	Locustella	lanceolata	Patari Futki		
466	Common Grasshopper-warbler	Locustella	naevia	Pati Foringfutki		
467	Baikal Grasshopper-warbler (Baikal Bush-warbler, David's Bush-warbler)	Locustella	davidi	Baikal Foringfutki		
468	Spotted Grasshopper-warbler (Bush-warbler)	Locustella	thoracica	Tila Foringfutki		
469	Striated Grassbird	Megalurus	palustris	Dagi Ghashpakhi		
470	Bristled Grassbird	Chaetornis	striata	Shotodagi Ghashpakhi	VU	EN
471	Asian House Martin	Delichon	dasypus	Aesio Ghornakuti		
472	Nepal House Martin	Delichon	nipalense	Nepali Ghornakuti		
473	Streak-throated Swallow	Petrochelidon	fluvicola	Dagigola Ababeel		
474	Red-rumped Swallow	Cecropis	daurica	Laalkomor Ababeel		
475	Wire-tailed Swallow	Hirundo	smithii	Tarleja Ababeel		
476	Barn Swallow	Hirundo	rustica	Pati Ababeel		
477	Asian Plain (Brown-throated) Martin	Riparia	chinensis	Nirol Nakuti		
478	Collared Sand Martin	Riparia	riparia	Bali Nakuti		
479	Pale Sand Martin	Riparia	diluta	Mlan Nakuti		
480	White-throated Bulbul	Alophoixus	flaveolus	Dholagola Bulbul		
481	Olive (Cachar) Bulbul	Iole	virescens	Jolpai Bulbul		
482	Ashy Bulbul	Hemixos	flavala	Kalchey Bulbul		
483	Black Bulbul	Hypsipetes	leucocephalus	Kala Bulbul		
484	Black-crested Bulbul	Pycnonotus	flaviventris	Kalajhuti Bulbul		
485	Red-whiskered Bulbul	Pycnonotus	jocosus	Sipahi Bulbul		
486	Red-vented Bulbul	Pycnonotus	cafer	Bangla Bulbul		
487	Flavescent Bulbul	Pycnonotus	flavescens	Metey Bulbul		DD
488	Black-headed Bulbul	Brachypodius	atriceps	Kalamatha Bulbul		
489	Yellow-browed Warbler	Phylloscopus	inornatus	Holdeyvru Futki		

	English Name	Genus	Species	Bangla Name	Th	BDTh
490	Hume's Leaf-warbler	*Phylloscopus*	*humei*	Humer Patafutki		
491	Lemon-rumped Leaf-warbler (Warbler)	*Phylloscopus*	*chloronotus*	Lebukomor Patafutki		
492	Dusky Warbler	*Phylloscopus*	*fuscatus*	Kalchey Futki		
493	Smoky Warbler	*Phylloscopus*	*fuligiventer*	Metey Futki		
494	Siberian Chiffchaff	*Phylloscopus*	*tristis*	Siberio Chiffchaff		
495	Tickell's Leaf-warbler	*Phylloscopus*	*affinis*	Tickeller Patafutki		
496	White-spectacled Warbler	*Phylloscopus*	*intermedius*	Dholachoshma Futki		
497	Green-crowned Warbler	*Phylloscopus*	*burki*	Shobujchadi Futki		DD
498	Grey-crowned Warbler	*Phylloscopus*	*tephrocephalus*	Meteychadi Futki		DD
499	Whistler's Warbler	*Phylloscopus*	*whistleri*	Whiistlerer Futki		DD
500	Chestnut-crowned Warbler	*Phylloscopus*	*castaniceps*	Khoyrachadi Futki		
501	Green Warbler	*Phylloscopus*	*nitidus*	Shobooj Futki		
502	Greenish Warbler	*Phylloscopus*	*trochiloides*	Shobjey Futki		
503	Pale-legged Leaf-warbler	*Phylloscopus*	*tenellipes*	Dholapa Patafutki		
504	Large-billed Leaf-warbler	*Phylloscopus*	*magnirostris*	Borothot Patafutki		
505	Yellow-vented Warbler	*Phylloscopus*	*cantator*	Holdeytola Futki		
506	Blyth's Leaf-warbler	*Phylloscopus*	*reguloides*	Blyther Patafutki		
507	Western Crowned Leaf-warbler	*Phylloscopus*	*occipitalis*	Poshchima Mathafutki		DD
508	Grey-hooded Warbler	*Phylloscopus*	*xanthoschistos*	Meteymukhosh Futki		
509	Slaty-bellied Tesia	*Tesia*	*olivea*	Slatepet Tesia		
510	Grey-bellied Tesia	*Tesia*	*cyaniventer*	Meteypet Tesia		
511	Chestnut-crowned Bush-warbler	*Cettia*	*major*	Khoyrachadi Jharfutki		
512	Grey-sided Bush-warbler	*Cettia*	*brunnifrons*	Meteypash Jharfutki		
513	Chestnut-headed Tesia	*Cettia*	*castaneocoronata*	Khoyramatha Tesia		
514	Asian Stubtail	*Urosphena*	*squameiceps*	Aesio Bhotalej		
515	Yellow-bellied Warbler	*Abroscopus*	*superciliaris*	Holdeypet Futki		NT
516	Mountain Tailorbird	*Phyllergates*	*cucullatus*	Pahari Tuntuni		
517	Brownish-flanked Bush-warbler	*Horornis*	*fortipes*	Meteypash Futki		
518	Aberrant Bush-warbler	*Horornis*	*flavolivaceus*	Pashua Jharfutki		
519	Eastern Orphean Warbler	*Sylvia*	*crassirostris*	Puber Shurelafutki		
520	Lesser Whitethroat	*Sylvia*	*curruca*	Choto Dholagola		
521	Yellow-eyed Babbler	*Chrysomma*	*sinense*	Holdeychokh Satarey		VU
522	Black-breasted Parrotbill	*Paradoxornis*	*flavirostris*	Kalabook Tiathuti	VU	RE
523	Spot-breasted Parrotbill	*Paradoxornis*	*guttaticollis*	Tilabook Tiathuti		RE
524	Rufous-headed (Greater Rufous-headed) Parrotbill	*Psittiparus*	*bakeri*	Boro Laalmatha Tiathuti		RE
525	Striated Yuhina	*Yuhina*	*castaniceps*	Dagi Yuhina		DD
526	Indian (Oriental) White-eye	*Zosterops*	*palpebrosus*	Udoyee Dholachokh		
527	Red-billed Scimitar-babbler	*Pomatorhinus*	*ochraceiceps*	Laalthot Kasteysatarey		DD
528	White-browed Scimitar-babbler	*Pomatorhinus*	*schisticeps*	Dholavru Kasteysatarey		NT
529	Large Scimitar-babbler	*Erythrogenys*	*hypoleucos*	Boro Kasteysatarey		
530	Spot-breasted Scimitar-babbler	*Erythrogenys*	*mcclellandi*	Tilabook Kasteysatarey		DD
531	Grey-throated Babbler	*Stachyris*	*nigriceps*	Meteygola Satarey		
532	Chestnut-capped Babbler	*Timalia*	*pileata*	Laaltupi Satarey		
533	Pin-striped (Striped) Tit-babbler	*Mixornis*	*gularis*	Dagi Titsatarey		
534	Rufous-fronted Babbler	*Cyanoderma*	*rufifrons*	Laalkopal Satarey		NT

English Name	Genus	Species	Bangla Name	Th	BDTh
535	White-hooded Babbler	*Gampsorhynchus*	*rufulus*	Dholamookhosh Satarey		EN
536	Swamp Grass-babbler (Prinia)	*Laticilla*	*cinerascens*	Bada Ghashsatarey	EN	
537	Puff-throated Babbler	*Pellorneum*	*ruficeps*	Golafola Satarey		
538	Marsh Babbler	*Pellorneum*	*palustre*	Bada Satarey	VU	DD
539	Spot-throated Babbler	*Pellorneum*	*albiventre*	Tilabook Satarey		DD
540	Buff-breasted Babbler	*Trichastoma*	*tickelli*	Khoyrabook Satarey		EN
541	Abbott's Babbler	*Malacocincla*	*abbotti*	Abbotter Satarey		
542	Streaked Wren-babbler	*Turdinus*	*brevicaudatus*	Dagi Tunisatarey		
543	Indian Grass-babbler (Rufous-rumped Grassbird)	*Graminicola*	*bengalensis*	Bangla Ghashpakhi	NT	EN
544	Brown-cheeked Fulvetta	*Alcippe*	*poioicephala*	Khoyragal Fulvetta		NT
545	Nepal Fulvetta	*Alcippe*	*nipalensis*	Nepali Fulvetta		NT
546	Striated Babbler	*Argya*	*earlei*	Dagi Satarey		
547	Common Babbler	*Argya*	*caudata*	Pati Satarey		
548	Jungle Babbler	*Turdoides*	*striata*	Bon Satarey		
549	Lesser Necklaced Laughingthrush	*Garrulax*	*monileger*	Choto Malapenga		
550	White-crested Laughingthrush	*Garrulax*	*leucolophus*	Dholamookhosh Penga		
551	Greater Necklaced Laughingthrush	*Garrulax*	*pectoralis*	Boro Malapenga		
552	Rufous-necked Laughingthrush	*Garrulax*	*ruficollis*	Laalghar Penga		
553	Yellow-throated Laughingthrush	*Garrulax*	*galbanus*	Holdeygola Penga		DD
554	Long-tailed Sibia	*Heterophasia*	*picaoides*	Lanja Sibia		
555	Rusty-fronted Barwing	*Actinodura*	*egertoni*	Laalmukh Dagidana		RE
556	Bar-tailed Treecreeper	*Certhia*	*himalayana*	Dagilej Gachachra		
557	Indian Nuthatch	*Sitta*	*castanea*	Deshi Bonomali		
558	Chestnut-bellied Nuthatch	*Sitta*	*cinnamoventris*	Khoyrapet Bonomali		
559	Velvet-fronted Nuthatch	*Sitta*	*frontalis*	Kalakopal Bonomali		
560	Common Starling	*Sturnus*	*vulgaris*	Pati Kathshalik		
561	Rosy Starling	*Pastor*	*roseus*	Golapi Kathshalik		
562	Purple-backed (Daurian) Starling	*Agropsar*	*sturninus*	Laalpith Kathshalik		
563	Chestnut-cheeked Starling	*Agropsar*	*philippensis*	Khoyragal Kathshalik		
564	Asian Pied Starling (Pied Myna)	*Gracupica*	*contra*	Pakra Shalik		
565	Brahminy Starling	*Sturnia*	*pagodarum*	Bamuni Kathshalik		
566	Chestnut-tailed Starling	*Sturnia*	*malabaricus*	Khoyralej Kathshalik		
567	Common Myna	*Acridotheres*	*tristis*	Bhat Shalik		
568	Bank Myna	*Acridotheres*	*ginginianus*	Gang Shalik		
569	Jungle Myna	*Acridotheres*	*fuscus*	Jhuti Shalik		
570	Great (White-vented) Myna	*Acridotheres*	*grandis*	Dhlatola Shalik		
571	Spot-winged Starling	*Saroglossa*	*spilopterus*	Tila Telshalik		
572	Common Hill Myna	*Gracula*	*religiosa*	Pati Myna		
573	Asian Glossy Starling	*Aplonis*	*panayensis*	Aesio Telshalik		
574	'Plain Backed' Thrush complex (possibly Alpine)	*Zoothera*	*mollissima/salimalii*	(no name)		
575	Dark-sided Thrush	*Zoothera*	*marginata*	Kalapash Dama		DD
576	Long-billed Thrush	*Zoothera*	*monticola*	Lombathot Dama		
577	Scaly Thrush	*Zoothera*	*dauma*	Ashtey Dama		
578	Purple Cochoa	*Cochoa*	*purpurea*	Beguni Cochoa		
579	Orange-headed Thrush	*Geokichla*	*citrina*	Komla Dama		

	English Name	Genus	Species	Bangla Name	Th	BDTh
580	Grey-winged Blackbird	*Turdus*	*boulboul*	Dholapakh Kalidama		
581	Black-breasted Thrush	*Turdus*	*dissimilis*	Kalabook Dama		
582	Tickell's Thrush	*Turdus*	*unicolor*	Tickeller Dama		
583	Eyebrowed Thrush	*Turdus*	*obscurus*	Vrulekha Dama		
584	Tibetan Blackbird	*Turdus*	*maximus*	Tibeti Kalidama		
585	White-collared Blackbird	*Turdus*	*albocinctus*	Dholaghar Kalidama		
586	Dusky Thrush	*Turdus*	*eunomus*	Kalchey Dama		
587	Black- (Dark-) throated Thrush	*Turdus*	*atrogularis*	Kalagola Dama		
588	Rufous-throated (Red-throated) Thrush	*Turdus*	*ruficollis*	Laalgola Dama		
589	Oriental Magpie-robin	*Copsychus*	*saularis*	Udoyee Doel		
590	White-rumped Shama	*Kittacincla*	*malabarica*	Dholakomor Shama		
591	Dark-sided Flycatcher	*Muscicapa*	*sibirica*	Kalpash Chutki		
592	Ferruginous Flycatcher	*Muscicapa*	*ferruginea*	Morcheyrong Chutki		
593	Brown-breasted Flycatcher	*Muscicapa*	*muttui*	Meteybook Chutki		
594	Asian Brown Flycatcher	*Muscicapa*	*dauurica*	Aesio Khoyrachutki		
595	Spotted Flycatcher	*Muscicapa*	*striata*	Tila Chutki		
596	Rufous-bellied Niltava	*Niltava*	*sundara*	Laalpet Neelmoni		
597	Small Niltava	*Niltava*	*macgrigoriae*	Choto Neelmoni		
598	Large Niltava	*Niltava*	*grandis*	Boro Neelmoni		
599	Verditer Flycatcher	*Eumyias*	*thalassinus*	Ombor Chutki		
600	Pale Blue-flycatcher	*Cyornis*	*unicolor*	Neelchey Chutki		
601	Pale-chinned Flycatcher (Blue-flycatcher)	*Cyornis*	*poliogenys*	Dholagola Chutki		
602	Large Blue-flycatcher	*Cyornis*	*magnirostris*	Boro Neelchutki		
603	Hill Blue-flycatcher	*Cyornis*	*banyumas*	Pahari Neelchutki		
604	Tickell's Blue-flycatcher	*Cyornis*	*tickelliae*	Tickeller Neelchutki		
605	Blue-throated Blue-flycatcher	*Cyornis*	*rubeculoides*	Neelgola Neelchutki		
606	Lesser Shortwing	*Brachypteryx*	*leucophris*	Choto Khatodana		
607	Indian Blue Robin	*Larvivora*	*brunnea*	Deshi Neelrobin		
608	Siberian Blue Robin	*Larvivora*	*cyane*	Siberio Neelrobin		
609	Bluethroat	*Cyanecula*	*svecica*	Neelgola Fidda		
610	Firethroat	*Calliope*	*pectardens*	Laalgola Fidda	NT	NT
611	Siberian Rubythroat	*Calliope*	*calliope*	Siberio Chunikonthi		
612	Chinese (White-tailed) Rubythroat	*Calliope*	*tschebaiewi*	Dholalej Chunikonthi		NT
613	White-tailed Blue Robin	*Myiomela*	*leucura*	Dholalej Robin		
614	Rufous-breasted Bush-robin	*Tarsiger*	*hyperythrus*	Lalbook Bonrobin		
615	Himalayan (Orange-flanked) Bush-robin (Red-flanked Bluetail)	*Tarsiger*	*rufilatus*	Komlapash Bonrobin		
616	White-browed Bush-robin	*Tarsiger*	*indicus*	Dholavru Bonrobin		
617	Slaty-backed Forktail	*Enicurus*	*schistaceus*	Slatepith Cheralej		
618	Black-backed Forktail	*Enicurus*	*immaculatus*	Kalapith Cheralej		
619	White-crowned Forktail	*Enicurus*	*leschenaulti*	Dholachadi Cheralej		
620	Blue Whistling-thrush	*Myophonus*	*caeruleus*	Neel Sheeshdama		
621	Slaty-backed Flycatcher	*Ficedula*	*erithacus*	Slatepith Chutki		
622	Slaty-blue Flycatcher	*Ficedula*	*tricolor*	Kalcheyneel Chutki		

	English Name	Genus	Species	Bangla Name	Th	BDTh
623	Snowy-browed Flycatcher	Ficedula	hyperythra	Dholavru Chutki		
624	Rufous-gorgeted Flycatcher	Ficedula	strophiata	Laalmala Chutki		
625	Sapphire Flycatcher	Ficedula	sapphira	Neelkanto Chutki		
626	Ultramarine Flycatcher	Ficedula	superciliaris	Ghononeel Chutki		
627	Little Pied Flycatcher	Ficedula	westermanni	Choto Pakrachutki		
628	Rusty-tailed Flycatcher	Ficedula	ruficauda	Koyralej Chutki		
629	Red-breasted Flycatcher	Ficedula	parva	Laalbook Chutki		
630	Red-throated (Taiga) Flycatcher	Ficedula	albicilla	Taiga Chutki		
631	Blue-fronted Redstart	Phoenicurus	frontalis	Neelkopali Girdi		
632	White-capped Water-redstart	Phoenicurus	leucocephalus	Dholatupi Paangirdi		
633	Plumbeous Water-redstart	Phoenicurus	fuliginosus	Neel Paangirdi		
634	Black Redstart	Phoenicurus	ochruros	Kala Girdi		
635	Daurian Redstart	Phoenicurus	auroreus	Daurian Girdi		
636	Blue-capped Rock-thrush	Monticola	cinclorhyncha	Neeltupi Shiladama		
637	Chestnut-bellied Rock-thrush	Monticola	rufiventris	Khoyrapet Shiladama		
638	Blue Rock-thrush	Monticola	solitarius	Neel Shiladama		
639	Jerdon's Bushchat	Saxicola	jerdoni	Jerdoner Jharfidda		DD
640	Grey Bushchat	Saxicola	ferreus	Metey Jharfidda		
641	White-throated (Hodgson's) Bushchat	Saxicola	insignis	Dholagola Jharfidda	VU	DD
642	Pied Bushchat	Saxicola	caprata	Pakra Jharfidda		
643	White-tailed Stonechat	Saxicola	leucurus	Dholalej Shilafidda		
644	Common Stonechat	Saxicola	torquatus	Pati Shilafidda		
645	Desert Wheatear	Oenanthe	deserti	Ooshor Kankali		
646	Asian Fairy Bluebird	Irena	puella	Aesio Neelpori		
647	Golden-fronted Leafbird	Chloropsis	aurifrons	Sonakopali Horbola		
648	Jerdon's Leafbird	Chloropsis	jerdoni	Jerdon Horbola		
649	Orange-bellied Leafbird	Chloropsis	hardwickii	Komlapet Horbola		
650	Blue-winged Leafbird	Chloropsis	moluccensis	Neeldana Horbola		
651	Yellow-bellied Flowerpecker	Dicaeum	melanozanthum	Holdeypet Phooljhuri		
652	Yellow-vented Flowerpecker	Dicaeum	chrysorrheum	Holeytola Phooljhuri		
653	Thick-billed Flowerpecker	Dicaeum	agile	Thotmota Phooljhuri		
654	Orange-bellied Flowerpecker	Dicaeum	trigonostigma	Komlapet Phooljhuri		
655	Pale-billed Flowerpecker	Dicaeum	erythrorhynchos	Meteythot Phooljhuri		
656	Plain Flowerpecker	Dicaeum	minullum	Nirol Phooljhuri		
657	Scarlet-backed Flowerpecker	Dicaeum	cruentatum	Laalpith Phooljhuri		
658	Fire-breasted Flowerpecker	Dicaeum	ignipectus	Laalbook Phooljhuri		
659	Little Spiderhunter	Arachnothera	longirostra	Choto Makormar		
660	Streaked Spiderhunter	Arachnothera	magna	Boro Makormar		
661	Ruby-cheeked Sunbird	Chalcoparia	singalensis	Chunimookhi Moutushi		
662	Purple-rumped Sunbird	Leptocoma	zeylonica	Begunikomor Moutushi		
663	Maroon-bellied Sunbird	Leptocoma	brasiliana	Begunigola Moutushi		
664	Purple Sunbird	Cinnyris	asiaticus	Beguni Moutushi		
665	Olive-backed Sunbird	Cinnyris	jugularis	Jolpaipith Moutushi		
666	Fire-tailed Sunbird	Aethopyga	ignicauda	Laal-lej Moutushi		
667	Green-tailed Sunbird	Aethopyga	nipalensis	Shobujlej Moutushi		
668	(Mrs) Gould's Sunbird	Aethopyga	gouldiae	Begumgolder Moutushi		

	English Name	Genus	Species	Bangla Name	Th	BDTh
669	Crimson Sunbird	*Aethopyga*	*siparaja*	Sidurey Moutushi		
670	Black-breasted Weaver	*Ploceus*	*benghalensis*	Bangla Babui		
671	Streaked Weaver	*Ploceus*	*manyar*	Dagi Babui		
672	Baya Weaver	*Ploceus*	*philippinus*	Deshi Babui		
673	Red Avadavat	*Amandava*	*amandava*	Laal Mamunia		
674	Indian Silverbill	*Euodice*	*malabarica*	Deshi Chadithot		
675	White-rumped Munia	*Lonchura*	*striata*	Dholakomor Munia		
676	Scaly-breasted Munia	*Lonchura*	*punctulata*	Tila Munia		
677	Tricoloured (Black-headed) Munia	*Lonchura*	*malacca*	Kalamatha Munia		
678	Chestnut Munia	*Lonchura*	*atricapilla*	Khoyra Munia		
679	House Sparrow	*Passer*	*domesticus*	Pati Chorui		
680	Eurasian Tree Sparrow	*Passer*	*montanus*	Euresio Gaas-chorui		
681	Forest Wagtail	*Dendronanthus*	*indicus*	Bon Khonjon		
682	Tree Pipit	*Anthus*	*trivialis*	Geso Tulika		
683	Olive-backed Pipit	*Anthus*	*hodgsoni*	Jolpaipith Tulika		
684	Red-throated Pipit	*Anthus*	*cervinus*	Laalgola Tulika		
685	Rosy Pipit	*Anthus*	*roseatus*	Golapi Tulika		
686	Richard's Pipit	*Anthus*	*richardi*	Richarder Tulika		
687	Paddyfield Pipit	*Anthus*	*rufulus*	Dhani Tulika		
688	Blyth's Pipit	*Anthus*	*godlewskii*	Blyther Tulika		
689	Tawny Pipit	*Anthus*	*campestris*	Tamatey Tulika		
690	Long-billed Pipit	*Anthus*	*similis*	Lombathot Tulika		
691	Western Yellow Wagtail	*Motacilla*	*flava*	Poshchima Holdeykhonjon		
692	Grey Wagtail	*Motacilla*	*cinerea*	Metey Khonjon		
693	Citrine Wagtail	*Motacilla*	*citreola*	Citrine Khonjon		
694	Eastern Yellow Wagtail	*Motacilla*	*tschutschensis*	Puber Holdeykhonjon		
695	White-browed Wagtail	*Motacilla*	*maderaspatensis*	Dholavru Khonjon		
696	White Wagtail	*Motacilla*	*alba*	Dhola Khonjon		
697	Common Rosefinch	*Carpodacus*	*erythrinus*	Pati Tuti		
698	Crested Bunting	*Emberiza*	*lathami*	Jhutial Chatak		
699	Black-headed Bunting	*Emberiza*	*melanocephala*	Kalamatha Chatak		
700	Chestnut-eared Bunting	*Emberiza*	*fucata*	Lalkan Chatak		
701	Grey-necked Bunting	*Emberiza*	*buchanani*	Meteyghar Chatak		
702	Yellow-breasted Bunting	*Emberiza*	*aureola*	Holdeybook Chatak	CR	VU
703	Little Bunting	*Emberiza*	*pusilla*	Khudey Chatak		
704	Black-faced Bunting	*Emberiza*	*spodocephala*	Kalamukh Chatak		
705	Tristram's Bunting	*Emberiza*	*tristrami*	Tristram Chatak		

APPENDIX 3: HYPOTHETICAL AND UNCONFIRMED BIRDS THAT MIGHT OCCUR IN BANGLADESH

There are published references to the following 162 species for Bangladesh, but the evidence, or lack of it, has rendered these birds to be treated here as hypothetical/unconfirmed. More than 30 species that were similarly assessed in the early 2000s had confirmed sightings in Bangladesh by December 2019. Species included here were considered possible by the authors, or were repeated in various documents and published lists without a verifiable or documented source record. In some cases, genuine records may have been relegated here because of identification challenges, or the lack of adequate identification information and related notes in the past, or the loss of specimens, or confusion over locations and lists (for example, Bangladesh was part of Pakistan between 1947 and 1971).

1. Japanese Quail *Coturnix japonica*
Listed as possible for Bangladesh by Rashid (1967); although a vagrant to NE India there is no confirmed record in Bangladesh.

2. See-see Partridge *Ammoperdix griseogularis*
Listed as a winter visitor in Mountfort (1969) with no further details, presumably an error as that expedition recorded this desert species in (then West) Pakistan.

3. Manipur Bush-quail *Perdicula manipurensis*
Although listed as possible for Bangladesh by several authors, these claims are probably all based on the statement in Baker (1922–30) quoted in BirdLife International (2001) 'quite possibly it is equally common in the similar immense grassy areas in Western Cachar and Sylhet'. Searches in the Chittagong Hill Tracts and any other tall grassland remain a priority for this globally threatened (EN) species.

4. Red Spurfowl *Galloperdix spadicea*
Although the online database ornisnet.org of November 2006 (ORNIS, 2006) lists a specimen in the Museum of Comparative Zoology, Harvard, there is no published record, and it seems unlikely to be from Bangladesh as this species is limited to India south of the Gangetic plains.

5. Chinese Francolin *Francolinus pintadeanus*
Listed as possible for Bangladesh by Rashid (1967), but there is no confirmed record; the nearest part of its range is in Manipur.

6. Mountain Bamboo-partridge *Bambuscicola fytchii*
Although included in Baker (1922–30), Rashid (1967) and Harvey (1990), Rasmussen & Anderton (2005) found no specimens for Bangladesh, nor for Meghalaya, so treated as unproven.

7. Blyth's Tragopan *Tragopan blythii*
Mentioned for Bangladesh in Olivier (1979) but there is no confirmed record.

8. Mrs Hume's Pheasant *Syrmaticus humiae*
Although listed for Bangladesh in Husain (1979) and Sarker & Sarker (1988), there is no confirmed record. BirdLife International (2001) concluded that this species was unlikely to occur at the low altitudes of the Bangladesh hills.

9. Red-breasted Goose *Branta ruficollis*
Listed as possible by Rashid (1967) and Sarker & Sarker (1988), but there are no confirmed records and this species was treated as hypothetical in South Asia by Rasmussen and Anderton (2005).

10. Bean Goose *Anser fabalis*
There are no details of either a possible record (Mountfort 1969) or a report from Rangamati (Husain 1975; Khan 1982), so this species is considered as unconfirmed. There are very few records from South Asia (Ali & Ripley 1987; Rasmussen & Anderton 2005).

11. Pink-footed Goose *Anser brachyrhynchus*
Listed as possible by Rashid (1967) but there are no confirmed records from South Asia (Ali & Ripley 1987; Rasmussen & Anderton 2005).

12. Marbled Teal *Marmaronetta angustirostris*
Although listed in Khan (1982), no location or details of this record are available; it is only a rare vagrant to E India, and this is treated as unconfirmed.

13. White-tailed Tropicbird *Phaethon lepturus*
Listed by Rashid (1967) but there is no evidence that this pelagic species has occurred in Bangladesh and its territorial waters, although there is one record from Assam (Rasmussen & Anderton 2005).

14. Speckled Woodpigeon *Columba hodgsonii*
Listed by Sarker & Sarker (1988) but without details so it is considered as unconfirmed.

15. Painted Sandgrouse *Pterocles indicus*
Although listed as possibly occurring by Mountfort & Poore (1968), there is no confirmed record of this dry-country species; the closest records are from West Bengal, India (Rasmussen & Anderton 2005).

16. White-throated Needletail *Hirundapus caudacutus*
Listed for Bangladesh by Rashid (1967) and Ali & Ripley (1987), but no specimens or detailed records have been traced (Rasmussen & Anderton 2005).

17. Alpine Swift *Tachymarptis melba*
Although listed for Bangladesh by Rashid (1967) and Harvey (1990) on the basis of unspecified records in 1982–83, no details were forthcoming. Rasmussen & Anderton (2005) found no specimens from Bangladesh.

18. Dark-rumped Swift *Apus acuticauda*
Listed as hypothetical by Rashid (1967), and it nests in Meghalaya on cliffs overlooking Bangladesh, but there are no definite records.

19. Common Swift *Apus apus*
Listed as possible by Rashid (1967) but there is no confirmed record.

20. Red-legged Crake *Rallina fasciata*
Although listed by Rashid (1967) and Sarker & Sarker (1986),

no detailed evidence was given. Although it is partly migratory, Rasmussen & Anderton (2005) also considered old records from nearby Cachar in Assam to be unconfirmed.

21. Little Crake *Zapornia parva*

The observers of the record published in Thompson *et al.* (1993) subsequently withdrew this sighting as possible confusion with Baillon's Crake could not be eliminated (Thompson & Johnson 2003).

22. Black-tailed Crake *Zapornia bicolor*

Although listed by Rashid (1967), there are no confirmed records.

23. Siberian Crane *Leucogeranus leucogeranus*

Although listed by Olivier (1979), there are no other claims of this globally threatened (CR) species, whose closest records were in Bihar in the 19th century (BirdLife International 2001).

24. Wilson's Storm-petrel *Oceanites oceanicus*

Although listed by Sarker (1984), Ali & Ripley (1987) noted it was not recorded in the N Bay of Bengal. Rasmussen and Anderton (2005) mention reports from West Bengal, but it is considered unproven in Bangladesh territorial waters.

25. Black-bellied Storm-petrel *Fregetta tropica*

Included in several lists (Rashid 1967; Ripley 1982; Sarker 1984), presumably on the basis of one specimen from the 'Bay of Bengal'. However, Rasmussen & Anderton (2005) found that this specimen 'lacks an original label and provenance may be unreliable', so it is treated as unproven.

26. Oriental Stork *Ciconia boyciana*

Although included in some lists of threatened birds of Bangladesh (Oliver 1979; Rahman & Akonda 1987), there is no evidence that this globally threatened (EN) species has occurred, and Rasmussen & Anderton (2005) treated it as hypothetical in South Asia.

27. Common Little Bittern *Ixobrychus minutus*

Although listed by Rashid (1967) and Ripley (1982), there are no confirmed records further east than Uttar Pradesh in South Asia (Rasmussen & Anderton 2005) and confusion with Yellow Bittern *I. sinensis* may have occurred.

28. Western Reef-egret *Egretta gularis*

Thompson and Johnson (2003) reviewed past records including published lists and their own observations, and concluded that without photographic evidence past claims should be treated as probable or hypothetical, in part because of possible confusion with Pacific Reef-egret *E. sacra*.

29. Dalmatian Pelican *Pelecanus crispus*

Although listed for Bengal by Ali & Ripley (1987), there are no details of any records from Bangladesh.

30. Red-footed Booby *Sula sula*

Although listed by Ripley (1982) and Sarker (1984), there seems no reason to be certain that the sight records and specimen reported in Ali & Ripley (1987) from the 'Bay of Bengal' came from Bangladesh territorial waters and so this species is treated as hypothetical.

31. Brown Booby *Sula leucogaster*

Although included in the list of Sarker & Sarker (1988) there are no details or other claims of this seabird so it is treated as hypothetical.

32. Eurasian Golden Plover *Pluvalis apricaria*

Listed as possible by Rashid (1967) but there are no confirmed records; the nearest vagrant report is from Assam, India (Rasmussen & Anderton 2005).

33. Sociable Lapwing *Vanellus gregarious*

Included in several lists for Bangladesh, including Rashid (1967) and Husain *et al.* (1974), but no details are available of any observation. Given a lack of 19th-century records it is unlikely that this now globally threatened (CR) species has occurred; the closest records are from Bihar, India (Rasmussen & Anderton 2005).

34. Solitary Snipe *Gallinago solitaria*

Although listed by Rashid (1967) and in subsequent works (e.g. Harvey 1990), there are no detailed records of this high-altitude species. It has been recorded in winter from nearby Meghalaya, but most records are from above 2,000m altitude (Rasmussen & Anderton 2005).

35. Great Snipe *Gallinago media*

Although reported by Mountfort & Poore (1968), no details are available. This scarce species is a vagrant to S India so is unlikely but could occur in Bangladesh.

36. Indian Courser *Cursorius coromandelicus*

Although included in several Bangladesh lists, e.g. Rashid (1967) and Husain *et al.* (1974), no details are available for any observation, so it is regarded as unconfirmed; the closest records are from West Bengal (Rasmussen & Anderton 2005).

37. Black Noddy *Anous minutus*

Rasmussen & Anderton (2005) note that a specimen of this species was collected from the 'mouth of Ganges' as reported by Hume (1888), and confirmed it as this species rather than Lesser Noddy *A. tenuirostris* which had been reported by Ali & Ripley (1987) and Harvey (1990). While the majority of what was known as the mouth of the Ganges lies in what is now Bangladesh, part of this area lies in West Bengal. Therefore, without a more precise location, it is only probable that this genuine record came from Bangladesh.

38. Common White Tern *Gygis alba*

An old specimen from the Bay of Bengal is referred to in several works (Ripley 1982; Sarker 1984; Harvey 1990), but it cannot be confirmed that this was in what are now Bangladesh territorial waters.

39. Sooty Tern *Onychoprion fuscatus*

Although listed by Husain (1979), Khan (1982) and others, there is no evidence that this pelagic species has been recorded.

40. Roseate Tern *Sterna dougallii*

Although included in some lists (Rashid 1967; Husain *et al.* 1983), there is no confirmed evidence that this species has been recorded.

41. Oriental Bay-owl *Phodilus badius*

Listed by Rashid (1967) and consequently by others, e.g. Rahman & Akonda (1987), but there is no confirmed record.

42. Eastern Grass-owl *Tyto longimembris*
Although listed by, for example, Rashid (1967), there are no recent records nor any detailed historical accounts or specimens (Rasmussen & Anderton 2005). While it is likely to have occurred historically when suitable habitat existed, there is no proof of this.

43. Jungle Owlet *Glaucidium radiatum*
Mentioned for Bangladesh by Lister (1951) and Harvey (1990) but without details. Thompson & Johnson (2003), the sources of most recent claimed records, concluded that these records were not certain as they were based on calls now known to be made by Asian Barred Owlet *G. cuculoides*.

44. Mottled Wood-owl *Strix ocellata*
Mentioned for Bangladesh by Lister (1951) and Rashid (1967) but without details. This species has been recorded in West Bengal (Rasmussen & Anderton (2005) but is treated as unproven in Bangladesh.

45. Tawny Owl *Strix aluco*
Listed as possible by Rashid (1967), but there is no confirmed record.

46. Rock Eagle-owl *Bubo bengalensis*
Although listed for Bangladesh by Ali & Ripley (1987) and reported once from the Sundarbans (Khan 2005), there are no details or evidence supporting its occurrence.

47. Lesser Fish-eagle *Ichthyophaga humilis*
Although listed for the Sundarbans (Sarker & Sarker 1985, 1986) there are no details or descriptions. Considering the possibility of confusion with the widespread Grey-headed Fish-eagle *I. ichthyaetus* (Lethaby 2005) and that Lesser Fish-eagle is a species of hill-rivers and pools, its occurrence in Bangladesh is considered unproven.

48. Red Kite *Milvus milvus*
Although reported from Dhaka by Sarker & Sarker (1985) no description or details were given. Past records of this potential vagrant species throughout South Asia were treated as hypothetical by Rasmussen & Anderton (2005), so we consider it unconfirmed.

49. Rufous-necked Hornbill *Aceros nipalensis*
Included in several lists for Bangladesh (Rashid 1967; Khan 1982; Ripley 1982), but there appears to be no detailed record.

50. Brown-headed Barbet *Psilopogon zeylanica*
Although reported by Husain (1968) from the Chittagong Hill Tracts, he did not include it in subsequent publications; Lister (1951) and Rashid (1967) also reported it but without details. Its occurrence is considered unproven.

51. Golden-throated Barbet *Psilopogon franklinii*
Although listed by Rashid (1967) and Ripley (1982), there are no confirmed records of this montane species.

52. White-naped Woodpecker *Chrysocolaptes festivus*
Listed by Sarker & Sarker (1988) but there are no confirmed records; however, it has been recorded near Calcutta (Rasmussen & Anderton 2005).

53. Laced Woodpecker *Picus vittatus*
Reported from the Sundarbans based on one specimen collected in 1958 (Paynter 1970). However, Rasmussen (2000) concluded that this was misidentified and that Streak-breasted Woodpecker *P. viridanus* is found in the Sundarbans, as has subsequently been confirmed in the field.

54. Rufous-bellied Woodpecker
Dendrocopos hyperythrus
Although included in several lists (Husain 1979; Khan 1982; Ripley 1982; Harvey 1990), there appear to be no documented records.

55. Stripe-breasted Woodpecker *Dendrocopos atratus*
Listed by Husain (1979) and Khan (1982), no detailed evidence was presented so it might easily have been mistaken with Fulvous-breasted Woodpecker.

56. Collared Falconet *Microhierax caerulescens*
Although included in Rashid (1967), there is no evidence of any confirmed records. This species is recorded from the adjacent Garo Hills of India (Rasmussen & Anderton 2005) so there is a chance it could wander to Bangladesh.

57. Pied Falconet *Microhierax melanoleucos*
Although listed for Bangladesh by Rashid (1967) and Ripley (1982), Rasmussen & Anderton (2005) failed to trace any specimens and there are no recent records. This species is recorded from the nearby hill states of India so could occur in Bangladesh.

58. Slaty-headed Parakeet *Psittacula himalayana*
Listed by Rashid (1967), but there is no evidence that this Himalayan species has ever occurred.

59. Eared Pitta *Hydrornis phayrei*
Reported from Lawachara NP by Vestergaard (1998). However, Thompson & Johnson (2003), and Rasmussen & Anderton (2005) considered that it was not certainly identified and best treated as unproven.

60. White-browed Shrike-babbler *Pteruthius aeralatus*
Listed as possible by Rashid (1967) and Sarker & Sarker (1988), but there are no certain records, although it occurs in Meghalaya.

61. Black-eared Shrike-babbler *Pteruthius melanotis*
Although listed for Bangladesh by Rashid (1967), Husain (1979), Khan (1982), Ripley (1982) and Harvey (1990), there appear to be no detailed records, although it occurs in Meghalaya.

62. White-bellied Minivet *Pericrocotus erythropygius*
Listed for the Sundarbans by Sarker & Sarker (1986) but without details, and this species would be unusual in mangroves away from its preferred dry open woodland. Treated as unconfirmed although it has wandered to West Bengal, India.

63. Grey-chinned Minivet *Pericrocotus solaris*
Listed for Bangladesh by Rashid (1967), Husain (1979) and Khan (1982), but without details. Although recorded from Meghalaya, we consider it unconfirmed in Bangladesh.

64. Short-billed Minivet *Pericrocotus brevirostris*
Listed for Bangladesh by Rashid (1967), Mountfort & Poore (1968), Husain (1979) and Vestergaard (1998) but no details are given. Although recorded from hills in neighbouring Indian states, we consider it unconfirmed in Bangladesh

considering the ease of confusion with Long-tailed Minivet *P. ethologus*, itself a vagrant.

65. White-browed Fantail *Rhipidura aureola*
Listed by several authors for Bangladesh including Lister (1951), Rashid (1967) and Husain *et al.* (1983); and implied by Ali & Ripley (1987) to be common, but we are not aware of confirmed records of this species.

66. Collared Treepie *Denrocitta frontalis*
Although listed by Rashid (1967) as possible, there is no confirmed record.

67. Yellow-billed Blue Magpie *Urocissa flavirostris*
Although listed by Rashid (1967) as possible, there is no confirmed record.

68. Plain-crowned Jay *Garrulus bispecularis*
Although Rashid (1967) listed Eurasian Jay *G. glandarius* (before splitting) as possible, there is no confirmed record.

69. Yellow-bellied Fairy-fantail *Chelidorhynx hypoxanthus*
Although listed for Bangladesh by Rashid (1967), Husain (1979), Khan (1982) and Ripley (1982), there appear to be no detailed records.

70. Fire-capped Tit *Cephalopyrus flammiceps*
Although listed as possible by Rashid (1967), there have been no confirmed records of this species, which occurs at its closest in Assam.

71. Yellow-browed Tit *Sylviparus modestus*
Although listed as possible by Rashid (1967), there have been no confirmed records of this species, which occurs at its closest in Assam.

72. Marsh Tit *Poecile palustris*
Although listed as possible by Rashid (1967), this Palearctic species has not been recorded in South Asia.

73. Yellow-cheeked Tit *Machlolophus spilonotus*
Although listed as possible by Rashid (1967), there have been no records of this species, which occurs at its closest in the neighbouring hills of India.

74. Indian Bushlark *Mirafra erythroptera*
Although listed by Mountfort (1969) and Khan (1982), there is no detailed record of this species, which is recorded as close as S West Bengal.

75. Striated Prinia *Prinia crinigera*
Although listed for Bangladesh by Rashid (1967), Khan (1982), Ripley (1982) and Harvey (1990), there is no documented record, although it occurs in the neighbouring hills of India.

76. Black-throated Prinia *Prinia atrogularis*
Although listed for Bangladesh by Rashid (1967), Ripley (1982) and Harvey (1990), there is no documented record, although it occurs in the neighbouring hills of India.

77. Grey-crowned Prinia *Prinia cinereocapilla*
Although listed as possible for Bangladesh by Rashid (1967), there is no documented record of this globally threatened (VU) species, which occurs at its closest in Bhutan.

78. Jungle Prinia *Prinia sylvatica*
Although listed for Bangladesh by Rashid (1967), Khan

(1982), Ripley (1982) and Harvey (1990), there is no documented record; this is a species of drier habitats, which occurs at its closest in Bihar, India.

79. Scaly-breasted Cupwing *Pnoepyga albiventer*
Listed as possible by Rashid (1967), but there have been no records of this species, which occurs at its closest in the hills of Manipur.

80. Chinese Grasshopper-warbler *Locustella tacsanowskia*
Listed for Bangladesh by Rashid (1967), but there is no documented record of this species, although it has occurred in the neighbouring hills of India.

81. Mountain Bulbul *Ixos mcclellandii*
Although listed for Bangladesh by Rashid (1967), Husain (1979), Khan (1982), Ripley (1982) and Harvey (1990), there is no documented record of this species, which occurs in the neighbouring hills of India.

82. Crested Finchbill *Spizixos canifrons*
Although listed for Bangladesh by Rashid (1967), Husain (1979) and Khan (1982), there are no documented records of this species, which occurs in the neighbouring hills of India.

83. Himalayan Bulbul *Pycnonotus leucogenys*
Although listed as possible by Husain (1979), there are no records of this Himalayan species, which occurs in the foothills and plains in winter.

84. Buff-barred Warbler *Phylloscopus pulcher*
Listed as possible by Rashid (1967), but there have been no records of this Himalayan species.

85. Ashy-throated Warbler *Phylloscopus maculipennis*
Listed as possible by Rashid (1967), but there have been no records of this mountain species, although it occurs in Meghalaya.

86. Sulphur-bellied Leaf-warbler *Phylloscopus griseolus*
Although listed for Bangladesh by Law (1954), Rashid (1967), Husain (1979) and Khan (1982), there are no details and, as this species winters at its closest in C India, it is treated as hypothetical.

87. Yellow-streaked Warbler *Phylloscopus armandii*
Listed as possible by Rashid (1967), but there are no confirmed records for South Asia (Rasmussen & Anderton 2005).

88. Radde's Warbler *Phylloscopus schwarzi*
Three sightings were reported in Thompson *et al.* (1993), but in a review of past records/claims of this species in South Asia, and after consulting with the Bangladesh observers, it was concluded that these could not eliminate the very similar Yellow-streaked Warbler, and so they are treated as unconfirmed (Praveen *et al.* 2019).

89. Grey-cheeked Warbler *Phylloscopus poliogenys*
Although listed for Bangladesh by Rashid (1967), Ripley (1982) and Harvey (1990), there are no detailed records of this species, which occurs in S Assam.

90. Eastern Crowned Warbler *Phylloscopus coronatus*
Although listed for Bangladesh by Ripley (1982) and Harvey (1990), possible sightings in forests in NE Bangladesh in the

1980s–1990s are best treated as unconfirmed given the lack of detailed descriptions and possible confusion with similar species. Rasmussen & Anderton (2005) treated this species as hypothetical in South Asia.

91. Two-barred Warbler *Phylloscopus plumbeitarsus*
Listed as possible by Rashid (1967), but there have been no records, although one specimen exists from Assam (Rasmussen & Anderton 2005).

92. Pale-footed Bush-warbler *Hemitesia pallidipes*
Listed for Bangladesh by Rashid (1967), but there is no documented record of this species, although it occurs in the neighbouring hills of India in winter.

93. Rufous-faced Warbler *Abroscopus albogularis*
Although listed for Bangladesh by Rashid (1967), Ripley (1982) and Harvey (1990), there are no detailed records of this species, which occurs in Meghalaya.

94. Black-faced Warbler *Abroscopus schisticeps*
Listed as possible by Rashid (1967), but there are no records of this mountain species, which occurs at its closest in Nagaland.

95. Korean Bush-warbler *Horornis canturians*
Although listed as possible for Bangladesh by Rashid (1967), there is no documented record of this species, which has occurred as a vagrant in nearby states of India.

96. Black-throated Tit *Aegithalos concinnus*
Although listed as possible by Rashid (1967), there have been no records of this species, which occurs at its closest in NE India.

97. Jerdon's Babbler *Chrysomma altirostre*
After extensive review of the literature, BirdLife International (2001) concluded that past claims from Bangladesh were based on its possible occurrence and a reference to its inhabiting the Sylhet plains in Baker (1922–30). Although this area held habitats similar to those found in Cachar (where it occurs) at that time, there is no definite evidence that this globally threatened (VU) species ever occurred in Bangladesh.

98. Grey-headed Parrotbill *Psittiparus gularis*
Although listed for Bangladesh by Rashid (1967), Ripley (1982) and Harvey (1990), there appear to be no detailed records. At its closest it occurs in S Assam (Rasmussen & Anderton 2005).

99. Pale-billed Parrotbill *Chleuasicus atrosuperciliaris*
Although listed for Bangladesh by Rashid (1967) and Sarker & Sarker (1988), there appear to be no detailed records. At its closest it occurs in Cachar (Rasmussen & Anderton 2005).

100. Stripe-throated Yuhina *Yuhina gularis*
Although listed for Bangladesh by Rashid (1967) and Sarker & Sarker (1988), there appear to be no detailed records. At its closest it occurs in Manipur (Rasmussen & Anderton 2005).

101. Whiskered Yuhina *Yuhina flavicollis*
Although listed for Bangladesh by Rashid (1967), Ripley (1982) and Harvey (1990), there appear to be no detailed records. At its closest it occurs in Meghalaya.

102. White-naped Yuhina *Yuhina bakeri*
Although listed for Bangladesh by Baker (1922–30), Rashid (1967), Ripley (1982) and Harvey (1990), there appear to be no detailed records. At its closest it occurs in Cachar (Rasmussen & Anderton 2005).

103. Coral-billed Scimitar-babbler *Pomatorhinus ferruginosus*
Listed as possible by Rashid (1967), but there have been no records of this species, which occurs at its closest in the hills of Meghalaya.

104. Slender-billed Scimitar-babbler *Pomatorhinus superciliaris*
Although listed for Bangladesh by Rashid (1967), Husain (1979), Khan (1982), Ripley (1982) and Harvey (1990), there appear to be no detailed records. At its closest it occurs in Nagaland (Rasmussen & Anderton 2005).

105. Streak-breasted Scimitar-babbler *Pomatorhinus ruficollis*
Although listed for Bangladesh by Rashid (1967), Husain (1979), Khan (1982), Ripley (1982), Sarker & Sarker (1988), Harvey (1990) and Rasmussen & Anderton (2005), there are no detailed records, so it may best be treated as probable. It occurs nearby in the hills of Meghalaya.

106. Tawny-bellied Babbler *Dumetia hyperythra*
Although listed for Bangladesh by Rashid (1967), Mountfort (1969), Khan (1982), Ripley (1982) and Harvey (1990), there appear to be no records. At its closest the species occurs in the dry forests of Bihar.

107. Golden Babbler *Cyanoderma chrysaeum*
Although listed by Rashid (1967), Husain (1979), Khan (1982), Ripley (1982) and Harvey (1990), there appear to be no detailed records. At its closest the species occurs in Meghalaya.

108. Rufous-capped Babbler *Cyanoderma ruficeps*
Although reported from Lawachara NP by Vestergaard (1998), this is the only claim, no description was given and confusion with Rufous-fronted Babbler *C. rufifrons* was possible, so it is treated as unconfirmed.

109. Rufous-throated Fulvetta *Schoeniparus rufogularis*
Although listed for Bangladesh by Baker (1922–30), Rashid (1967), Husain (1979), Ripley (1982) and Harvey (1990), there appear to be no detailed records. At its closest the species occurs in Meghalaya.

110. Yellow-throated Fulvetta *Schoeniparus cinereus*
Although listed for Bangladesh by Rashid (1967), Ripley (1982) and Harvey (1990), there appear to be no detailed records. At its closest it occurs in S Assam.

111. Rufous-winged Fulvetta *Schoeniparus castaneceps*
Although listed for Bangladesh by Rashid (1967), Ripley (1982) and Harvey (1990), there appear to be no detailed records. At its closest the species occurs in Meghalaya.

112. Eyebrowed Wren-babbler *Napothera epilepidota*
Listed as possible by Rashid (1967), but there have been no records of this species, which occurs at its closest in the hills of Meghalaya.

113. Long-billed Wren-babbler *Rimator malacoptilus*
Although listed for Bangladesh by Rashid (1967), Husain (1979), Khan (1982) and Harvey (1990), there appear to be no detailed records. At its closest the species occurs in Meghalaya (Rasmussen & Anderton 2005).

114. Himalayan Cutia *Cutia nipalensis*
Listed as possible by Rashid (1967), but there are no records of this species, which occurs at its closest in the hills of Nagaland.

115. Large Grey Babbler *Argya malcolmi*
Although listed by Husain (1979), there are no details and this must be treated as hypothetical since the species has not yet been recorded east of Bihar, India.

116. Slender-billed Babbler *Chatarrhaea longirostris*
After an extensive literature review, BirdLife International (2001) concluded that past claims from Bangladesh were based on its possible occurrence and unsubstantiated comments in Baker (1922–30). There is no definite evidence that this globally threatened (VU) species ever occurred in Bangladesh.

117. Spot-breasted Laughingthrush *Garrulax merulinus*
Listed as possible by Rashid (1967), but there have been no records of this species, which occurs at its closest in Meghalaya.

118. Rufous-chinned Laughingthrush
Garrulax rufogularis
Although listed by Baker (1922–30), Rashid (1967) and Harvey (1990), there are no detailed records and the nearest population is in Meghalaya.

119. White-browed Laughingthrush *Garrulax sannio*
Although listed by Husain (1979), there are no detailed records and the nearest populations are in the hills of Manipur.

120. Rufous-vented Laughingthrush *Garrulax gularis*
Although listed for Bangladesh by Rashid (1967), Husain (1979), Khan (1982) and Harvey (1990), there are no detailed records of this species, which occurs in Meghalaya. The only published sighting (Thompson *et al.* 1993) was subsequently withdrawn by the observer as a probable misidentification (Thompson & Johnson 2003).

121. Blue-winged Laughingthrush
Trochalopteron squamatum
Listed as possible by Rashid (1967), but there have been no records of this species, which occurs in Meghalaya.

122. Striped Laughingthrush *Trochalopteron virgatum*
Listed for Bangladesh by Husain (1979), Khan (1982) and Sarker & Sarker (1988), but there are no detailed records of this species, which is resident in Manipur.

123. Assam Laughingthrush
Trochalopteron chrysopterum
Listed for Bangladesh by Husain (1979), but there are no details of this species, which occurs in Meghalaya.

124. Grey Sibia *Heterophasia gracilis*
Listed as possible by Rashid (1967), but there have been no records of this species, which occurs at its closest in Meghalaya.

125. Silver-eared Mesia *Leiothrix argentauris*
Although listed for Bangladesh by Rashid (1967), Husain (1979), Khan (1982), Ripley (1982) and Harvey (1990), there appear to be no detailed records. At its closest the species occurs in Meghalaya.

126. Red-billed Leiothrix *Leiothrix lutea*
Listed as possible by Rashid (1967), but there have been no records of this species, which occurs in Meghalaya.

127. Rufous-backed Sibia *Leioptila annectens*
Listed as possible by Rashid (1967), but there have been no records of this species, which occurs at its closest in Meghalaya.

128. Red-tailed Minla *Minla ignotincta*
Although listed for Bangladesh by Rashid (1967), Ripley (1982) and Harvey (1990), there appear to be no detailed records. At its closest the species occurs in Nagaland.

129. Red-faced Liocichla *Liocichla phoenicea*
Although listed for Bangladesh by Rashid (1967), Husain (1979), Khan (1982) and Harvey (1990), there appear to be no detailed records. At its closest the species occurs in Manipur.

130. Blue-winged Minla *Siva cyanouroptera*
Although listed for Bangladesh by Baker (1922–30), Rashid (1967), Ripley (1982) and Harvey (1990), there appear to be no detailed records. At its closest the species occurs in Meghalaya.

131. Sikkim Treecreeper *Certhia discolour*
Although listed as possible by Rashid (1967), there are no records of this species, which occurs at its closest in Meghalaya.

132. White-tailed Nuthatch *Sitta himalayensis*
Although listed as possible by Rashid (1967), there have been no records of this species, which occurs at its closest in Nagaland.

133. Beautiful Nuthatch *Sitta formosa*
Although listed for Bangladesh by Rashid (1967), Husain (1979) and Khan (1982), there is no documented record. One reported in Harvey (1990) was subsequently treated as possible (Thompson *et al.* 1993), although it occurs in Assam.

134. Brown Dipper *Cinclus pallasii*
Although listed for Bangladesh by Rashid (1967) and Ali & Ripley (1987), there is no documented record.

135. Golden-crested Myna *Ampeliceps coronatus*
Although listed as possible by Rashid (1967), there have been no records of this species, which occurs at its closest in NE Assam.

136. Green Cochoa *Cochoa viridis*
Listed as possible by Rashid (1967), but there have been no records of this species, which occurs in the neighbouring hills of India.

137. Pied Thrush *Geokichla wardii*
Although listed by Rashid (1967) there are no confirmed records. One specimen exists from the neighbouring Garo Hills of Meghalaya, but other records from NE India have been treated as doubtful (Rasmussen & Anderton 2005).

138. Chestnut Thrush *Turdus rubrocanus*
Listed as possible by Rashid (1967), there are no confirmed records although it has occurred in winter in Meghalaya.

139. Indian Robin *Saxicoloides fulicatus*
Although listed by Baker (1922–30), Rashid (1967), Mountfort (1969), Khan (1982), Ali & Ripley (1987) and Harvey (1990), there appear to be no confirmed records.

140. White-gorgeted Flycatcher *Anthipes monileger*
Although listed by Rashid (1967) and it is resident in Meghalaya, there are no confirmed records.

141. Himalayan Shortwing *Brachypteryx cruralis*
Although listed in Rashid (1967), there are no confirmed records. The two sightings reported as this species in Crosby (1995) were subsequently identified as White-browed Bushrobin *Tarsiger indicus* (Thompson & Johnson 2003).

142. Golden Bush-robin *Tarsiger chrysaeus*
Although listed as possible by Rashid (1967), there are no confirmed records; this species is a winter visitor in the nearby hills of India.

143. Little Forktail *Enicurus scouleri*
Although listed by Rashid (1967), Husain (1979) and Ripley (1982), there are no documented records; it occurs in the neighbouring hills of India.

144. Spotted Forktail *Enicurus maculatus*
Although listed by Rashid (1967), Husain (1979), Khan (1982) and Ripley (1982), there appear to be no documented records; it occurs in the neighbouring hills of India.

145. Malabar Whistling-thrush *Myophonus horsfieldii*
Although reported from Pablakhali Wildlife Sanctuary in the Chittagong Hill Tracts by Husain (1975), this is the only claim of a species that is resident in and endemic to peninsular India (at its closest in the hills of Orissa), and is considered erroneous.

146. Pygmy Blue-flycatcher *Ficedula hodgsoni*
Although listed by Rashid (1967), there are no confirmed records of this species, which occurs in NE India.

147. White-throated Redstart *Phoenicurus schisticeps*
Although listed for what is now Bangladesh in Baker (1922–30), there are no other claims of this mountain species, which is recorded at its closest in Bhutan.

148. White-winged Redstart
Phoenicurus erythogastrus
Although listed as possible by Rashid (1967), there have been no records of this species of high mountains.

149. Hodgson's Redstart *Phoenicurus hodgsoni*
Although listed by Rashid (1967), Husain (1979) and Khan (1982), there are no documented records of this winter visitor to the neighbouring hills of India.

150. Spotted Elachura *Elachura formosa*
Although listed by Rashid (1967), Khan (1982), Ripley (1982), Harvey (1990) and Rasmussen & Anderton (2005), there appear to be no detailed records. At its closest the species occurs in Meghalaya.

151. Purple-naped Spiderhunter
Arachnothera hypogrammica
Listed as possible by Rashid (1967), but there have been no records, nor has it been recorded anywhere in South Asia. At its closest it occurs in N Myanmar.

152. Brown-throated Sunbird *Anthreptes malacensis*
Listed as possible by Rashid (1967), but there have been no records, nor has it been recorded anywhere in South Asia. At its closest it occurs on the Arakan coast of Myanmar.

153. Black-throated Sunbird *Aethopyga saturata*
Although listed by Rashid (1967), Husain (1979), Ripley (1982) and Harvey (1990), there appear to be no detailed records. At its closest the species occurs in Meghalaya.

154. Finn's Weaver *Ploceus megarhynchus*
Although listed by Rashid (1967), there have been no confirmed records of this globally threatened (VU) species, which occurs in the grasslands of NE India.

155. Russet Sparrow *Passer cinnamomeus*
Although included in lists for the Bangladesh Sundarbans by Mountfort (1969), Hendrichs (1975), Sarker & Sarker (1986), no details are available and this seems unlikely for a species of the hills that occurs at its closest in Manipur, so it is treated as hypothetical.

156. Chestnut-shouldered Bush-sparrow
Gymnoris xanthocollis
Listed as possible by Rashid (1967), but there have been no records, although at its closest the species occurs in West Bengal.

157. Buff-bellied Pipit *Anthus rubescens*
Listed as possible by Rashid (1967), but there have been no confirmed records of this species, although it winters in Nepal.

158. Water Pipit *Anthus spinoletta*
Listed for Bangladesh in Mountfort (1969), but without details this is presumably a mistake since this species is a winter visitor to Pakistan (also covered in the same book).

159. Spot-winged Grosbeak *Mycerobas melanozanthos*
Although listed by Rashid (1967), there have been no confirmed records of this species, which occurs at its closest in Assam.

160. Red-headed Bunting *Emberiza bruniceps*
Although listed by Rashid (1967), Ripley (1982) and Harvey (1990), there appear to be no detailed records. At its closest the species occurs in much of peninsular India.

161. Yellowhammer *Emberiza citrinella*
Listed as possible by Rashid (1967), but there have been no confirmed records of this species, which occurs at its closest as a vagrant to Nepal.

162. Chestnut Bunting *Emberiza rutila*
Listed as possible by Rashid (1967), but there have been no confirmed records of this species, which occurs at its closest in Manipur.

References for Appendix 3

Ali, S. & Ripley, S.D. (1987) *The Compact Handbook of the Birds of India and Pakistan.* New Delhi: Oxford University Press.

Baker, E.C.S. (1922–30) *Fauna of British India. Aves.* Second edition. 8 vols. London: Taylor & Francis.

BirdLife International (2001) *Threatened Birds of Asia: the BirdLife International Red Data Book.* 2 vols. Cambridge UK: BirdLife International.

Crosby, M. (1995) From the field. *Oriental Bird Club Bulletin* 21: 68.

Harvey, W.G. (1990) *Birds in Bangladesh.* Dhaka: University Press Ltd.

Hendrichs, H.B. (1975) The status of the Tiger *Panthera tigris* (Linné, 1758) in the Sundarbans mangrove forests (Bay of Bengal). *Saugetierkundliche Mitteilungen* 23: 161–199.

Husain, K.Z. (1968) Field-notes on the birds of the Chittagong Hill-Tracts. (Notes based on the records of the Asiatic Society of Pakistan Expedition to the Chittagong Hill-Tracts, 1965), Paper V: Non-passerine birds. *J. Asiatic Soc. Pakistan* 13: 91–101.

Husain, K.Z. (1975) Birds of Pablakhali Wildlife Sanctuary (the Chittagong Hill-Tracts). *Bangladesh J. Zool.* 3: 155–157.

Husain, K.Z. (1979) *Birds of Bangladesh.* Dhaka: Government of Bangladesh.

Husain, K.Z., Sarker, S.U. & Khan, A.R. (1974) Birds of Dacca with some notes on their present status. *Bangladesh J. Zool.* 2: 153–170.

Husain, K.Z., Sarker, S.U. & Rahman, M. (1983) Summer birds of the Sundarbans, 'Nilkamal Sanctuary'. *Bangladesh J. Zool.* 11: 48-51.

Khan, M.A.R. (1982) *Wildlife of Bangladesh: a Checklist.* Dhaka: University of Dhaka.

Khan, M.M.H. (2005) Species diversity, relative abundance and habitat use of the birds in the Sundarbans East Wildlife Sanctuary, Bangladesh. *Forktail* 21: 79–86.

Lethaby, N. (2005) The occurrence of Lesser Fish Eagle *Ichthyophaga humilis* on the Cauvery River, Karnataka, India and some notes on the identification of this species. *Birding ASIA* 4: 33–38.

Lister, M.D. (1951) Some bird associations of Bengal. *J. Bombay Nat. Hist. Soc.* 49: 695–728.

Mountfort, G. (1969) *The Vanishing Jungle.* London: Collins.

Mountfort, G. & Poore, D. (1968) Report on the second World Wildlife Fund Expedition to Pakistan. (WWF Project 311). Unpublished report. Gland, Switzerland: WWF.

Olivier, R.C.D. (1979) *Wildlife Conservation and Management in Bangladesh.* FAO Field Doc. No 10. Rome: FAO.

ORNIS (2006) Downloaded 1 November 2006.

Paynter, R.A. (1970) Species with Malaysian affinities in the Sundarbans, East Pakistan. *Bull. Brit. Orn. Club* 90: 118–119.

Praveen, J., Inskipp, T. & Thompson, P.M. (2019) Radde's Warbler *Phylloscopus schwarzi* from the Indian Subcontinent – withdrawal of records. *Indian Birds* 15(3): 96A.

Rahman, S.A. & Akonda, A.W. (1987) Bangladesh National Conservation Strategy: Wildlife and Protected Areas.

Rashid, H. (1967) *Systematic List of the Birds of East Pakistan.* Publication no. 20. Dacca: The Asiatic Society of Pakistan.

Rasmussen, P.C. & Anderton, J.C. (2005) *Birds of South Asia: The Ripley Guide.* 2 vols. Smithsonian Institution and Lynx Edicions: Washington, D.C. and Barcelona.

Ripley, S. D. (1982) *A Synopsis of the Birds of India and Pakistan together with those of Nepal, Bhutan, Bangladesh and Sri Lanka.* Bombay: Bombay Natural History Society.

Sarker, S.U. (1984) Seabirds of the Bay of Bengal of Bangladesh coast and their conservation. *Tigerpaper* 11(4): 9–13.

Sarker, S.U. & Sarker, N.J. (1985) Birds of prey and their conservation in the Sundarbans mangrove forests, Khulna, Bangladesh. pp 205–209 in Chancellor R.D. & Newton I. *Conservation Studies on Raptors.* ICBP Technical Publication No 5. Cambridge: International Council for Bird Preservation.

Sarker, S.U. & Sarker, N.J. (1986) Status and distribution of birds of the Sundarbans, Bangladesh. *J. Noami* 3: 19–33.

Sarker, S.U. & Sarker, N.J. (1988) *Wildlife of Bangladesh: a Systematic List.* Dhaka: Rico Printers.

Thompson, P.M., Harvey, W.G., Johnson, D.L., Millin, D.J., Rashid, S.M.A., Scott, D.A., Stanford, C. & Woolner, J.D. (1993) Recent notable bird records from Bangladesh. *Forktail* 9: 13–44.

Thompson, P.M. & Johnson, D.L. (2003) Further notable bird records from Bangladesh. *Forktail* 19: 85–102.

Vestergaard, M. (1998) Eared Pitta *Pitta phayrei*: a new species for Bangladesh and the Indian subcontinent. *Forktail* 14: 69–70.

REFERENCES

BirdLife International (2003) *Saving Asia's Threatened Birds: a Guide for Government and Civil Society.* BirdLife International, Cambridge.

BirdLife International (2004) *Important Bird Areas in Asia: Key Sites for Conservation.* BirdLife International, Cambridge.

Chowdhury, S.U. (2020) Birds of the Bangladesh Sundarbans: Status, Threats and Conservation Recommendations. *Forktail* 36: 35–46.

Collar, N.J., Andreev, A.V., Chan, S., Crosby, M., Subramanya, S. & Tobias, J.A. (2001) *Threatened Birds of Asia.* BirdLife International, Cambridge.

Dickinson, E.C. & Remsen, J.V., eds. (2013) *The Howard and Moore Complete Checklist of the Birds of the World.* 4th edition. Vol. 1 Non-passerines. Aves Press, Eastbourne.

Dickinson, E.C. & Christidis, L., eds. (2014) *The Howard and Moore Complete Checklist of the Birds of the World.* 4th edition. Vol. 2 Passerines. Aves Press, Eastbourne.

Gill, F. & Wright, M. (2006) *Birds of the World: Recommended English Names.* Christopher Helm, London.

Grimmett, R., Inskipp, C. and Inskipp, T. (1998) *Birds of the Indian Subcontinent.* Christopher Helm, London.

Grimmett, R., Inskipp, C. & Inskipp, T. (2011) *Birds of the Indian Subcontinent.* 2nd edition. Christopher Helm, London.

Halder, R.R. (2010) *A Photographic Guide to the Birds of Bangladesh.* Baikal Teal Production, Dhaka.

Haque, E.U. (2014) *Feathered Splendours: Birds of Bangladesh.* The University Press Ltd., Dhaka.

HBW and BirdLife International (2019) *Handbook of the Birds of the World and BirdLife International digital checklist of the birds of the world.* Version 4. http://datazone.birdlife.org/userfiles/file/Species/Taxonomy/HBW-BirdLife_Checklist_v4_Dec19.

del Hoyo, J. & Collar, N. (2014–2016) *HBW and BirdLife International Illustrated Checklist of the Birds of the World.* 2 vols.

Inskipp, T., Lindsey, N. & Duckworth, W. (1996) *An Annotated Checklist of the Birds of the Oriental Region.* Oriental Bird Club.

IUCN Bangladesh (2015) *Red List of Bangladesh*, Vol 3, Birds. IUCN, International Union for Conservation of Nature, Bangladesh Country Office, Dhaka.

Khan, M.M.H. (2008) *Protected Areas of Bangladesh – a Guide to Wildlife.* Nishorgo Program, Bangladesh Forest Department, Dhaka.

Khan, M.M.H. (2018) *Photographic Guide to the Wildlife of Bangladesh.* Arannayk Foundation, Dhaka.

Rasmussen, P.C. & Anderton, J.C. (2012) *Birds of South Asia: The Ripley Guide.* 2nd edition. Lynx Edicions, Barcelona.

Round, P.D., Haque, E.U., Dymond, N., Pierce, A.J. & Thompson, P.M. (2014) Ringing and Ornithological Exploration in Northeast Bangladesh Wetlands. *Forktail* 30: 115–127.

Siddiqui, K.U., Islam, M.A., Kabir, S.M.H., Ahmad, M., Ahmed, A.T.A., Rahman, A.K.A., Haque, E.U., Ahmed, Z.U., Begum, Z.N.T., Hassan, M.A., Khondker, M. & Rahman, M.M., eds. (2008) *Encyclopedia of Flora and Fauna of Bangladesh*, 26. Birds. Asiatic Society of Bangladesh, Dhaka.

Thompson, P.M. & Chowdhury, S.U. (2020) A checklist of birds of Bangladesh. Retrieved from www.facebook.com/groups/2403154788/permalink/10158218941829789/

Thompson, P.M., Chowdhury, S.U., Haque, E.U., Khan, M.M.H. & Halder, R. (2014) Notable Bird records from Bangladesh from July 2002 to July 2013. *Forktail* 30: 50–65.

Thompson, P.M., Harvey, W.G., Johnson, D.L., Millin, D.J., Rashid, S.M.A., Scott, D.A., Stanford, C. & Woolner, J.D. (1993) Recent Notable Bird Records from Bangladesh. *Forktail* 9: 13–44.

Thompson, P.M. & Johnson, D.L. (2003) Further Notable Bird Records from Bangladesh. *Forktail* 19: 85–102.

INDEX

QUICK INDEX TO THE MAIN GROUPS OF BIRDS

Figures in **bold** refer to plate numbers